endgame

VOLUME I

THE PROBLEM
OF CIVILIZATION

endgame

VOLUME I

THE PROBLEM
OF CIVILIZATION

derrick jensen

SEVEN STORIES PRESS

new york • london • melbourne • toronto

A Seven Stories Press First Edition

Seven Stories Press
140 Watts Street
New York, NY 10013
www.sevenstories.com

In Canada:
Publishers Group Canada, 559 College Street, Toronto, ON M6G 1A9

In the UK:
Turnaround Publisher Services Ltd., Unit 3, Olympia Trading Estate, Coburg Road, Wood Green, London N22 6TZ

In Australia:
Palgrave Macmillan, 15-19 Claremont Street, South Yarra VIC 3141

Library of Congress Cataloging-in-Publication Data

Jensen, Derrick, 1960–
    Endgame / by Derrick Jensen.
        p. cm.
    Vol. 1 has subtitle: The end of civilization; vol. 2: Resistance.
    Includes bibliographical references.
    ISBN: 978-1-58322-730-5 (pbk. : alk. paper)
    ISBN: 978-1-58322-724-4 (pbk. : alk. paper)
    1. Nature--Effect of human beings on. 2. Human ecology. 3. Civilization. I. Title.

GF75.J45 2006
304.2--dc22
                    2006003645

College professors may order examination copies of Seven Stories Press titles for a free six-month trial period. To order, visit www.sevenstories.com/textbook or fax on school letterhead to (212) 226-1411.

Book design by Jon Gilbert

Printed in the USA

9 8

Contents

Volume I: The Problem of Civilization

Volume II: Resistance

Premises

PREMISE ONE: Civilization is not and can never be sustainable. This is especially true for industrial civilization.

PREMISE TWO: Traditional communities do not often voluntarily give up or sell the resources on which their communities are based until their communities have been destroyed. They also do not willingly allow their landbases to be damaged so that other resources—gold, oil, and so on—can be extracted. It follows that those who want the resources will do what they can to destroy traditional communities.

PREMISE THREE: Our way of living—industrial civilization—is based on, requires, and would collapse very quickly without persistent and widespread violence.

PREMISE FOUR: Civilization is based on a clearly defined and widely accepted yet often unarticulated hierarchy. Violence done by those higher on the hierarchy to those lower is nearly always invisible, that is, unnoticed. When it is noticed, it is fully rationalized. Violence done by those lower on the hierarchy to those higher is unthinkable, and when it does occur is regarded with shock, horror, and the fetishization of the victims.

PREMISE FIVE: The property of those higher on the hierarchy is more valuable than the lives of those below. It is acceptable for those above to increase the amount of property they control—in everyday language, to make money—by destroying or taking the lives of those below. This is called *production*. If those below damage the property of those above, those above may kill or otherwise destroy the lives of those below. This is called *justice*.

PREMISE SIX: Civilization is not redeemable. This culture will not undergo any sort of voluntary transformation to a sane and sustainable way of living. If we do not put a halt to it, civilization will continue to immiserate the vast majority of humans and to degrade the planet until it (civilization, and probably the planet) collapses. The effects of this degradation will continue to harm humans and nonhumans for a very long time.

PREMISE SEVEN: The longer we wait for civilization to crash—or the longer we wait before we ourselves bring it down—the messier the crash will be, and the worse things will be for those humans and nonhumans who live during it, and for those who come after.

PREMISE EIGHT: The needs of the natural world are more important than the needs of the economic system.

Another way to put Premise Eight: Any economic or social system that does not benefit the natural communities on which it is based is unsustainable, immoral, and stupid. Sustainability, morality, and intelligence (as well as justice) require the dismantling of any such economic or social system, or at the very least disallowing it from damaging your landbase.

PREMISE NINE: Although there will clearly someday be far fewer humans than there are at present, there are many ways this reduction in population may occur (or be achieved, depending on the passivity or activity with which we choose to approach this transformation). Some will be characterized by extreme violence and privation: nuclear Armageddon, for example, would reduce both population and consumption, yet do so horrifically; the same would be true for a continuation of overshoot, followed by a crash. Other ways could be characterized by less violence. Given the current levels of violence by this culture against both humans and the natural world, however, it's not possible to speak of reductions in population and consumption that do not involve violence and privation, not because the reductions themselves would necessarily involve violence, but because violence and privation have become the default of our culture. Yet some ways of reducing population and consumption, while still violent, would *consist* of decreasing the current levels of violence—required and caused by the (often forced) movement of resources from the poor to the rich—and would of course be marked by a reduction in current violence against the natural world. Personally and collectively we may be able to both reduce the amount and soften the character of violence that occurs during this ongoing and perhaps long-term shift. Or we may not. But this much is certain: if we do not approach it actively—if we do not talk about our predicament and what we are going to do about it—the violence will almost undoubtedly be far more severe, the privation more extreme.

PREMISE TEN: The culture as a whole and most of its members are insane. The culture is driven by a death urge, an urge to destroy life.

PREMISE ELEVEN: From the beginning, this culture—civilization—has been a culture of occupation.

PREMISE TWELVE: There are no rich people in the world, and there are no poor people. There are just people. The rich may have lots of pieces of green paper that many pretend are worth something—or their presumed riches may be even more abstract: numbers on hard drives at banks—and the poor may not. These "rich" claim they own land, and the "poor" are often denied the right to make that same claim. A primary purpose of the police is to enforce the delusions of those with lots of pieces of green paper. Those without the green papers generally buy into these delusions almost as quickly and completely as those with. These delusions carry with them extreme consequences in the real world.

PREMISE THIRTEEN: Those in power rule by force, and the sooner we break ourselves of illusions to the contrary, the sooner we can at least begin to make reasonable decisions about whether, when, and how we are going to resist.

PREMISE FOURTEEN: From birth on—and probably from conception, but I'm not sure how I'd make the case—we are individually and collectively enculturated to hate life, hate the natural world, hate the wild, hate wild animals, hate women, hate children, hate our bodies, hate and fear our emotions, hate ourselves. If we did not hate the world, we could not allow it to be destroyed before our eyes. If we did not hate ourselves, we could not allow our homes—and our bodies—to be poisoned.

PREMISE FIFTEEN: Love does not imply pacifism.

PREMISE SIXTEEN: The material world is primary. This does not mean that the spirit does not exist, nor that the material world is all there is. It means that spirit mixes with flesh. It means also that real world actions have real world consequences. It means we cannot rely on Jesus, Santa Claus, the Great Mother, or even the Easter Bunny to get us out of this mess. It means this mess really is a mess, and not just the movement of God's eyebrows. It means we have to face this mess ourselves. It means that for the time we are here on Earth—whether or not we end up somewhere else after we die, and whether we are condemned or privileged to live here—the Earth is the point. It is primary. It is our home. It is everything. It is silly to think or act or be as though this world is not real and primary. It is silly and pathetic to not live our lives as though our lives are real.

PREMISE SEVENTEEN: It is a mistake (or more likely, denial) to base our decisions on whether actions arising from them will or won't frighten fence-sitters, or the mass of Americans.

PREMISE EIGHTEEN: Our current sense of self is no more sustainable than our current use of energy or technology.

PREMISE NINETEEN: The culture's problem lies above all in the belief that controlling and abusing the natural world is justifiable.

PREMISE TWENTY: Within this culture, economics—not community well-being, not morals, not ethics, not justice, not life itself—drives social decisions.

Modification of Premise Twenty: Social decisions are determined primarily (and often exclusively) on the basis of whether these decisions will increase the monetary fortunes of the decision-makers and those they serve.

Re-modification of Premise Twenty: Social decisions are determined primarily (and often exclusively) on the basis of whether these decisions will increase the power of the decision-makers and those they serve.

Re-modification of Premise Twenty: Social decisions are founded primarily (and often exclusively) on the almost entirely unexamined belief that the decision-makers and those they serve are entitled to magnify their power and/or financial fortunes at the expense of those below.

Re-modification of Premise Twenty: If you dig to the heart of it—if there is any heart left—you will find that social decisions are determined primarily on the basis of how well these decisions serve the ends of controlling or destroying wild nature.

For Tecumseh

We have spent too much time in thinking, supposing that if we weigh in advance the possibilities of any action, it will happen automatically. We have learnt, rather too late, that action comes not from thought, but from a readiness for responsibility.

Dietrich Bonhoeffer, written while in prison for resisting the Nazis[1]

Cowardice asks the question, "Is it safe?" Expediency asks the question, "Is it politic?" And Vanity comes along and asks the question, "Is it popular?" But Conscience asks the question, "Is it right?" And there comes a time when one must take a position that is neither safe, nor politic, nor popular, but he must do it because Conscience tells him it is right.

Martin Luther King, Jr.

APOCALYPSE

When a white man kills an Indian in a fair fight it is called
honorable, but when an Indian kills a white man in a fair fight
it is called murder.[2] When a white army battles Indians and
wins it is called a great victory, but if they lose it is called a
massacre and bigger armies are raised. If the Indian flees
before the advance of such armies, when he tries to return he
finds that white men are living where he lived. If he tries to
fight off such armies, he is killed and the land is taken anyway. When an Indian is killed, it is a great loss which leaves a
gap in our people and a sorrow in our heart; when a white is
killed three or four others step up to take his place and there
is no end to it. The white man seeks to conquer nature, to
bend it to his will and to use it wastefully until it is all gone and
then he simply moves on, leaving the waste behind him and
looking for new places to take. The whole white race is a monster who is always hungry and what he eats is land.

Chiksika[3]

AS A LONGTIME GRASSROOTS ENVIRONMENTAL ACTIVIST, AND AS a creature living in the thrashing endgame of civilization, I am intimately acquainted with the landscape of loss, and have grown accustomed to carrying the daily weight of despair. I have walked clearcuts that wrap around mountains, drop into valleys, then climb ridges to fragment watershed after watershed, and I've sat silent near empty streams that two generations ago were "lashed into whiteness" by uncountable salmon coming home to spawn and die.

A few years ago I began to feel pretty apocalyptic. But I hesitated to use that word, in part because of those drawings I've seen of crazy penitents carrying "The End is Near" signs, and in part because of the power of the word itself. Apocalypse. I didn't want to use it lightly.

But then a friend and fellow activist said, "What will it take for you to finally call it an apocalypse? The death of the salmon? Global warming? The ozone hole? The reduction of krill populations off Antarctica by 90 percent, the turning of the sea off San Diego into a dead zone, the same for the Gulf of Mexico? How about the end of the great coral reefs? The extirpation of two hundred species per day? Four hundred? Six hundred? Give me a specific threshold, Derrick, a specific point at which you'll finally use that word."

❆ ❆ ❆

Do you believe that our culture will undergo a voluntary transformation to a sane and sustainable way of living?

For the last several years I've taken to asking people this question, at talks and rallies, in libraries, on buses, in airplanes, at the grocery store, the hardware store. Everywhere. The answers range from emphatic *nos* to laughter. No one answers in the affirmative. One fellow at one talk did raise his hand, and when everyone looked at him, he dropped his hand, then said, sheepishly, "Oh, voluntary? No, of course not." My next question: how will this understanding—that this culture will not voluntarily stop destroying the natural world, eliminating indigenous cultures, exploiting the poor, and killing those who resist—shift our strategy and tactics? The answer? Nobody knows, because we never talk about it: we're too busy pretending the culture will undergo a magical transformation.

This book is about that shift in strategy, and in tactics.

❨ ❨ ❨

I just got home from talking to a new friend, another longtime activist. She told me of a campaign she participated in a few years ago to try to stop the government and transnational timber corporations from spraying Agent Orange, a potent defoliant and teratogen, in the forests of Oregon. Whenever activists learned a hillside was going to be sprayed, they assembled there, hoping their presence would stop the poisoning. But each time, like clockwork, helicopters appeared, and each time, like clockwork, helicopters dumped loads of Agent Orange onto the hillside and onto protesting activists. The campaign did not succeed.

"But," she said to me, "I'll tell you what did. A bunch of Vietnam vets lived in those hills, and they sent messages to the Bureau of Land Management and to Weyerhaeuser, Boise Cascade, and the other timber companies saying, 'We know the names of your helicopter pilots, and we know their addresses.'"

I waited for her to finish.

"You know what happened next?" she asked.

"I think I do," I responded.

"Exactly," she said. "The spraying stopped."

FIVE STORIES

Nations and peoples are largely the stories they feed themselves. If they tell themselves stories that are lies, they will suffer the future consequences of those lies. If they tell themselves
stories that face their own truths, they will free their histories
for future flowerings.

Ben Okri [4]

Unquestioned beliefs are the real authorities of a culture.
Therefore, if an individual can express what is undeniably real
to him without invoking any authority beyond his own experience, he is transcending the belief systems of his culture.

Robert Combs [5]

LAST TUESDAY THE TWIN TOWERS OF THE WORLD TRADE CENTER collapsed, killing thousands of people. That same day a portion of the Pentagon also collapsed, killing more than a hundred. In addition, a jet airliner crashed in Pennsylvania.

Let's tell this story again: Last Tuesday nineteen Arab terrorists unleashed their fanaticism on the United States by hijacking four planes, each containing scores of innocent victims. These terrorists, who do not value life the way we Americans do, slammed two of the planes into the World Trade Center and a third into the Pentagon. Courageous men and women in the fourth plane wrestled with their attackers and drove the plane into the ground, sacrificing themselves rather than allowing the killers to attack the headquarters of the CIA or any other crucial target. Our government will find and punish those who masterminded the attack. This will be difficult because, as President George W. Bush said, "This enemy hides in shadows and has no regard for human life. This is an enemy that [*sic*] preys on innocent and unsuspecting people and then runs for cover."[6] When we find them, we must kill them. This killing will not be easy on us. We must steel ourselves against the possibility—inevitability—that we may be forced to kill even those whose guilt we cannot finally establish. As former Secretary of State Lawrence Eagleburger said, "There is only one way to begin to deal with people like this, and that is you have to kill some of them even if they are not immediately directly involved in this thing."[7] Many politicians and journalists have spoken yet more directly. "This is no time," syndicated columnist (and bestselling author) Ann Coulter wrote, "to be precious about locating the exact individuals directly involved in this particular terrorist attack. . . . We should invade their countries, kill their leaders and convert them to Christianity."[8]

Here is another version of the same story: Last Tuesday nineteen young men made their mothers proud. They gave their lives to strike a blow against the United States, the greatest terrorist state ever to exist. This blow was struck in response to U.S. support for the dispossession and murder of Palestinians, to the forced installation of pro-Western governments in Saudi Arabia, Egypt, and many other countries, to the hundreds of thousands of Iraqi civilians killed by U.S. bombs, to the nine thousand babies who die every month as a direct result of U.S. sanctions on Iraq, and to the irradiation of Iraq with depleted uranium. More broadly, it was a response to the CIA-backed murder of 650,000 people in Indonesia, and to the hundreds of thousands murdered by U.S.-backed death squads in Central and South America. To the four million civilians killed in North Korea. To the theft of American Indian land and the killings of millions of Indians. To the imposition of business-friendly dictators like Mobutu Sese Seko, Augusto Pinochet, the Shah, Suharto, or Ferdinand Marcos. (As Secretary of Defense William Cohen said to a group of *Fortune* 500 leaders, "Business follows the flag. . . . We provide the security. You provide the investment."[9]) It was a response to an American foreign policy driven by the needs of industrial production—as manifested through the unnatural logic of the bottom line—not life. This was a blow delivered not only against the United States but against a murderous global economy—a half a million babies die each year as a direct result of so-called debt repayment[10]—that is a continuation of the same old colonialism under which those who exploit get rich and the rest get killed. The poor of the world would all be better off if the global economy—run by transnational corporations backed by the military power of the United States—disappeared tomorrow. When a country, an economy, and a culture are all based on the systematic violent exploitation of humans and nonhumans the world over, it should come as no surprise when at long last someone fights back. We can only hope and pray that the organizations behind this have the resources and stamina to keep at it until they bring down the global economy.

Here's another version: Last Tuesday was a tragedy for the planet, and at least a temporary victory for rage and hatred. But let us not seek to pinpoint blame, nor meet negativity with negativity. The terrorists were wrong to act as they did, but to meet their violence with our own would be just as wrong. Violence never solves anything. As Gandhi said, "An eye for an eye only ends up making the whole world blind." Even if you believe the United States and the global economy are fundamentally destructive, you cannot use the master's tools to dismantle the master's house. The most important thing any of us can do is eradicate the anger that lies within our own hearts, that wounds the world as surely as do all the hijackers in Arabia and all the bombs in the United States. If I wish to experience peace, I must provide peace for another. If I wish to heal another's anger, I must first heal my own. I know that all of the terrorists of the world are, beneath it all, searching for love. It is the task of those of us who've been granted this understanding to teach them this, simply by loving them, and then by loving them more. For love is the only cure. I deplore violence, and if the United States goes to war, I will oppose that war in whatever peaceful ways I can, with love in my heart. And I will love and support our brave troops.

❨ ❨ ❨

Or how about this: It should be clear to everyone by now—even those with a vested interest in ignorance—that industrial civilization is killing the planet. It's causing unprecedented human privation and suffering. Unless it's stopped, or somehow stops itself, or most likely collapses under the weight of its inherent ecological and human destructiveness, it will kill every living being on earth. It should be equally clear that the efforts of those of us working to stop or slow the destruction are insufficient. We file our lawsuits; write our books; send letters to editors, representatives, CEOs; carry signs and placards; restore natural communities; and not only do we not stop or slow the destruction, but it actually continues to accelerate. Rates of deforestation continue to rise, rates of extinction do the same, global warming proceeds apace, the rich get richer, the poor starve to death, and the world burns.

At the same time that we so often find ourselves seemingly helpless in facing down civilization's speeding train of destruction, we find that there's a huge gap in our discourse. We speak much of the tactics of civil disobedience, much of the spiritual politics of cultural transformation, much of the sciences of biotechnology, toxicology, biology, and psychology. We talk of law. We also talk often of despair, frustration, and sorrow.

Yet our discourse remains firmly embedded in that which is sanctioned by the very overarching structures that govern the destruction in the first place. We do not often speak of the tactics of sabotage, and even less do we speak of violence. We avoid them, or pretend they should not be allowed to enter even the realm of possibility, or that they simply do not exist, like disinherited relatives who show up at a family reunion.

Several years ago I interviewed a long-term and well-respected Gandhian activist. I asked him, "What if those in power are murderous? What if they're not willing to listen to reason at all? Should we continue to approach them nonviolently?"

He responded, reasonably enough, "When a house is on fire, and has gone far beyond the point where you can do anything about it, all you can do is bring lots of water to try to stop its spread. But you can't save the house. Nonviolence is a precautionary principle. Before the house is on fire you have to make sure you have a fire hydrant, clearly marked escape routes, emergency exits. The same is true in society. You educate your children in nonviolence. You educate your media in nonviolence. And when someone has a grievance, you don't ignore or suppress it, but you listen to that person, and ask, 'What is your concern?' You say, 'Let's sit down and solve it.'"

I agreed with what he said, so far as it went, but that didn't stop me from understanding that he'd sidestepped the question.

Before I could bring him back, he continued, "Say a father beats his children. Once he has already reached that stage, you have to say, 'What kind of a childhood did he have? How did he not learn the skills of coping with adverse situations in a calm, compassionate, composed way?'"

This Gandhian's compassion, I thought, was entirely misplaced. Where was his compassion for the children being beaten? I responded that I believed the first question we need to ask is how we can get the children to a safe place. Once safety has been established, by any means possible, I said, and once the emotional needs of the children are being met, only then do we have the luxury of asking about the father's emotional needs, and his history.

What happened next is really the point of this story. I asked this devoted adherent of nonviolence if in his mind it would ever be acceptable to commit an act of violence were it determined to be the only way to save the children. His answer was revealing, and symbolizes the hole in our discourse: he changed the subject.

After I transcribed and edited the interview, I sent it to him with a new question inserted, attempting once again to pin him down. What did he do this time? He deleted my question.

Too often this is the response of all of us when faced with this most difficult of questions: when is violence an appropriate means to stop injustice? But with the world dying—or rather being killed—we no longer have the luxury to change the subject or delete the question. It's a question that won't go away.

❪ ❪ ❪

I had two reasons for telling the four versions of the World Trade Center bombing.[11] The first was to point out that all writers are propagandists. Writers who claim differently, or who otherwise do not understand this, have succumbed to the extremely dangerous propaganda that narrative can be divorced from value. This is not true. All descriptions carry with them weighty presumptions of value. This is as true for wordless descriptions such as mathematical formulae—which value the quantifiable and ignore everything else—as it is for the descriptions I gave above. The first version, by giving only current actions—"the twin towers of the World Trade Center collapsed, killing hundreds of people"—devalues (by their absence) cause and context. Why did the towers collapse? What were the events surrounding the collapse? This neat excision of both cause and context is the standard now in journalism, where, for example, we often hear of devastating mudslides in the colonies killing thousands of people who, seemingly unaccountably, were foolish enough to build villages beneath unstable slopes; toward the end of these articles we sometimes see sidelong references to "illegal logging," but nowhere do we see mention of Weyerhaeuser, Hyundai, Daishowa, or other transnational timber companies, which cut the steep slopes over the objections—and sometimes dead bodies—of the villagers. Or we may read of the rebel group UNITA slaughtering civilians in Angola, with no mention of two decades of U.S. financial and moral support for this group. So far as the bombing of the World Trade Center, despite yard after column yard of ink and paper devoted to the attacks, analyses of potential reasons for hatred of the United States rarely venture beyond, "They're fanatics," or "They're jealous of our lifestyle," or even, and I'm not making this up, "They want our resources."

The second, patriotic version carries with it the inherent presumption that the United States did nothing to deserve or even lead to the attack: if the United States kills citizens of other countries, and survivors of that violence respond by killing United States citizens—even if the casualty counts of the counter-strikes are by any realistic assessment much smaller—the United States is then justified in killing yet more citizens of those other countries. As Thomas Jefferson put it, "In war, they will kill some of us; we shall destroy all of them."[12] Another presumption of the patriotic version is that the lives of people killed by foreign terrorists are more worthy of notice, vengeance, and future protection than those killed, for example, by unsafe working conditions, or by the turning of our total environment into a carcinogenic stew. Let's say that three thousand people died in those attacks. In no way do I mean to demean these lives once presumably full of love, friendship, drama, sorrow, and so on, but more Americans die each month from toxins and other workplace hazards, and more Americans die each *week* from *preventable* cancers that are for the most part direct results of the activities of large corporations, and certainly the results of the industrial economy.[13] The lack of outrage over these deaths commensurate to the outrage expressed over the deaths in the 9/11 bombings reveals much—if we care to reflect on it—about the values and presumptions of our culture.

The third version, from the perspective of the bombers or their supporters, presumes that there are conditions under which it is morally acceptable to kill noncombatants, to kill those who themselves have done you no direct harm.[14] It also presumes that to kill people within the United States (by bombs, of course, since carcinogens spewed in the service of production evidently do not count as causes of atrocity) may cause those who run the governments of the United States—both nominal, that is, political, and de facto, that is, economic—to re-think their position of violently dominating the rest of the planet.

The fourth version presumes it is possible to halt or significantly slow violence through nonviolent means.

❨ ❨ ❨

Here's a question I've been asking: can the same action seem immoral from one perspective and moral from another? From the perspective, for example, of salmon or other creatures, including humans, whose lives depend on free-flowing rivers, dams are murderous and immoral. To remove dams would, from this perspective, be extremely moral. Of course the most moral thing would have been to not build these or any other large dams in the first place. But they're built, and they continue to be built the world over, to the consistent short-term fiscal benefit of huge corporations and over the determined yet usually unsuccessful resistance of the poor. The second most moral thing would be to let the water out slowly, and then breach the dams more or less gently, taking the survival needs (as opposed to the more abstract requirements of the dominant economic system) of all humans and nonhumans into account as we let rivers once again run free. But the dams are there, they're killing rivers—because of dams in the Northwest, for example, salmon and sturgeon are fast disappearing, and in the Southwest, I'm not sure what more I need to say except that the Colorado River no longer even reaches the ocean—and the current political, economic, and social systems have shown themselves to be consistently unresponsive to and irredeemably detrimental to human and nonhuman needs. Faced with a choice between healthy functioning natural communities on one hand and profits on the other (or behind those profits, and motivating them, the centralization of power) of course those in power always choose the latter. What, then, becomes the moral thing to do? Do we stand by and watch the last of the salmon die? Do we write letters and file lawsuits that we know in our hearts will ultimately not make much difference? Do we take out the dams ourselves?

Here's another question: What would the rivers themselves want?

I'm aiming at a far bigger and more profound target than the nearly twelve million cubic yards of cement that went into the Grand Coulee Dam. I want in this book to examine the morality and feasibility of intentionally taking down not just dams but all of civilization. I aim to examine this as unflinchingly and honestly as I can, even, or especially, at the risk of examining topics normally considered off-limits to discourse.

I am not the first to make the case that the industrial economy, indeed, civilization (which underpins and gives rise to it), is incompatible with human and nonhuman freedoms, and in fact with human and nonhuman life.[15] If you accept that the industrial economy—and beneath it, civilization—is destroying the planet and creating unprecedented human suffering among the poor (and if you don't accept this, go ahead and put this book down, back away slowly, turn on the television, and take some more *soma*: the drug should kick in soon enough, your agitation will disappear, you'll forget everything I've written, and then everything will be perfect again, just like the voices from the television tell you over and over), then it becomes clear that the best thing that can happen, from the perspective of essentially all nonhumans as well as the vast majority of humans, is for the industrial economy (and civilization) to go away or, in the shorter run, for it to be slowed as much as humanly possible during the time we await its final collapse. But here's the problem: this slowing of the industrial economy will inconvenience many of those who benefit from it, including nearly everyone in the United States. Many of those who will be inconvenienced identify so much more with their role as participants in the industrial economy than they do with being human that they may very well consider this inconvenience to be a threat to their very lives. Those people will not allow themselves to be inconvenienced without a fight. What, then, is the right thing to do? Is it possible to talk about fundamental social change without asking ourselves the question the Gandhian refused to answer?

CIVILIZATION

Civilization originates in conquest abroad and repression at home.

Stanley Diamond[16]

IF I'M GOING TO CONTEMPLATE THE COLLAPSE OF CIVILIZATION, I need to define what it is. I looked in some dictionaries. *Webster's* calls civilization "a high stage of social and cultural development."[17] *The Oxford English Dictionary* describes it as "a developed or advanced state of human society."[18] All the other dictionaries I checked were similarly laudatory. These definitions, no matter how broadly shared, helped me not in the slightest. They seemed to me hopelessly sloppy. After reading them, I still had no idea what the hell a civilization is: define *high*, *developed*, or *advanced*, please. The definitions, it struck me, are also extremely self-serving: can you imagine writers of dictionaries willingly classifying themselves as members of "a low, undeveloped, or backward state of human society"?

I suddenly remembered that all writers, including writers of dictionaries, are propagandists, and I realized that these definitions are, in fact, bite-sized chunks of propaganda, concise articulations of the arrogance that has led those who believe they are living in the most advanced—and best—culture to attempt to impose by force this way of being on all others.

I would define a civilization much more precisely, and I believe more usefully, as a culture—that is, a complex of stories, institutions, and artifacts—that both leads to and emerges from the growth of cities (*civilization*, see *civil*: from *civis*, meaning *citizen*, from Latin *civitatis*, meaning *city-state*), with cities being defined—so as to distinguish them from camps, villages, and so on—as people living more or less permanently in one place in densities high enough to require the routine importation of food and other necessities of life. Thus a Tolowa village five hundred years ago where I live in Tu'nes (*meadow long* in the Tolowa tongue), now called Crescent City, California, would not have been a city, since the Tolowa ate native salmon, clams, deer, huckleberries, and so on, and had no need to bring in food from outside. Thus, under my definition, the Tolowa, because their way of living was not characterized by the growth of city-states, would not have been civilized. On the other hand, the Aztecs were. Their social structure led inevitably to great city-states like Iztapalapa and Tenochtitlán, the latter of which was, when Europeans first encountered it, far larger than any city in Europe, with a population five times that of London or Seville.[19] Shortly before razing Tenochtitlán and slaughtering or enslaving its inhabitants, the explorer and conquistador Hernando Cortés remarked that it was easily the most beautiful city on earth.[20] Beautiful or not, Tenochtitlán required, as do all cities, the (often forced) importation of food and other resources. The story of any civilization is the story of the rise of city-states, which means it is the story of the funneling of resources toward these centers (in order to sustain them and cause them to grow), which means it is the story of an increasing region of unsustainability surrounded by an increasingly exploited countryside.

German Reichskanzler Paul von Hindenburg described the relationship perfectly: "Without colonies no security regarding the acquisition of raw materials, without raw materials no industry, without industry no adequate standard of living and wealth. Therefore, Germans, do we need colonies."[21]

Of course someone already *lives* in the colonies, although that is evidently not of any importance.

But there's more. Cities don't arise in political, social, and ecological vacuums. Lewis Mumford, in the second book of his extraordinary two-volume *Myth of the Machine*, uses the term *civilization* "to denote the group of institutions that first took form under kingship. Its chief features, constant in varying proportions throughout history, are the centralization of political power, the separation of classes, the lifetime division of labor, the mechanization of production, the magnification of military power, the economic exploitation of the weak, and the universal introduction of slavery and forced labor for both industrial and military purposes."[22] (The anthropologist and philosopher Stanley Diamond put this a bit more succinctly when he noted, "Civilization originates in conquest abroad and repression at home."[23]) These attributes, which inhere not just in this culture but in all civilizations, make civilization sound pretty bad. But, according to Mumford, civilization has another, more benign face as well. He continues, "These institutions would have completely discredited both the primal myth of divine kingship and the derivative myth of the machine had they not been accompanied by another set of collective traits that deservedly claim admiration: the invention and keeping of the written record, the growth of visual and musical arts, the effort to widen the circle of communication and economic intercourse far beyond the range of any local community: ultimately the purpose to make available to all men [sic] the discoveries and inventions and creations, the works of art and thought, the values and purposes that any single group has discovered."[24]

Much as I admire and have been influenced by Mumford's work, I fear that when he began discussing civilization's admirable face he fell under the spell of the same propaganda promulgated by the lexicographers whose work I consulted: that this culture really is "advanced," or "higher." But if we dig beneath this second, smiling mask of civilization—the belief that civilization's visual or musical arts, for example, are more developed than those of noncivilized peoples—we find a mirror image of civilization's other face, that of power. For example, it wouldn't be the whole truth to say that visual and musical arts have simply *grown* or become more highly advanced under this system; it's more true that they have long ago succumbed to the same division of labor that characterizes this culture's economics and politics. Where among traditional indigenous people—the "uncivilized"—songs are sung by everyone as a means to bond members of the community and celebrate each other and their land-base, within civilization songs are written and performed by experts, those with "talent," those whose lives are devoted to the production of these arts. There's no reason for me to listen to my neighbor sing (probably off-key) some amateurish song of her own invention when I can pop in a CD of Beethoven, Mozart, or Lou Reed (okay, so Lou Reed sings off-key, too, but I like it). I'm not certain I'd characterize the conversion of human beings from participants in the ongoing creation of communal arts to more passive consumers of artistic products manufactured by distant experts—even if these distant experts are *really* talented—as a good thing.

I could make a similar argument about writing, but Stanley Diamond beat me to it: "Writing was one of the original mysteries of civilization, and it reduced the complexities of experience to the written word. Moreover, writing provides the ruling classes with an ideological instrument of incalculable power. The word of God becomes an invincible law, mediated by priests; therefore, respond the Iroquois, confronting the European: 'Scripture was written by the Devil.' With the advent of writing, symbols became explicit; they lost a certain richness. Man's word was no longer an endless exploration of reality, but a sign that could be used against him. . . . For writing splits consciousness in two ways—it becomes more authoritative than talking, thus degrading the meaning of speech and eroding oral tradition; and it makes it possible to use words for the political manipulation and control of others. Written signs supplant memory; an official, fixed, and permanent version of events can be made. If it is written, in early civilizations [and I would suggest, now], it is bound to be true."[25]

I have two problems, also, with Mumford's claim that the widening of communication and economic intercourse under civilization benefits people as a whole. The first is that it presumes that uncivilized people do not communicate or participate in economic transactions beyond their local communities. Many do. Shells from the Northwest Coast found their way into the hands of Plains Indians, and buffalo robes often ended up on the coast. (And let's not even mention noncivilized people communicating with their nonhuman neighbors, something rarely practiced by the civilized: talk about restricting yourself to your own community!) In any case, I'm not certain that the ability to send emails back and forth to Spain or to watch television programs beamed out of Los Angeles makes my life particularly richer. It's far more important, useful, and enriching, I think, to get to know my neighbors. I'm frequently amazed to find myself sitting in a room full of fellow human beings, all of us staring at a box watching and listening to a story concocted and enacted by people far away. I have friends who know Seinfeld's neighbors better than their own. I, too, can get lost in valuing the unreality of the distant over that which surrounds me every day. I have to confess I can navigate the mazes of the computer game Doom 2: Hell on Earth far better than I can find my way along the labyrinthine game trails beneath the trees outside my window, and I understand the intricacies of Microsoft Word far better than I do the complex dance of rain, sun, predators, prey, scavengers, plants, and soil in the creek twenty yards away. The other night, I wrote till late, and finally turned off my computer to step outside and say goodnight to the dogs. I realized, then, that the wind was blowing hard through the tops of the redwood trees, and the trees were sighing and whispering. Branches were clashing, and in the distance I heard them cracking. Until that moment I had not realized such a symphony was taking place so near, much less had I gone out to participate in it, to feel the wind blow my hair and to feel the tossed rain hit me in the face. All of the sounds of the night had been drowned out by the monotone whine of my computer's fan. Just yesterday I saw a pair of hooded mergansers playing on the pond outside my bedroom. Then last night I saw a television program in which yet another lion chased yet another zebra. Which of those two scenes makes me richer? This perceived widening of communication is just another replication of the problem of the visual and musical arts, because given the impulse for centralized control that motivates civilization, widening communication in this case really means reducing us from active participants in our own lives and in the lives of those around us to consumers sucking words and images from some distant sugar tit.

I have another problem with Mumford's statement. In claiming that the widening of communication and economic intercourse are admirable, he seems to have forgotten—and this is strange, considering the sophistication of the rest of his analysis—that this widening can only be universally beneficial when all parties act voluntarily and under circumstances of relatively equivalent power. I'd hate to have to make the case, for example, that the people of Africa—perhaps 100 million of whom died because of the slave trade, and many more of whom find themselves dispossessed and/or impoverished today—have benefited from their "economic intercourse" with Europeans. The same can be said for Aborigines, Indians, the people of pre-colonial India, and so on.

I want to re-examine one other thing Mumford wrote, in part because he makes an argument for civilization I've seen replicated so many times elsewhere, and that actually leads, I think, to some of the very serious problems we face today. He concluded the section I quoted above, and I reproduce it here just so you don't have to flip back a couple of pages: "ultimately the purpose [is] to make available to all men [sic] the discoveries and inventions and creations, the works of art and thought, the values and purposes that any single group has discovered." But just as a widening of economic intercourse is only beneficial to everyone when all exchanges are voluntary, so, too, the imposition of one group's values and purposes onto another, or its appropriation of the other's discoveries, can lead only to the exploitation and diminution of the latter in favor of the former. That this "exchange" helps all was commonly argued by early Europeans in America, as when Captain John Chester wrote that the Indians were to gain "the knowledge of our faith," while the Europeans would harvest "such ritches as the country hath."[26] It was argued as well by American slave owners in the nineteenth century: philosopher George Fitzhugh stated that "slavery educates, refines, and moralizes the masses by bringing them into continual intercourse with masters of superior minds, information, and morality."[27] And it's just as commonly argued today by those who would teach the virtues of blue jeans, Big Macs™, Coca-Cola™, Capitalism™, and Jesus Christ™ to the world's poor in exchange for dispossessing them of their landbases and forcing them to work in sweatshops.

Another problem is that Mumford's statement reinforces a mindset that leads inevitably to unsustainability, because it presumes that discoveries, inventions, creations, works of art and thought, and values and purposes are transposable over space, that is, that they are separable from both the human context and landbase that created them. Mumford's statement unintentionally reveals perhaps more than anything else the power of the stories that hold us in thrall to the machine, as he put it, that is civilization: even in brilliantly dissecting the myth of this machine, Mumford fell back into that very same myth by seeming to implicitly accept the notion that ideas or works of art or discoveries are like tools in a toolbox, and can be meaningfully and without negative consequence used out of their original context: thoughts, ideas, and art as tools rather than as tapestries inextricably woven from and into a community of human and nonhuman neighbors. But discoveries, works of thought, and purposes that may work very well in the Great Plains may be harmful in the Pacific Northwest, and even moreso in Hawai'i. To believe that this potential transposition is positive is the same old substitution of what is distant for what is near: if I really want to know how to live in Tu'nes, I should pay attention to Tu'nes.

There's another problem, though, that trumps all of these others. It has to do with a characteristic of this civilization unshared even by other civilizations. It is the deeply and most-often-invisibly held beliefs that there is really only one way to live, and that we are the one-and-only possessors of that way. It becomes our job then to propagate this way, by force when necessary, until there are no other ways to be. Far from being a loss, the eradication of these other ways to be, these other cultures, is instead an actual gain, since Western Civilization is the only way worth being anyway: we're doing ourselves a favor by getting rid of not only obstacles blocking our access to resources but reminders that other ways to be exist, allowing our fantasy to sidle that much closer to reality; and we're doing the heathens a favor when we raise them from their degraded state to join the highest, most advanced, most developed state of society. If they don't want to join us, simple: we kill them. Another way to say all of this is that something grimly alchemical happens when we combine the arrogance of the dictionary definition, which holds this civilization superior to all other cultural forms; hypermilitarism, which allows civilization to expand and exploit essentially at will; and a belief, held even by such powerful and relentless critics of civilization as Lewis Mumford, in the desirability of cosmopolitanism, that is, the transposability of discoveries, values, modes of thought, and so on over time and space. The twentieth-century name for that grimly alchemical transmutation is genocide: the eradication of cultural difference, its sacrifice on the altar of the one true way, on the altar of the centralization of perception, the conversion of a multiplicity of moralities all dependent on location and circumstance to one morality based on the precepts of the ever-expanding machine, the surrender of individual perception (as through writing and through the conversion of that and other arts to consumables) to predigested perceptions, ideas, and values imposed by external authorities who with all their hearts—or what's left of them—believe in, and who benefit by, the centralization of power. Ultimately, then, the story of this civilization is the story of the reduction of the world's tapestry of stories to only one story, the best story, the real story, the most advanced story, the most developed story, the story of the power and the glory that is Western Civilization.

CLEAN WATER

A sense of *place* is critical. For people who live *with* the land, the land becomes the center of their universe. It's a marriage. We are in a symbiotic relationship with the land where we live, and the notion that this relationship should or even can be transcended is central to many of our problems, and to many of the problems we've created for others. Land is something to be respected, and this respect for land makes respect for self and others possible.

Richard Drinnon[28]

OR MAYBE I SHOULD RESTATE THAT. THE STORY OF WESTERN CIVILIZATION is not the story of that reduction, but of its *attempted* reduction. Certainly it has already succeeded in eliminating many of the stories—the stories of great auks, passenger pigeons, many of the indigenous of Europe, North America, Africa, and elsewhere, the great herds of bison, the stories of free-flowing rivers—but it will never succeed in reducing all stories to one. The world won't let it. And, to the very best of my abilities, neither will I.

(((

An action's morality—or at the very least its perceived morality—can shift depending not only on one's perspective, but also of course on circumstance. For one example of this, let's talk for a moment about sex.

I have a friend who was a virgin into his thirties, mainly because he was terrified of women, terrified of life, terrified of himself. One day he somehow got hooked up on a blind date—the first date of his life—with a woman, also in her thirties, who had one child. This woman, too, was frightened, but of something else. She was afraid of raising a child by herself, of growing old with no one at her side.

That first night they had sex, at her instigation. My friend, who had never before spent any real private time with *any* woman, was hooked. It felt good to have someone want him. She, in her desperation and loneliness, I thought, took advantage of his naïveté and fear to quickly reel him in.

That's how I saw it at the time. And though I felt protective of my friend, I didn't say anything because I didn't feel it was my place. I'm glad now that I didn't because I was wrong. Meeting her, and having sex with her, and entering into a relationship with her, was the best thing that ever happened to him. Their love became the centerpost of his life, and he is a far better and happier man for it. She, too, is happier than she would otherwise have been.

I knew someone else who, by the time he was thirty, had long lost count of his sexual partners. They had to number in the hundreds. The number really doesn't matter. His compulsion does. He knew only one way to relate to women, which was, to use his word, to "bone" them. I didn't know him very long, but in that short time he told me, or attempted to tell me before I'd leave the room, of his sexual encounters on the presidential yacht (he was a lobbyist, with access to the president, or at least his yacht), in elevators, in the bathroom of an air-plane, in the back seats of enough cars to make his own parade. His sexuality was objectifying and harmful. It was clear from his accounts—although equally clear was the fact that he did not allow himself to become consciously aware of this—that his sexual use of women hurt many of them: he kept asking me why I thought so many of these women insisted he never under any circumstances contact them again.

The point is simple: life—and morality—is far too complex to allow us to say that sex is either good or bad. Sometimes it's good, sometimes not. It is possi-ble for a sexual act to be profoundly moral and beautiful, and it is equally pos-sible for other circumstances, participants, motivations, to cause it to be just as profoundly immoral and/or ugly. Sometimes sexuality can be neither moral nor immoral, and carry no particular moral weight. Of course sex isn't the point. Remaining open to one's current experience is. Life is circumstantial. Morality is circumstantial. An action that may be moral to the point of obliga-tion in one circumstance could as easily be just as immoral in another. This is true of any action with moral implications.

None of this is to say that there are no moral absolutes. It's merely to say we have become confused as to what they are, how we can discern them, and hav-ing discerned them, how we can make sense of them and allow them to guide our lives.

❨ ❨ ❨

Years ago I got into an argument with a woman over whether rape is a bad thing. I said it was. She—and I need to say that she was dating a philosopher at the time, and has since regained her sanity—responded, "No, we can *say* that rape is a bad thing. But since humans assign all value"—and presumably both she and her philosopher boyfriend meant most especially those humans who have most fully internalized the messages of this culture, and who therefore receive its greatest social rewards—"humans can decide whether rape is good or bad. There is nothing inherently good or bad about it. It just is. Now, we can cer-tainly tell ourselves a series of stories that cause us to believe that rape is bad, that is, we can construct a set of narratives reinforcing the notion that rape is harmful, but we could just as easily construct a set of narratives that tell us quite the opposite."

There are two ways in which she was absolutely correct. The first has to do with the importance of stories in telling us how to live. If the stories you heard from birth on repeat to you in one way or another the messages that industrial civilization benefits human beings; that "civilized" people do not commit atrocities (they are, so to speak, civil); that violence is "barbaric," and that "barbarians" are violent; that someone who is violent is an "animal," a "brute"; that only the most successful at dominating survive; that nonhumans (and many humans) are here for us to use; that nonhumans (and many humans) have no desires of their own; that sorrow, anger, frustration, loneliness will somehow dissipate if only you buy something;[29] that the government of the United States (or Nazi Germany, or the Soviet Union, or Luxembourg, for that matter) has your best interests at heart; that working in the wage economy, i.e., having a job, is natural, normal, desirable, or necessary; that the world is a vale of tears and that you will go to a better place when you die; that the morality of violence or any other action is simple; that those in power are too strong—or perhaps they rule by divine right or its modern equivalent, historical inevitability— to be brought down; that we would all suffer if civilization were taken out; that there have been no other ways to live that have been more peaceful, sustainable, and just plain happy than civilization; that those in power have the right to destroy the planet, and that there is little or nothing we can do to stop them, then of course you will come to believe all of that. If, on the other hand, the stories you are told are different, you will grow to believe and act far differently.

The second way she was correct is that it's clearly possible to construct stories teaching us that rape is acceptable. The Bible certainly stands out as an example of this. Further, given that 25 percent of women in this culture are raped during their lifetimes, and another 19 percent have to fend off rape attempts,[30] it seems pretty obvious that a lot of men have learned well the lessons that women are objects to be used, and that men have the right to do whatever violence they would like to women. These stories are told to us by people like, to choose just one egregious example among an entire culture's worth, Brian De Palma, director of such films as *Dressed to Kill*, *Carrie*, and *The Untouchables*, who said, "I'm always attacked for having an erotic, sexist approach—chopping up women, putting women in peril. I'm making suspense movies! What else is going to happen to them?"[31] Even more to the point, he also said that "using women in situations where they are killed or sexually attacked" is simply a "genre convention . . . like using violins when people look at each other."[32]

Similarly, we can create a series of stories that cause us to believe it makes sense to deforest the planet, vacuum the oceans, impoverish the majority of humans. If the stories are good enough—effective enough at convincing us the stories are more important than physical reality—it will not only make sense to destroy the world, but we will feel good about it, and we will feel good about killing anyone who tries to stop us.

One of the problems with all of this is that not all narratives are equal. Imagine, to take a silly example, that someone told you story after story extolling the virtues of eating dog shit. You've been told these stories since you were a child. You believe them. You eat dog-shit hot dogs, dog-shit ice cream, General Tso's dog shit. Sooner or later, if you are exposed to some other foods, you might figure out that dog shit really doesn't taste that good.[33] Or if you cling too tightly to these stories you've been told about eating dog shit (or if your enculturation is so strong that dog shit actually does taste good to you), the diet might make you sick or kill you. To make the example a little less silly, substitute the word *pesticides* for *dog shit*. (Who was the genius who decided [for us] that it was a good idea to put poisons on our own food?). Or, for that matter, substitute *Big Mac*™, *Whopper*™, or *Coca Cola*™. Physical reality eventually trumps narrative. It has to. It just can take a long time. In the case of civilization, it has so far taken some six thousand years (considerably less, of course, for its victims).

It took me a couple of years to articulate a response to my friend. One afternoon I called her up. We went to dinner.

She said, "Well?"

"Water," I said.

"Water?"

"Water."

"That's it?" she asked.

"It's everything," I responded.

She didn't understand.

"Your basic point was that nothing is inherently good or bad . . ."

"Right." She nodded.

"And that the stories we tell ourselves determine not only whether we *perceive* something as good or bad, but whether it in fact is . . ."

"Yes," she said. "Because humans are the sole definers of value . . ."

We'd been through this before, and I'd been through this with so many other people, too. Once I had an office at a university next to that of a philosophy professor. I wandered in sometimes to chat, but was always quickly repulsed by his relentless strangeness and illogic. "Because humans are the sole definers of value, nothing in the world has any value unless we decide it does," he said time and again, as though by repeating his starting assumption enough times he would force me to accept it, just as the possibility of failing his class forced his students to do the same. I fled the room each time in disarray, but I've always wished I would have returned with a hammer. He would have asked about it, and I would have replied, "If I hit your thumb, you won't *decide* cognitively that getting hit by a hammer hurts. Not getting hit by a hammer has inherent value, no matter what you decide about it."

Unfortunately, this form of narcissism—that only humans (and more specifically some very special humans, and even more specifically the disembodied thoughts of these very special humans) matter—is central to this culture. It pervades everything from this culture's religion to its economics to its philosophy, literature, medicine, politics, and so on. And it certainly pervades our relationships with nonhuman members of the natural world. If it did not, we could not cause clearcuts nor construct dams. I once read a book on zoos and wildlife in which the authors asked why wildlife should be preserved, and then answered their own question in a way that makes this arrogance and stupidity especially clear: "Our answer is that the human world would be impoverished, for animals are preserved solely for human benefit, because human beings have decided they want them to exist for human pleasure. The notion that they are preserved for *their sakes* is a peculiar one, for it implies that animals might wish a certain condition to endure. It is, however, nonsensical for humans to imagine that animals might want to continue the existence of their species."[34]

I told my friend this story.

"Were they serious, or ironic?" she asked.

"Dead fucking serious."

She replied, "They're full of shit. The arguments are unfounded."

I raised my eyebrows.

She said, "I'm not so hard-core as I used to be." She'd long-since dumped the philosopher, and started making sense again. "If the stories we live by are going to mean anything, they have to be grounded, anchored. We have to have a reference point we can rely on."

I said, "I can name for you something that is good, no matter what stories we tell ourselves."

"And it is . . ."

I held up my glass. "Drinkable quantities of clean water."

"I don't understand."

"Drinkable quantities of clean water are unqualifiedly a good thing, no matter the stories we tell ourselves."

She got it. She smiled before saying, "And breathable clean air."

We both nodded.

She continued, "Without them you die."

"Exactly," I said. "Without them, everyone dies."

Now she was excited. "That's the anchor," she said. "We can build an entire morality from there."

It was my turn to get excited. "Exactly," I said again.

We spent the rest of the evening sitting at the restaurant discussing—fleshing out—what an embodied morality would look like, feel like, be. If the foundation for my morality consists not of commandments from a God whose home is not primarily of this Earth and whose adherents have committed uncountable atrocities, nor of laws created by those in political power to serve those in political power, nor even the perceived wisdom—the common law—of a culture that has led us to ecological apocalypse, but if instead the foundation consists of the knowledge that I am an animal who requires habitat—including but not limited to clean water, clean air, non-toxic food—what does my consequent morality suggest about the rightness or wrongness of, say, pesticide production? If I understand that as human animals we require healthy landbases for not only physical but emotional health, how will I perceive the morality of mass extinction? How does the understanding that humans and salmon thrived here together in Tu'nes for at least twelve thousand years affect my perception of the morality of the existence of dams, deforestation, or anything else that destroys this long-term symbiosis by destroying salmon?

Although we both enjoyed our talk, we each knew we were leaving something unsaid. Not until we were outside the restaurant, returning to our respective cars, did either one of us mention it. She said, "I understand the immorality of poisoning our bodies and toxifying landbases, and of course I *know* that rape is immoral, but how does the fact that we have bodies, the fact that we have needs, the fact that we are animals, *make* rape immoral?"

I took a deep breath. The answer was right there. I could see it, taste it. I almost had it. I opened my mouth to say it. But then it was gone. I lost it, almost had it again, then lost it entirely. My mind was fried from all the thinking and talking.

"It's late," she said. "We'll talk again soon."

"Soon," I said.[35]

CATASTROPHE

Modern man likes to pretend that his thinking is wide-awake. But this wide-awake thinking has led us into the mazes of a nightmare in which the torture chambers are endlessly repeated in the mirrors of reason. When we emerge, perhaps we will realize that we have been dreaming with our eyes open, and that the dreams of reason are intolerable. And then, perhaps, we will begin to dream once more with our eyes closed.

Octavio Paz[36]

IT IS CUSTOMARY WHEN WRITING TO HIDE ONE'S PRESUMPTIONS. The hope is that readers will flow along with the narrative and get swept up by the language until by the end they've reached roughly the same conclusions as the author, never realizing that oftentimes the unstated starting point was far more important to the conclusion than the arguments themselves. For example, you hear some talking head on television ask, "How are we going to best make the U.S. economy grow?" Premise one: We want the U.S. economy to grow. Premise two: We want the U.S. economy to exist. Premise three: Who the hell is *we*?

I'm going to try to not slide premises by you. I want to lay them out as clearly as I can, for you to accept or reject. Part of the reason I want to do this is that the questions I'm exploring regarding civilization are the most important questions we as a culture and as individuals have ever been forced to face. I don't want to cheat. I want to convince neither you nor me unfairly (nor, for that matter, do I want to convince either of us at all), but instead to help us both better understand what to do (or not do) and how to do it (or why not). This goal will be best served by as much transparency—and honesty—as I can muster.

Some of the assertions undergirding this book are self-evident, some I've shown elsewhere, some I will support here. Of course I cannot list every one of my premises, since many of them are hidden even from me, or far more fundamentally are inherent in English, or the written word (books, for example, presume a beginning, middle, and end). Nonetheless, I'll try my best.

The first premise I want to mention is so obvious I'm embarrassed to have to write it down, as silly in its way as having to state that clean air or clean water are good and necessary, and as self-evident as the polluted air we breathe and water we drink. But our capacity and propensity for self-delusion—indeed the *necessity* of self-delusion if we're to continue to propagate this culture—means I need to be explicit. The first premise is: *Civilization is not and can never be sustainable. This is especially true for industrial civilization.*

Years ago I was riding in a car with friend and fellow activist George Draffan. He has influenced my thinking as much as any other one person. It was a hot day in Spokane. Traffic was slow. A long line waited at a stoplight. I asked, "If you could live at any level of technology, what would it be?"

As well as being a friend and an activist, George can be a curmudgeon. He was in one of those moods. He said, "That's a stupid question. We can fantasize about living however we want, but the only sustainable level of technology is the Stone Age. What we have now is the merest blip—we're one of only six or seven generations who ever have to hear the awful sound of internal combustion engines (especially two-cycle)—and in time we'll return to the way humans have lived for most of their existence. Within a few hundred years at most. The only question will be what's left of the world when we get there."

He's right, of course. It doesn't take a rocket scientist to figure out that any social system based on the use of nonrenewable resources is by definition unsustainable: in fact it probably takes anyone *but* a rocket scientist to figure this one out. The hope of those who wish to perpetuate this culture is something called "resource substitution," whereby as one resource is depleted another is substituted for it (I suppose there is at least one hope more prevalent than this, which is that if we ignore the consequences of these actions they will not exist). Of course on a finite planet this merely puts off the inevitable, ignores the damage caused in the meantime, and begs the question of what will be left of life when the last substitution has been made. Question: When oil runs out, what resource will be substituted in order to keep the industrial economy running? Unstated premises: a) equally effective substitutes exist; b) we want to keep the industrial economy running; and c) keeping it running is worth more to us (or rather to those who make the decisions) than the human and nonhuman lives destroyed by the extraction, processing, and utilization of this resource.

Similarly, any culture based on the nonrenewable use of renewable resources is just as unsustainable: if fewer salmon return each year than the year before, sooner or later none will return. If fewer ancient forests stand each year than the year before, sooner or later none will stand. Once again, the substitution of other resources for depleted ones will, some say, save civilization for another day. But at most this merely holds off the inevitable while it further damages the planet. This is what we see, for example, in the collapse of fishery after fishery worldwide: having long-since fished out the more economically valuable fish, now even so-called trash fish are being extirpated, disappearing into civilization's literally insatiable maw.

Another way to put all of this is that any group of beings (human or nonhuman, plant or animal) who take more from their surroundings than they give back will, obviously, deplete their surroundings, after which they will either have to move, or their population will crash (which, by the way, is a one sentence disproof of the notion that competition drives natural selection: if you hyperexploit your surroundings you will deplete them and die; the only way to survive in the long run is to give back more than you take. Duh). This culture—Western Civilization—has been depleting its surroundings for six thousand years, beginning in the Middle East and expanding now to deplete the entire planet. Why else do you think this culture has to continually expand? And why else, coincident with this, do you think it has developed a rhetoric—a series of stories that teach us how to live—making plain not only the necessity but desirability and even morality of continual expansion—causing us to boldly go where no man has gone before—as a premise so fundamental as to become invisible? Cities, the defining feature of civilization, have always relied on taking resources from the surrounding countryside, meaning, first, that no city has ever been or ever will be sustainable on its own, and second, that in order to continue their ceaseless expansion cities must ceaselessly expand the areas they must ceaselessly hyperexploit. I'm sure you can see the problems this presents and the end point it must reach on a finite planet. If you cannot or will not see these problems, then I wish you the best of luck in your career in politics or business. Our collective studied-to-the-point-of-obsessive avoidance of acknowledging and acting on the surety of this end point is, especially given the consequences, more than passing strange.

Yet another way to say that this way of living is unsustainable is to point out that because ultimately the only real source of energy for the planet is the sun (the energy locked in oil, for example, having come from the sun long ago; and I'm excluding nuclear power from consideration here because only a fool would intentionally fabricate and/or refine materials that are deadly poisonous for tens or hundreds of thousands of years, especially to serve the frivolous, banal, and anti-life uses to which electricity is put: think retractable stadium roofs, supercolliders, and aluminum beer cans), any way of being that uses more energy than that currently coming from the sun will not last, because the non-current energy—stored in oil that could be burned, stored in trees that could be burned (stored, for that matter, in human bodies that could be burned)—will in time be used up. As we see.

I am more or less constantly amazed at the number of intelligent and well-meaning people who consistently conjure up magical means to maintain this current disconnected way of living. Just last night I received an email from a very smart woman who wrote, "I don't think we can go backward. I don't think Hunter/Gatherer is going to be it. But is it possible to go forward in a way that will bring us around the circle back to sustainability?"

It's a measure of the dysfunction of civilization that no longer do very many people of integrity believe we can or should go forward with it because it serves us well, but rather the most common argument in its favor (and this is true also for many of its particular manifestations, such as the global economy and high technology) seems to be that we're stuck with it, so we may as well make the best of a very bad situation. "We're here," the argument goes, "We've lost sustainability and sanity, so now we have no choice but to continue on this self- and other-destructive path." It's as though we've already boarded the train to Treblinka, so we might as well stay on for the ride. Perhaps by chance or by choice (someone else's) we'll somehow end up somewhere besides the gas chambers.

The good news, however, is that we don't need to go "backward" to anything, because humans and their immediate evolutionary predecessors lived sustainably for at least a million years (cut off the word *immediate* and we can go back billions). It is not "human nature" to destroy one's habitat. If it were, we would have done so long before now, and long-since disappeared. Nor is it the case that stupidity kept (and keeps) noncivilized peoples from ordering their lives in such a manner as to destroy their habitat, nor from developing technologies (for example, oil refineries, electrical grids, and factories) that facilitate this process. Indeed, were we to attempt a cross-cultural comparison of intelligence, maintenance of one's habitat would seem to me a first-rate measure with which to begin. In any case, when civilized people arrived in North America, the continent was rich with humans and nonhumans alike, living in relative equilibrium and sustainability. I've shown this elsewhere, as have many others,[37] most especially the Indians themselves.

Because we as a species haven't fundamentally changed in the last several thousand years, since well before the dawn of civilization, each new child is still a human being, with the potential to become the sort of adult who can live sustainably on a particular piece of ground, if only the child is allowed to grow up within a culture that values sustainability, that lives by sustainability, that rewards sustainability, that tells itself stories reinforcing sustainability, and strictly disallows the sort of exploitation that would lead to unsustainability. This is natural. This is who we are.

In order to continue moving "forward," each child must be made to forget what it means to be human and to learn instead what it means to be civilized. As psychiatrist and philosopher R. D. Laing put it, "From the moment of birth, when the Stone Age baby confronts the twentieth-century mother, the baby is subject to these forces of violence . . . as its mother and father, and their parents and their parents before them, have been. These forces are mainly concerned with destroying most of its potentialities, and on the whole this enterprise is successful. By the time the new human being is fifteen or so, we are left with a being like ourselves, a half-crazed creature more or less adjusted to a mad world. This is normality in our present age."[38]

Another problem with the idea that we cannot abandon or eliminate civilization, because to do so would be to go backwards, is that the idea emerges from a belief that history is natural—like water flowing downhill, like spring following winter—and that social (including technological) "progress" is as inevitable as personal aging. But history is a product of a specific way of looking at the world, a way that is, in fact, influenced by, among other things, environmental degradation.

I used to be offended by the World History classes I took in school, which seemed almost Biblical in the pretension that the world began six thousand years ago. Oh, sure, teachers and writers of books made vague allowances for the Age of the Dinosaurs, and moved quickly—literally in a sentence or two—through the tens or hundreds of thousands of years of human existence constituting "prehistory," preferring to avert their eyes from such obviously dead subjects. These few moments were always the briefest prelude to the only human tale that has ever really mattered: Western Civilization. Similarly short shrift was always given to cultures that have existed (or for now still exist) coterminous with Western Civilization, as other civilizations such as the Aztec, Incan, Chinese, and so on were given nothing more than a cousinly nod, and ahistorical cultures were mentioned only when it was time for their members to be enslaved or exterminated. It was always clear that the real action started in the Middle East with the "rise" of civilization, shifted its locus to the Mediterranean, to northern and western Europe, sailed across the ocean blue with Christopher Columbus and the boys, and now shimmers between the two towns struck by the September 11, 2001, attacks in New York and DC (and to a lesser extent, Tinseltown). Everything, everyone, and everywhere else matters only in relation to this primary story.

I was bothered not only by the obvious narcissism and arrogance of relegating all of these other stories to the periphery (I'd like to call it racism as well as arrogance, but the white-skinned indigenous of Europe were ignored in these histories as steadfastly as everyone else), and by the just-as-obvious stupidity and unsustainability of not making one's habitat the central figure of one's stories, but also by the language itself. History, I was told time and again, in classes and in books, began six thousand years ago. Before that, there was no history. It was prehistory. Nothing much happened in this long dark time of people grunting in caves (never mind that extant indigenous languages are often richer, more subtle, more complex than English).

But the truth is that history *did* begin six thousand years ago. Before then there were personal histories, but there were no significant social histories of the type we're used to thinking about, in part because the cultures were cyclical (based on cycles of nature) instead of linear (based on the changes brought about by this social group on the world surrounding them).

I have to admit that I still don't like the word *pre*history, because it imputes to history an inaccurate inevitability. For the truth is that history didn't have to happen. I'm not merely saying that any *particular* history isn't inevitable,[39] but instead that history itself—the existence of any social history whatsoever—was not always inevitable. It is inevitable for now, but at one point it did not exist, and at some point it will again cease to be.

History is predicated on at least two things, the first physical, the second perceptual. As always, the physical and the perceptual are intertwined. So far as the former, history is marked by change. An individual's history can be seen as a series of welcomings and leavetakings, a growth in physical stature and abilities followed by a tailing off, a gradual exchange of these abilities for memories, experiences, and wisdom. Fragments of my history. I went to college. I was a high jumper. I remember the eerie, erotic smoothness of laying out over the bar, higher than my head. I lost my springs in my late twenties. I was still a fast runner, chopping the softball toward short and beating out the throw every time. In my thirties arthritis stole my speed, until now I run like a pitching coach, or like an extra in an Akira Kurasawa movie. Twenty years ago I was an engineer. Eighteen years ago a beekeeper. Sixteen years ago I became an environmental activist. Now I'm writing a book about the problem of civilization. I do not know what my future history will look like.

Social histories are similarly marked by change. The deforestation of the Middle East to build the first cities. The first written laws of civilization, which had to do with the ownership of human and nonhuman slaves. The fabrication of bronze, then iron, the ores mined by slaves, the metals used to conquer. The first empires. Greece and its attempts to take over the world. Rome and its attempts. The conquest of Europe. The conquest of Africa. The conquest of the Americas. The conquest of Australia, India, much of Asia. The deforestation of the planet.

Just as with my own future history, I do not know what the future history of our society will be, nor of the land that lies beneath it. I do not know when the Grand Coulee Dam will come down, nor whether there will still be salmon to reinhabit the Upper Columbia. I do not know when the Colorado will again reach the sea, nor do I know whether civilization will collapse before grizzly bears go extinct, or prairie dogs, gorillas, tuna, great white sharks, sea turtles, chimpanzees, orangutans, spotted owls, California red-legged frogs, tiger sala-manders, tigers, pandas, koalas, abalones, and so many others on the brink.

The point is that history is marked by change. No change, no history.

And some day history will come to an end. When the last bit of iron from the last skyscraper rusts into nothingness, when eventually the earth, and humans on the earth, presuming we still survive, find some sort of new dynamic equi-librium, there will no longer be any history. People will live once again in the cycles of the earth, the cycles of the sun and moon, the seasons. And longer cycles, too, of fish who slip into seas then return to rivers full of new life, of insects who sleep for years to awaken on hot summer afternoons, of martens who make massive migrations once every several human generations, of the rise and fall of populations of snowshoe hare and the lynx who eat them. And longer cycles still, the birth, growth, death, and decay of great trees, the swaying of rivers in their courses, the rise and fall of mountains. All these cycles, these cir-cles great and small.

That's looking at history from an ecological level. From a social or percep-tual level, history started when certain groups or classes of people for whatever reason gained the ability to tell the story of what was going on. Monopolizing the story allowed them to set up a worldview to which they could then get other people to subscribe. History is *always* told by the people in control. The lower classes—and other species—may or may not subscribe to an academic or upper class description of events, but to some degree most of us do buy into it.

And buying into it carries a series of perceptual consequences, not the least of which is the inability to envision living ahistorically, which means living sustain-ably, because a sustainable way of living would not be marked, obviously, by changes in the larger landscape. Another way to say all of this is that to perceive history as inevitable or natural is to render impossible the belief that we can go "back" to being nonindustrialized, indeed noncivilized, and to create the notion that to do either of these isn't, in a larger sense, backwards at all. To perceive his-tory as inevitable is to make sustainability impossible. The opposite is true as well. To the degree that we can liberate ourselves from the historical perspective that holds us captive and fall again into the cyclical patterns that characterize the natural world—including natural human communities—we'll find that the notions of forward and backward will likewise lose their primacy. At that point we will once again simply be living. We will learn to not make those markers on the earth that *cause* history, markers of environmental degradation, and both we and the rest of the world will at long last be able to heave a huge sigh of relief.

❰ ❰ ❰

A few years ago, I had an interesting conversation with George Draffan. We were talking about civilization, power, history, discourse, propaganda, and how and why we all buy into the current unsustainable system. George said he really likes the social and political model called "the three faces of power." He said, "The first face is the myth of American democracy, that everyone has equal power, and society or politics is just the give and take of different interest groups that come together and participate, with the best ideas and most active participants winning. This face says that the losers are basically lazy. The second face says it's more complex than that, that some groups have more power than others, and actually control the agenda, so that some things, like the distribution of property, never get discussed. The third face of power is operating when we stop noticing that some things aren't on the agenda, and start believing that unequal power and starvation and certain economic and social decisions aren't actually decisions, they're 'just the way things are.' At this point even the powerless perceive unjust social relations as the natural order." He paused before he said something that has haunted me ever since: "Conspiracy's unnecessary when everyone thinks the same."

❰ ❰ ❰

George also said, "The three faces of power were developed as conflicting descriptions of reality but I'm starting to see them as a progression over time, as the story of history.

"At some point we were all equal. The social structures of many indigenous cultures were set up to guarantee that power remained fluid. But then within some cultures as power began to be centralized, the powerful created a discourse—in religion, philosophy, science, economics—that rationalized injustice and institutionalized it into a group projection. At first the powerless might not have believed in this discourse, but by now, many thousands of years later, we're all deluded to some extent and believe that these differentials in power are natural. Some of us may want to change the agenda a little bit, but there's no seeing through the whole matrix. Power, like property, like land and water, has become privatized and concentrated. And it's been that way for so long and we believe it to such an extent that we think that's the natural order of things."

❰ ❰ ❰

It's not.

〈 〈 〈

Just today I came across an article in *Nature* magazine with the title "Catastrophic Shifts in Ecosystems." Conventional scientific thought, it seems, has generally held that ecosystems—natural communities like lakes, oceans, coral reefs, forests, deserts, and so on—respond slowly and steadily to climate change, nutrient pollution, habitat degradation, and the many other environmental impacts of industrial civilization. A new study suggests that instead, stressors like these can cause natural communities to shift almost overnight from apparently stable conditions to very different, diminished conditions. The lead author of the study, Marten Scheffer, an ecologist at the University of Wageningen in the Netherlands, said, "Models have predicted this, but only in recent years has enough evidence accumulated to tell us that resilience of many important ecosystems has become undermined to the point that even the slightest disturbance can make them collapse."

It's pretty scary. A co-author of the study, Jonathan Foley, a climatologist at the University of Wisconsin-Madison, added, "In approaching questions about deforestation or endangered species or global climate change, we work on the premise that an ounce of pollution equals an ounce of damage. It turns out that assumption is entirely incorrect. Ecosystems may go on for years exposed to pollution or climate changes without showing any change at all and then suddenly they may flip into an entirely different condition, with little warning or none at all."

For example, six thousand years ago, great parts of what is now the Sahara Desert were wet, featuring lakes and swamps that teemed with crocodiles, hippos, and fish. Foley said: "The lines of geologic evidence and evidence from computer models shows that it suddenly went from a pretty wet place to a pretty dry place. Nature isn't linear. Sometimes you can push on a system and push on a system and, finally, you have the straw that breaks the camel's back."

Once the camel's back is broken, it often cannot or will not heal the way it was before.

Another co-author, limnologist Stephen Carpenter, past president of the Ecological Society of America, said that this understanding—of the discontinuous nature of ecological change—is beginning to suffuse the scientific community, and then he continued, "We realize that there is a common pattern we're seeing in ecosystems around the world. Gradual changes in vulnerability accumulate and eventually you get a shock to the system, a flood or a drought, and boom, you're over into another regime. It becomes a self-sustaining collapse."[40]

After I read the article, I received a call from a friend, Roianne Ahn, a woman smart and persistent enough that even a Ph.D. in psychology hasn't clouded her insight into how people think and act. "It never ceases to amaze me," she said, "that it takes experts to convince us of what we already know."

That wasn't the response I'd been expecting.

She continued, "That's one of my roles as a therapist. I just listen and reflect back to clients things they know, but don't have the confidence to believe until they hear an outside expert say them."

"Do you think people will listen to these scientists?"

"It depends on how much denial they're in. But the bottom line is that what they're describing is no big surprise. It's what happens when a person is under stress: she can only take so much before she falls apart. This is what happens in relationships. It happens in families. It happens in communities. Naturally it will be true on this larger scale, too."

"What do you mean?"

"We work as hard as we can, even overextend ourselves, to maintain our stability, and when the pressure gets too much, something's got to give. We collapse. Sometimes that's bad, sometimes it's good."

There was silence while I thought about the fact that some collapses are unnecessary—the breaking down of prisoners under torture, the systematic dismantling of self-esteem under the grinding regime of an abusive parent or partner, ongoing ecological apocalypse—while others can be healing.

She continued, "It's obvious why people try to maintain healthy structures that make them happy. It's not always quite so obvious why we, and I include myself, seem to work just as hard to maintain structures and systems that make them miserable. We're all familiar with the notion that many addicts have to hit rock bottom before they change, even when their addiction is killing them."

I asked, "When do you think the culture will change?"

"This culture is clearly addicted to civilization," she said. "So I think the answer to that question is another one: how far down does it have to go before it hits bottom?"

❨ ❨ ❨

I talked to another friend about all of this. It was late at night. The wind blew outside. The computer was off. We heard the wind. This friend, an excellent thinker and writer, used to live in New York City, and carries with her a certain loyalty not only to that great city, but to cities in general. She was simultaneously sympathetic to and exasperated by me and what I said. After we'd been talking for hours, she asked, reasonably enough, "What right do you have to tell people they can't live in cities?"

"None at all. I couldn't care less where people live. But people who live in cities have no right to demand—much less steal—resources from everybody else."

"Do you have a problem if people in cities just buy them?"

"Buy resources, or people?" I was thinking of a line by Henry Adams: "We have a single system," he wrote, and in "that system the only question is the price at which the proletariat is to be bought and sold, the bread and circuses."[41]

She didn't laugh at my joke. She didn't think it was funny. Neither did I, but probably for a different reason.

I asked, "Buy them with what?"

"They give us food, we give them culture. Isn't that the way it works?"

Ah, I thought, she's following the Mumford line of thought. I asked, "What if the people in the country don't like opera, or Oprah, for that matter?"

"It's not just opera. Good food, books, ideas, the whole cultural ferment."

"And if people in the country like their own food, their own ideas, their own culture?"

"They're going to need protection."

"From whom?"

"Roving bands of marauders. Bandits who will steal their food."

"What if the only marauders are the people from the city?"

She hesitated before saying, "Manufactured goods, then. Because of economies of scale, people in the city can import raw materials from the countryside, work them into things people can use, and sell them back." Her first degree was in economics.

"What if people in the countryside also don't want manufactured goods?"

"Modern medicine then."

"And if they don't want that? I know plenty of Indians who to this day refuse all Western medicine."

She laughed and said, "So we go the opposite direction. Everybody wants Big Macs."

I shook my head, and more or less ignored her joke, as she'd ignored mine, for maybe the same reason. "People only want all this stuff after their own culture has been destroyed."

"I don't think it's necessary to destroy them. Much better to convince them. Modernity is good. Development is good. Technology is good. Consumer choice is good. What do you think advertising is for?"

Maybe both Henry Adams and the Roman satirist Juvenal should have mentioned advertising as well as bread and circuses. And maybe they should have mentioned the importance of dictionary definitions for keeping people in line. I stood my ground. "Intact cultures generally only open their doors wide to consumer goods at gunpoint. Sure, they might pick and choose, but not enough to counterbalance the loss of their resources. Think of what NAFTA and GATT have done to the poor in the Third World, or in the United States. Think of Perry opening Japan, or the Opium Wars, or—"

She cut me off: "I get your point." She thought a moment. "Instead of manufactured items, give them money. A fair price. No ripping them off. They can buy whatever they want with all their money, or rather our money."

"And what if they don't want money? What if they'd rather have their resources? What if they don't want to sell because they want or need the resources themselves? What if their whole way of life is dependent on these resources, and they'd rather have their way of life—for example, hunting and gathering—than money? Or what if they don't want to sell because they don't believe in buying and selling? What if they don't believe in economic transactions at all? Or even moreso, what if they don't believe in the whole idea of resources?"

She got a little annoyed. "They don't believe in trees? They don't believe fish exist? What do you think they catch when they go fishing? What are you telling me?"

"They believe in trees, and they believe in fish. It's just that trees and fish aren't resources."

"What are they, then?"

"Other beings. You can kill them to eat. That's part of the relationship. But you can't sell them."

She understood. "Like the Indians thought."

"Still think," I said. "Many traditional ones. And cities have gotten so large by now—the city mentality has grown to include the whole consumer culture— that people in the country certainly can't kill enough to feed the city without damaging their own landbase. By definition they never could. Which leads us back to the question: What if they don't want to sell? Do the people in the city have the right to take the resources anyway?"

"How else will they eat?"

We heard the wind again outside, and rain began to spatter against the windows. The rain often comes horizontally here in Crescent City, or Tu'nes.

She said, "If I were in charge of a city, and my people—*my people*, what an interesting phrase, as if I own them—are starving, I would take the food by force."

More wind, more rain. I said, "And what if you need slaves to run your industries? Would you take them, too? And if you need not just food and slaves, but if oil is the lifeblood of your economy, metal its bones? What if you need everything under the sun? Are you going to take it all?"

"If I need them—"

I cut her off: "Or perceive that you need them . . ."

She didn't seem to mind. "Yes," she said, thoughtfully. I could tell she was changing her mind. We were silent a moment, before she said, "And there's the land. Cities damage the land they're on."

I thought of pavement and asphalt. Steel. Skyscrapers. I thought of a five hundred-year-old oak I saw in New York City, on a slope overlooking the Hudson River. I thought of all that tree had experienced. As an acorn it fell in an ancient forest—except that back then there was no reason to call those forests *ancient*, or anything but *home*. It germinated in this diverse community, witnessed runs of fish up the Hudson so great they threatened to carry away the nets of those who would catch them, witnessed human communities living in these forests, the humans not depleting the forests, but rather enhancing them by their very presence, by what they gave back to their home. It witnessed the arrival of civilization, the building of a village, a town, a city, a metropolis, and from there, as Mumford put it, the "Parasitopolis turns into Patholopolis, the city of mental, moral, and bodily disorders, and finally terminates in Necropolis, the City of the Dead."[42] Along the way, the tree said good-bye to the wood bison, the passenger pigeon, the Eskimo curlew, the great American chestnuts, the wolverines who paced the shores of the Hudson. It said good-bye (at least for now) to humans living traditional ways. It said good-bye to the neighboring trees, to the forest where its life began. It witnessed the laying down of billions of tons of concrete, the erection of rigid steel structures and brick buildings topped with razor wire.

Unfortunately, it did not live long enough to witness all of this come back down. The tree, I learned last year, is no more. It was cut down by a landowner worried that its branches would fall on his roof. Environmentalists—doing what we seem to do best—gathered to say prayers over its stump.

I told her this story.

"Fuck," she said. "I get it." She shook her head. Pale brown hair fell to cover one eye. She pouted, as she often does when she thinks. Finally she said, "Damn it." Then she smiled just slightly, although I could tell from her eyes she was tired. Suddenly she said, "You know, if we're going to do this much damage, the least we can do is tell the truth."

VIOLENCE

A visitor from Mars could easily pick out the civilized nations.
They have the best implements of war.

Herbert V. Prochnow[43]

THE SECOND PREMISE OF THIS BOOK IS THAT, FOR OBVIOUS REASONS, *traditional communities do not often voluntarily give up or sell the resources on which their communities are based until their communities have been destroyed. They also do not willingly allow their landbases to be damaged so that other resources—gold, oil, and so on—can be extracted. It follows that those who want the resources will do what they can to destroy traditional communities.* This can be accomplished more or less physically, such as through the murder of the peoples and the land on which they depend, or more or less spiritually or psychologically, through the destruction of sacred sites, through aggressive and/or forceful proselytization, by forcefully addicting them to the aggressor's products, by kidnapping their children (most often legally), and through many other means all-too-familiar to those who attend to the relations between the civilized and noncivilized.

《 《 《

Resources for the civilized have always been more important than the lives of those in the colonies. A German colonial officer in South West Africa was more honest than many: "A right of the natives, which could only be realized at the expense of the development of the white race, does not exist. The idea is absurd that Bantus, Sudan-negroes, and Hottentots in Africa have the right to live and die as they please, even when by this uncounted people among the civilized peoples of Europe were forced to remain tied to a miserable proletarian existence instead of being able, by the full use of the productive capacities of our colonial possessions to rise to a richer level of existence themselves and also to help construct the whole body of human and national welfare."[44]

《 《 《

Following quickly on the heels of the second premise is the third, that *this way of living—industrial civilization—is based on, requires, and would collapse very quickly without persistent and widespread exploitation and degradation.* This includes exploitation and degradation of the natural world—for what is unsustainability except a fancy word for exploitation and degradation of nat-ural communities?—and it includes exploitation and degradation of those who do not want us to take their resources (or, to another way of thinking, to kill and sell their nonhuman neighbors). It also includes harming those humans and nonhumans who will come later, and who will inherit a pauper-ized world.

A few months ago I received an email from an activist who wrote, "I've been inspired by Bucky Fuller's vision for years. He says that we have enough of everything to give everyone on the planet a standard of living no one has known so far. But it will require taking all of our resources and technology off of weaponry and fully devoting them to 'livingry.' In other words, we can make it happen, but there's no room for greed in the equation. His whole thing was 'a world that works for everyone with no one left out.'"

Leaving aside the standard conceit that the civilized have higher standards of living than traditional hunter-gatherers (if you measure by some standards, such as the number of automobiles, yes; if you measure by others, such as leisure time, sustainability, social equality, and food security—meaning no one goes hungry—hunter-gatherers win hands down), Fuller's is a powerful—and powerfully dan-gerous—fantasy, and an odd statement coming from someone living on land taken by violence from its original inhabitants, and using the sorts of technolo-gies—for example, industrial forestry, mining, smelting—that violently shape the world to industrial ends. Just because Fuller designed groovy structures like geodesic domes (the one at Expo '67 in Montreal was way cool!) did not mean that violence was not done to the land—and to people—both there and elsewhere. Where, precisely, did Fuller believe these resources came from, and how did he believe he would get them without using force against both the "resources" them-selves and against the humans who live in close proximity to them?

I enjoy railing against the absurdity of the U.S. military budget as much as the next sane person. I often marvel at the extraordinary amounts of money that are spent seemingly for no other purpose than to kill people, and dream of what good could be accomplished if those who serve life had the same easy access to cash as those who serve death. Corporate Senators and Representatives are fond of complaining, for example, that it's too expensive to save species driven to the brink of extinction by the actions of the industrial economy, and that the corporations these men (and token women) represent must be allowed to continue their actions unimpeded. An industry front group calling itself the "Grassroots ESA Coalition" (a subgroup of the similarly deceivingly named industry front group "National Wilderness Institute") has stated that total costs for "the ten species covered by the most expensive endangered species recovery plans are: Atlantic Green Turtle $88,236,000; Loggerhead Turtle $85,947,000; Blunt-Nosed Leopard Lizard $70,252,000; Kemp's Ridley Sea Turtle $63,600,000; Colorado Squawfish $57,770,000; Humpback Chub $57,770,000; Bonytail Chub $57,770,000; Razorback Sucker $57,770,000; Black-Capped Vireo $53,538,000; Swamp Pink $29,026,000."[45] I'm not sure I trust their research, or, for that matter, their intelligence because even when trying to show how expensive implementation of the Endangered Species Act is, they left off more pricey efforts. Costs for projects aimed toward recovering salmon in the Northwest (or rather, projects aimed at providing the illusion of recovery while allowing business to continue as usual) were $119 million just in 1995. Not including land acquisition, annual expenditures for recovery efforts for all endangered species went from $43 million in 1989 to $312 million in 1995.[46] Recently, the federal government made big news when it granted more than $16 million to twenty-five states to promote the conservation of such varied species as marbled murrelets, salmon, bull trout, aplomado falcons, Karner blue butterflies, Florida scrub jays, and the Preble's meadow jumping mouse.[47] This may all seem like a lot of money, but in fiscal year 2001 the federal government spent more than $5.7 billion on the physical impossibility called the Ballistic Missile Defense System (a.k.a. Strategic Defense Initiative, a.k.a. Star Wars, and most especially a.k.a. a black hole into which money disappears, to conveniently reappear on the ledgers of favored corporations). It spent $3.9 billion on new F-22 fighters, $3 billion on new C-17 Transport aircraft, $1.7 billion on new V-22 Osprey aircraft (which seem capable so far only of killing their own crews), $4 billion as a partial payment on a new aircraft carrier, $3 billion as a partial payment on a new submarine. Even prior to the events of September 11, the military received nearly one billion dollars per day during fiscal year 2001.[48] Just in the last seven years, the military spent more than $100 million on airline tickets it did not use. The tickets were fully refundable, but the military never bothered to ask for a refund.[49] The United States government spends $44 billion per year on spying. I used to often fantasize about using all that that money used for harm—real money, not the crumbs tossed in the direction of wildlife—to help salmon, spotted owls, Carson wandering skipper butterflies (listed as endangered only after having been reduced to a few individuals), Columbia Basin pygmy rabbits (of whom only fifty remain in the wild), Mississippi gopher frogs (of whom only one hundred members remain, breeding in one pond), Tumbling Creek cavesnails (down to forty individuals), and so many others. But the truth is that this will never happen.

The reason that my fantasies are nothing more than fantasies, and the reason that the same is true for Buckminster Fuller's more well-known fantasies, is that the money *must* be spent on weaponry, and not on livingry. To believe the U.S. military does not serve an absolutely vital purpose is to have failed to pay any attention to the path of civilization for the past six thousand years. The importation of resources into cities has *always* required force, and always will. And that's why Fuller's fantasy is dangerous—as is my own, when I forget it is a fantasy—because it pretends that resource extraction can be accomplished without force and exploitation, thus diverting attention toward the outrageous and obscene military budgets and away from the social and technological processes that require them. If you need—or perceive yourself as needing—gold, wood, food, fur, land, or oil that resides in someone else's community, and if this other community does not want to hand these resources over to you—and why on God's green earth should they?—how are you going to get them? We have seen this process too many times to not know the answer.

❨ ❨ ❨

In late 2001, the United States military began bombing the people and landscape of Afghanistan, at a cost to the American public of approximately a billion dollars a day, or about four dollars for every man, woman, and child in this country (or more was spent in three hours than in all of 1995 ostensibly to save salmon). This amounts to about forty dollars per day for every one of the human targets, that is, every Afghan man, woman, and child. Based on Afghanistan's gross domestic product, forty dollars is about twenty times their average daily income.[50] It would, in a sense, be the equivalent of a government spending about eighteen hundred dollars per day per person in the United States to kill us here.

Much as I enjoy being the center of attention (I am, after all, a male), given the choice, I'd be willing to settle for a lot less attention were it given to me in the form of cash or foodstuffs, rather than bombs. No, thank you, I'd say politely, I don't really want a bomb, nor even a "bomblet," not even one as cool as a BLU-26 Sadeye, although I might be able to use a few of the hundreds of razor-sharp projectiles to cut some things around the house. Instead, a cow would be nice. And some chickens. And some native trees. You could buy all of that for me in one day. And then tomorrow we could talk about a bicycle, and then the day after that we could start thinking about a new well. Truth be told I wouldn't even know what to do with eighteen hundred dollars every day, or its equivalent in the Afghan community. I'd probably give most of it away. But I still think I'd rather have a cow and some chickens than a bomb.

On further reflection, do you know what I'd like even more? To simply be left alone.

《　《　《

The United States has historically spent about $70 million per year on humanitarian aid for Afghanistan (about four dollars per person per year, the equivalent of about a hundred and eighty dollars if it were given in the United States),[51] which is about how much the United States spent per Afghan on the bombing campaign every hour and forty minutes.

If United States citizens have paid four dollars a piece per day to support this war effort, the Afghan people have paid rather more. The bombs—such a nice, short word to describe inventions that have as their purpose destruction—include, fairly typically, the two-thousand pound MK-84, which was developed in the 1950s and has served its masters well in the time since. About twelve thousand were dropped on Iraq during the First Gulf War.[52] If the bomb detonates on contact with the ground, it creates a crater fifty feet in diameter, and thirty-six feet deep (Sorry, guys, this is not precisely what I had in mind when I asked for a new well). If it explodes above ground, it disperses shrapnel to a lethal radius of four hundred yards.[53]

A more commonly used incendiary device— incendiary device being even more abstract language than bomb—are cluster bombs. Instead of causing a single explosion, cluster bombs (or CBUs: Cluster Bomb Units, if we don't mind getting yet more abstract) contain dozens, hundreds, or even thousands of bomblets called BLUs (bomb live units). Each BLU then splits into hundreds of pieces of shrapnel. For example, the bomb called the CBU-75 may contain eighteen hundred BLU-26 Sadeyes. Each Sadeye contains six hundred sharp steel shards. A single CBU-75 will shoot these shards across an area of around 9.25 million square feet, which is about 212 acres, or more than 150 football fields, or nearly a third of a square mile. A single B-52 strategic bomber can carry forty of these cluster bombs, which could then blanket almost fourteen square miles at an average density of one shard every ten square feet. In just one day in the First Gulf War, twenty-eight B-52s dropped about four hundred and seventy tons of explosives on Iraq, enough to devastate approximately sixteen hundred square miles, an area about one-third the size of Connecticut.[54]

The United States military uses another type of bomb, this one "a terrific weapon" with "tremendous destructive power," according to U.S. General Wesley Clark.[55] It is the BLU-82, also known as the "Daisy Cutter." This fifteen-thousand-pound bomb, filled with an aqueous mixture of ammonium nitrate, aluminum powder, and polystyrene soap, is so large it can only be launched by rolling it out the rear door of a cargo aircraft, the MC-130 Hercules. The slowness of the cargo plane means Daisy Cutters can only be dropped when there are no defenses, in other words, only on those who are defenseless. (It must be stated that prior to the U.S. attack, the Afghans were not *precisely* defenseless: their Air Force did have two old planes, which might even have been jets.[56] It must also be stated that in the first days of the attack the Afghan military killed precisely one American soldier, and Afghan prisoners did manage to kill one CIA operative—who was probably "playing smacky face" with them, as the CIA has been known to put it—before they themselves were ultimately blown to bits. Far more U.S. military casualties were caused by so-called friendly fire and a plane wreck.) A parachute opens, then the Daisy Cutter floats toward the Earth. The parachute slows the descent enough to give the transport plane time to get away before the bomb explodes. The bomb detonates just above ground, producing what are called overpressures of one thousand pounds per square inch (overpressure is air pressure over and above normal air pressure: overpressures of just a few pounds are enough to kill people) disintegrating everything and everyone within hundreds of yards, and killing people (and nonhumans) at a range of up to three miles.[57] General Peter Pace, vice-chair of the U.S. joint chiefs of staff, put the purpose clearly: "As you would expect, they make a heck of a bang when they go off and the intent is to kill people." Marine Corps General Trainor was even more specific about the effect of Daisy Cutters on people in Afghanistan: "Besides the physical degradation, these—along with the regular ordinance dropped from B-52s—provide great psychological punishment, as victims begin to bleed from the eyes, nose, and ears, if they aren't killed outright, of course. It's a frightening, awesome assault they're suffering, and there's no doubt they're feeling our wrath."[58]

Even if the primary target of these bombs were members of the Afghan military (or terrorists, whatever or whomever they may be) those who were killed were mainly just people trying to survive. "We were farmers," said Kamal Huddin, after American planes made four passes over Kama Ado, his home village, killing more than half of the three hundred people who lived there. "We were poor people. And we didn't have any contact with any organizations."[59] It's no surprise that people like these—people living in mud huts with straw roofs, using wooden plows to till the soil exactly as their ancestors did—were killed. Colonel

John Warden, who planned the air campaign in Iraq, said that dropping any of these bombs I've mentioned "is like shooting skeet. Four hundred and ninety-nine out of five hundred pellets may miss the target, but that's irrelevant."[60]

So, who dies? I have seen pictures of the dead, dark-haired children laid out on mattresses, hands folded neatly above the last clothes they will ever wear by parents now standing looking downward, eyes red, in the background. The children's faces are bloated, and red, too, though not from tears but instead from blood which never seems to finally wash away. The parents' hands, too, are red where faint traces of their children's blood remains.

It is not acceptable in the United States to talk about these dead children. The official United States and capitalist media have declared it so. The Chair of CNN, Walter Isaacson, ordered journalists who work for CNN not to focus on the killing of Afghan citizens by the U.S. military, because it "seems perverse to focus too much on the casualties or hardship in Afghanistan." He went on to admonish his reporters who cover civilian deaths that they should never forget that it is "that country's leaders who are responsible for the situation Afghanistan is now in," perhaps forgetting that the same argument could just as easily be used to ignore the dead in this country. The head of standards [sic] for CNN, Rick Davis, followed up his boss's memo with some suggested language for newscasters to repeat, for example, "We must keep in mind, after seeing reports like this from Taliban-controlled areas, that these U.S. military actions are in response to a terrorist attack that killed close to 5,000 innocent people in the U.S.," or "We must keep in mind, after seeing reports like this, that the Taliban regime in Afghanistan continues to harbor terrorists who have praised the September 11 attacks that killed close to 5,000 innocent people in the U.S.," or "The Pentagon has repeatedly stressed that it is trying to minimize civilian casualties in Afghanistan, even as the Taliban regime continues to harbor terrorists who are connected to the September 11 attacks that claimed thousands of innocent lives in the U.S."[61] Each of these statements could of course be inverted: "We must keep in mind that the capitalist regime in Washington continues to harbor journalists, military leaders, politicians, and CEOs who have put in place and praised U.S. military and economic policies that kill millions of people annually."

Not to be outdone, Brit Hume of Fox News Channel recently wondered on air why journalists should bother to cover civilian deaths at all: "The question I have," Hume said, "is civilian casualties are historically, by definition, a part of war, really. [This is true only under a strict definition of history: as I've shown elsewhere, even for many of the warlike indigenous peoples—that is, those who are ahistorical, uncivilized—to kill noncombatants was unthinkable, and even killing combatants was a rarity, an event.] Should they be as big news as they've been?" One could, of course, ask the same question of civilian casualties in the United States. Mara Liasson of that bastion of liberal news National Public Radio answered Hume's question, and went right to the point: "No. Look, war is about killing people. Civilian casualties are unavoidable." Perhaps following the standards set down by Rick Davis, Liasson made sure to add that what she thought was missing from television coverage was "a message from the U.S. government that says we are trying to minimize them, but the Taliban isn't, and is putting their tanks in mosques, and themselves among women and children." *U.S. News & World Report* columnist and Fox commentator Michael Barone responded to Hume and Liasson, revealing the wide variety of opinion represented in the corporate media: "I think the real problem here is that this is poor news judgment on the part of some of these news organizations. Civilian casualties are not, as Mara says, news. The fact is that they accompany wars."[62]

As above, so below. The same avoidance of attention to those killed by the United States happens at smaller news outlets as well. A memo circulated at the Panama City, Florida, *News Herald* warned editors: "DO NOT USE photos on Page 1A showing civilian casualties from the U.S. war on Afghanistan. Our sister paper in Fort Walton Beach has done so and received hundreds and hundreds of threatening e-mails and the like.... DO NOT USE wire stories which lead with civilian casualties from the U.S. war on Afghanistan. They should be mentioned further down in the story. If the story needs rewriting to play down the civilian casualties, DO IT. The only exception is if the U.S. hits an orphanage, school or similar facility and kills scores or hundreds of children."[63]

❆ ❆ ❆

After 9/11, *The New York Times* took to publishing profiles of people killed in the attack on the World Trade Center. These profiles were syndicated through the country, letting us in on details of the lives of the dead. Thus we learn that one of the dead was an "efficient executive" who "never forgot the attention to spit and polish, in his work or play. 'It doesn't shine itself,' he'd reply when people admired his vintage car." We learn that another was "mad for Mantle," and "stubbornly stood by his Yankees, even when his two sons ... turned out to be Mets fans." A third, we learn, was a top stockbroker, and a "prankster with a heart" who "would pull up next to you in his Porsche—a 911—flip the bird, grin, and take off in the wind."[64] A friend from New York said of the profiles, "I smell a Pulitzer."

Here's my question: What is the premise (and purpose) of these profiles? The most basic answer is clear, that the dead are individuals worthy of consideration. Or, as someone put it in a letter to the editor, "I appreciate the efforts to humanize the victims. . . . They deserve to be remembered. They deserve justice."[65]

Here's another question that interests me even more: What is the premise (and purpose) of the silence surrounding victims of our way of life? That answer is clear as well, although we do not talk or even think about it.

Of course.

❨ ❨ ❨

Imagine how our discourse and actions would be different if people daily detailed for us the lives—the individuality, the small and large joys and fears and sorrows—of those whom this culture enslaves or kills. Imagine if we gave these victims that honor, that attention. Imagine if everyday newspapers carried an account of each child who starves to death because cities take the resources on which the child's traditional community has forever depended. *She never ran*, the article might read, *because she never had the energy, but she loved to be tickled, and loved to watch her mother, no matter what her mother did. When her mother carried her in a sling on her back, her large eyes took in every detail of her surroundings. She loved to smile at her neighbors, and smile also at little birds that landed on the ground near her mother's feet.* Imagine if we considered her life as valuable as that of the "efficient executive," and if we considered violence against her to be as heinous as we consider violence against him.

❨ ❨ ❨

Imagine, too, if our discourse included accounts of those nonhumans whose lives this culture makes unspeakably miserable: the billions of creatures bred for torture in feedlot, factory farm, or laboratory; the wild creatures worth money, who are pursued and destroyed no matter where they hide; the wild creatures unvalued by the economic system, who are eliminated because they are in the way of production. Imagine if we spoke of the threespine stickleback, the Miami blue butterfly, white abalone, spectacled eider, southwestern willow flycatcher, Holmgren's milkvetch, Pacific pocket mouse, individually and collectively. Imagine, finally, if we considered their lives as valuable as our own, and their contribution to the world and to our neighborhoods to be as valuable as that of a stockbroker—or even moreso—even if the stockbroker *does* drive a Porsche, flip us the bird, and take off in the wind.

❨ ❨ ❨

The fourth premise of this book is that *civilization is based on a clearly defined and widely accepted yet often unarticulated hierarchy. Violence done by those higher on the hierarchy to those lower is nearly always invisible, that is, unnoticed. When it is noticed, it is fully rationalized. Violence done by those lower on the hierarchy to those higher is unthinkable, and when it does occur it is regarded with shock, horror, and the fetishization of the victims.*

This is true when we talk about the acceptability—the expectedness, normality, necessity, even desirability (only when victims force their hand, of course)—of the U.S. military and its proxies killing civilians the world over and the unthinkability of counterattacks in kind. It is true when we talk about the acceptability of routine police violence against civilians and the fetishization of police officers killed on the job ("All gave some, some gave all," read the bumper stickers, but no one ever mentions, at the huge police funerals or elsewhere, that garbage collection is far more dangerous—with a far higher mortality rate—than police work; and don't hold your breath waiting for the next Bruce Willis or Tom Cruise action flick about courageous garbage collectors putting their lives on the line to clean up the mean streets of New York or L.A.). This is true when we talk about humans extirpating sharks and other species almost unnoticed while trumpeting the rare cases when sharks or others bite humans (usually when the humans have already either destroyed the creature's home, backed it into a corner, and/or physically tormented it): despite propaganda from books and movies like *Jaws*, the ratio of humans slaughtering sharks to sharks even attacking humans is approximately 20 million to one.[66] It is true when we talk about CEOs making decisions that lead to profits for the corporations they run and death for those humans (and nonhumans) they poison, and the victims of these CEOs for some reason refraining from similarly poisoning the CEOs, the politicians who protect them, and the families of both. And it's true when we talk about more intimate forms of violence, like those perpetrated en masse by men against women and children, and the relative rarity with which the women or children fight back. I wrote a book about the violence that took place within my family when I was a child. The violence was rigidly one-way: my father beat his wife and children with impunity. I remember the only time my brother defended himself by returning a single blow: he received the worst beating of his miserable childhood. Why? Because he had broken a fundamental unstated rule of our family (and of civilization): Violence flows in only one direction.

《 《 《

I've been thinking a lot lately about depleted uranium, in part because of some pictures I've seen. First the depleted uranium, then the pictures.

So-called depleted uranium is what's left of natural uranium after the "enriched uranium"—the fissionable isotope uranium 235—has been separated to produce fuel for nuclear reactors. The term *depleted uranium* is something of a misnomer in that it implies that the remaining uranium has become significantly less dangerous, more, well, depleted. But depleted uranium—99.8 percent uranium 238—is just as toxic and about 60 percent as radioactive as natural uranium. And with a half-life of 4.5 billion years, it will truly be one of this culture's trademark gifts that keeps on giving: it will kill essentially forever.[67]

The United States has made a lot of it, well over a billion pounds. Beginning in the 1950s, the feds started trying to figure out what they were going to do with all of this stuff. Providentially, uranium is extremely dense—about 1.7 times heavier than lead—and so can be used to make an artillery shell[68] that easily penetrates steel. Even better, it's pyrophoric, meaning heat from the impact causes it to vaporize, releasing huge amounts of energy. If you don't mind toxifying and irradiating the surrounding countryside and its human and nonhuman inhabitants, depleted uranium makes a tank-busting shell extraordinaire.

What this means in practice is that leaders of government and industry solved the problem of disposing of U-238 in typical win-win (for them) fashion by giving it away free to both national and foreign arms manufacturers (perhaps it never occurred to anyone in power that the planet had already come up with the best solution for storing uranium: keep it in its natural state underground). I suppose we should be thankful that the researchers didn't deem DU's most effective use to be in forks or the heating elements of toasters, or else we'd be up to our glowing eyeballs in it at home. But this gratitude is in truth unfounded, because that plan has long been floated by a committee of the National Academy of Sciences and many others as a way to get rid of various radioactive wastes. They want (note the use of present tense) to redefine certain forms of radioactive waste as "Below Regulatory Concern," recycle them (it's great to be green!), and thus give citizens "authorized doses" of radiation.[69] We should also be grateful, I guess, that they didn't just decide to put the DU in our water supplies and tell us it's good for our teeth. Oops, they've already done something like that, too. As is true for DU, fluoride is a toxic byproduct of this way of living (in this case the production of aluminum, fertilizer, cement, and weapons-grade plutonium and uranium). Also as is true for DU, fluoride is

extremely costly—if not impossible—to dispose of safely. The feds didn't know
what to do with it. Perhaps because fluoride didn't work very well either in
artillery shells or toaster ovens, those in power decided to get rid of it by adding
it to our municipal water supplies and toothpaste, which means that the old
John Birchers were right when they averred that fluoridation was a dangerous
plot ("to sap and impurify all of our precious bodily fluids," as General Jack D.
Ripper might have put it): they just had the wrong conspirators. Another sim-
ilarity between fluoride and DU is that both are dangerous: not only does fluoride derived from toxic waste contain impurities such as lead and arsenic, but
even at relatively small doses fluoride itself can cause cancer, osteoporosis, skele-
tal fluorosis, arthritis, and brain damage, among many other conditions.[70]
Here's another thought: just for grins, if you're ever in your grandparents' base-
ment, see if you can find an old container of rat poison. Check out the toxic
ingredient—the killer. Yep, you guessed it, sodium fluoride. Happy brushing.[71]

The list of countries using or purchasing weapons or shells made with
depleted uranium is long, and includes, among others, the United States, the
United Kingdom, France, Canada, Russia, Greece, Turkey, Israel, the monar-
chies in the Persian Gulf, Taiwan, South Korea, Pakistan, and Japan.[72] Spreading these toxic, radioactive materials around the world is bad enough, but the
real danger comes when the weapons are used. And they are used often. In
110,000 air raids against Iraq during the so-called First Gulf War ("so-called"
because my understanding is that for something to be called a war the other
side has to actually be able to fight back: casualties in the First Gulf Massacre corresponded closely to premise four of this book), U.S. A-10 Warthog aircraft fired
about 940,000 DU projectiles.[73] When a depleted uranium projectile hits a target, about 70 percent of the round vaporizes into (hot) dust as fine as talcum
powder, as does part of the target, which may also have been constructed of
depleted uranium. Three hundred tons of DU are estimated to be blowing in the
wind from this particular desert storm. An American soldier in charge of a crew
assigned to clean up DU around tanks destroyed by these shells said, "When we
climbed into vehicles after they'd been hit, no matter what time of day or night
it was, you couldn't see three feet in front of you. You breathed in that dust."[74]
Once the dust has been respirated, it can lodge in the lungs or make its way to
other organs, such as kidneys. In any case, you're in trouble. Uranium 238 and
the products from its decay—including other isotopes of uranium, thorium
234, and protactinium—release alpha and beta radiation that cause cancer and
genetic mutations in exposed individuals and their descendants more or less
into perpetuity. Two of that soldier's fifteen crew members are now dead, and

even the Department of Energy admits that this soldier's internal uranium con-
tamination is five thousand times that permissible. Ninety to one hundred
thousand American Gulf War veterans have reported medical problems associated with the "Gulf War Syndrome," and rates for malformations in their children approach 67 percent in some communities.

As well as affecting U.S. soldiers, DU has probably already harmed 250,000
Iraqis. The same can be said for residents of Bosnia, and soon we'll be saying the
same for the people of Afghanistan. Leukemias and cancers have gone up by 66
percent in recent years in southern Iraq, with some locales experiencing a 700
percent increase.[75] And there have been birth defects. Oh, how there have been
birth defects. One doctor began her report, "In August we had three babies born
with no heads. Four had abnormally large heads. In September we had six with
no heads, none with large heads, and two with short limbs. In October, one
with no head, four with big heads and four with deformed limbs or other types
of deformities."[76]

Which finally brings us to the pictures. There are two groups: pictures I have
not seen, and pictures I have. Here is what one person wrote about those I have
not (and of course I don't expect to soon see similar text in America's much-
vaunted and certainly uncensored capitalist "free press"™): "I thought I had a
strong stomach—toughened by the minefields and foul frontline hospitals of
Angola, by the handiwork of the death squads in Haiti and by the wholesale
butchery of Rwanda. But I nearly lost my breakfast last week at the Basrah
Maternity and Children's Hospital in southern Iraq. Dr. Amer, the hospital's
director, had invited me into a room in which were displayed colour photographs of what, in cold medical language, are called "congenital anomalies,"
but what you and I would better understand as horrific birth deformities. The
images of these babies were head-spinningly grotesque—and thank God they
didn't bring out the real thing, pickled in formaldehyde. At one point I had to
grab hold of the back of a chair to support my legs. I won't spare you the details.
You should know because—according to the Iraqis and in all likelihood the
World Health Organization, which is soon to publish its findings on the spiral-
ing birth defects in southern Iraq—we are responsible for these obscenities.
During the Gulf war, Britain and the United States pounded the city and its surroundings with 96,000 depleted-uranium shells. The wretched creatures in the
photographs—for they were scarcely human—are the result, Dr. Amer said. He
guided me past pictures of children born without eyes, without brains. Another
had arrived in the world with only half a head, nothing above the eyes. Then there
was a head with legs, babies without genitalia, a little girl born with her brain

outside her skull, and the whatever-it-was whose eyes were below the level of its nose. Then the chair-grabbing moment—a photograph of what I can only describe (inadequately) as a pair of buttocks with a face and two amphibian arms. Mercifully, none of these babies survived for long. Depleted uranium has an incubation period in humans of five years. In the four years from 1991 (the end of the Gulf War) until 1994, the Basrah Maternity Hospital saw 11 congenital anomalies. Last year there were 221."[77]

There are photographs, too, that I have seen, some of the worst of my life. There are infants with one large eye in the middle of the face; infants—still alive, huge eyes staring—with the exploded heads of the hydrocephalic; infants with translucent skin or skin covered with some unknown white substance or covered with welts or deep split-open fissures or with charred-looking skin or skin like dark glazed pottery; infants with ambiguous genitals (these are called, for some reason, "non-viable children"); infants—unfortunately alive—with no eyes, their bones fused and stunted; an infant—also unfortunately alive—with no anus, and with her bowel and urinary tract on the outside of her body.[78]

These pictures all lead me to ask, not rhetorically, but with all expectation of answers: What, precisely, is this culture's calculus of casualties? The lives of how many of these children are worth the life of one efficient executive, one prank-playing stockbroker? How many of these children's lives are worth one Porsche, or the gasoline it burns to take off in the wind? The lives of how many children add up to the value, to take a unit of modern currency, of a barrel of oil?

❨ ❨ ❨

The *San Francisco Chronicle* carried an article on page 3 entitled "Scientist's Urgent Warning of World's Failing Environment: Ailing Planet in Need of Mass Conservation." The article disturbed me for several reasons. First, of course, is that the planet doesn't so much need mass conservation as it needs to be relieved of that which is killing it: civilization. Next was the article's placement, on the same spread—implying equivalent importance—with an article, on page two, entitled "Suit Catches Psychic Line Off Guard: Miss Cleo Accused of Rampant Fraud." On page 1 of this day's paper, just below the masthead, implying far greater importance, was an article with the headline: "Silver Turns to Gold for Canadian Pair: Skating Union Makes Amends for Judge's Misconduct." Above the masthead was a teaser for the most important article of the day, even more important than the one about figure skaters getting ripped off in the Olympics, which was, "Britney Crosses Over: Spears Trods Well-worn Path from Pop Star to Movie Actress in 'Crossroads.'" And let's not *even* compare the importance of the article about the killing of the planet to, say, the entire sections of the newspaper devoted daily to business, travel, and sports (Go Giants!). It bothered me also, maybe even more than the placement, that three full paragraphs of even this meager coverage were devoted to a Danish statistician who has gained great fame by arguing that the global environment is in fact improving, revealing once again the truth behind the thesis of another of my books, that in order for us to maintain our way of living, we must tell lies to each other, and especially to ourselves.

It's important to note that the *Chron* followed up this article by giving the Danish statistician an article all to himself that was three times as large as the original (seventy column-inches versus twenty-four—yeah, I know, I've got to get a social life), covering an entire page (with the exception of two ads, one stating that larger Post-It notes give you "More yada yada per note," and one that reads "SEX FOR LIFE! Erection Problems? Premature Ejaculation? Immediate results after one consultation!"), complete with smiling photograph and statistical sidebar stating "it is not cost efficient to spend money on certain environmental problems" because "the cost per year of [human] life saved" is too high. Perhaps because this person's obscene calculations—his damn lies, or even worse, his *statistics*, as the saying goes—fit so well with the goals of civil society, he has been named to head a government-funded environmental monitoring agency in his native Denmark.[79]

I think, however, that what bothered me most about the original article was the pull-quote the editors chose to bold, which was, "We clearly will have an increasingly difficult time in maintaining our current levels of affluence."[80] The world is being killed before our eyes, and these editors are concerned primarily for the maintenance of their affluence?

That's a silly question. Of course the answer is yes.

But it makes me ask again: What is the calculus of casualties? There's no reason to confine this calculus to humans. How many baubles is life on the planet worth? How many salmon, how many generations of salmon, swimming upstream, spawning, dying, feeding humans, bears, eagles, their own offspring, entire forests, are worth the life of one politician, one executive, one lying statistician? The lives of how many species of salmon are worth the fortune of one politician, one executive? How many salmon are we willing to sacrifice so that an efficient executive can have a vintage car? How many rivers of fish—and how many rivers themselves, with their once-clean, free-flowing water—are worth sustaining a lifestyle based on exploitation, a lifestyle that will not last, and that will, we can only hope (the *we* in this case evidently not including the editors of the *San Francisco Chronicle*), end very soon.

☾ ☾ ☾

The fifth premise of this book is that *the property of those higher on the hierarchy is more valuable than the lives of those below. It is acceptable for those above to increase the amount of property they control—in everyday language, to make money—by destroying or taking the lives of those below. This is called* production. *If those below damage the property of those above, those above may kill or otherwise destroy the lives of those below. This is called* justice.

This is all certainly true of our intraspecies relations. Police can and routinely do bust up homeless camps, but homeless people are not allowed to dismantle police stations (or the homes of the police). Petrochemical companies are allowed to make people's homes uninhabitable by toxifying the surrounding landscape, but the residents of those homes are not allowed to destroy the refineries (or the homes of the owners). Whites could, should, and would systematically destroy the possessions of the Indians, but Indians were not allowed to return the favor. And it's true of our interspecies relations, as industrial production systematically devours the living planet, any nonhumans who threaten productivity must be destroyed. A functionary for the Canadian Department of Fisheries and Oceans expressed this perfectly—to present just one example among an entire planet full of them—in regards to the now-extinct Great Auk: "No matter how many there may have been, the Great Auk had to go. They must have consumed thousands of tons of marine life that commercial fish stocks depend on. There wasn't room for them in any properly managed fishery. Personally, I think we ought to be grateful to the old timers for handling the problem for us."[81] If we could change the culture such that this premise were no longer true, the calculations of the Danish statistician would be recognized for the insanity they represent, prisons would not be stocked with small-scale criminals, and civilization would collapse in a heartbeat.

IRREDEEMABLE

I think we must face the possibility that something is dreadfully wrong with society and that this is somehow connected to the bloody history of Western culture, a bloodiness that surpasses all others.

Deborah Root[82]

THE SIXTH PREMISE OF THIS BOOK, THE ONE ALLUDED TO EARLY ON, is that *civilization is not redeemable. This culture will not undergo any sort of voluntary transformation to a sane and sustainable way of living. If we do not put a halt to it, civilization will continue to immiserate the vast majority of humans and to degrade the planet until it (civilization, and probably the planet) collapses. The effects of this degradation will continue to harm humans and nonhumans for a very long time.*

Ever since I was a child, I've been asking: if this culture's destructive behavior isn't making us happy, why are we doing it?

I've come up with many answers so far. All of them, unfortunately, point toward the intractability of this culture's destructiveness. In my book *A Language Older Than Words*, part of my answer was that the entire culture suffers from what trauma expert Judith Herman calls *complex post-traumatic stress disorder*, or complex PTSD. By now most of us are familiar with normal PTSD, if not in our bodies then at least from having read about it. PTSD is an embodied response to extreme trauma, to extreme terror, to the loss of control, connection, and meaning that can happen at the moment of trauma, the moment when, as Herman puts it, "the victim is rendered helpless by overwhelming force."[83] This force may be nonhuman, as in an earthquake or fire; or inhuman, as in the violence on which this culture is based: the rape, assault, battery, and so on that characterizes so much of this culture's romantic and childrearing practices; the warfare that characterizes so much of this culture's politics; and the grinding coercion that makes up so much of the rest of this culture, such as its economics, schooling, and so on. Herman states, "Traumatic reactions occur when no action is of avail. When neither resistance nor escape is possible, the human [and the same is clearly true for the nonhuman] system of self-defense becomes overwhelmed and disorganized."[84] Traumatized people, she writes, "feel and act as though their nervous systems have been disconnected from the present."[85] They may experience hyperarousal, sensing danger everywhere. Certain triggers may stimulate "flashbacks," so that a child who was beaten by a parent while on a water skiing trip, for example, may even as an adult become terrified or full of rage when faced with this stimulus. The same may happen to a woman who was raped in a certain make and model of car. And the adult may wonder at the source of this sudden fear or anger. Those who have been traumatized may go into a state of surrender. Having been brought to the point of powerlessness, where any resistance was futile, this feeling may continue later into life. Faced with any emotionally threatening situation, these people may freeze, failing to resist even when resistance becomes feasible or necessary.

This entire culture is so violent, so traumatic, I argued in *Language*, as to render most all of us to one degree or another shell shocked, and therefore incapable of realizing or even imagining what it would be like to live a life not based on fear. This fear, in fact, runs so deep that it has become normalized in this culture, codified, made the basis of the entire society.

I am sure you can see these symptoms not only among those of your friends who may have been grotesquely and obviously traumatized, but in the culture at large: the culture is certainly disconnected from the present, else we could not possibly kill the planet (and each other) for the sake of production; it certainly sees danger everywhere, even when there is none (the culture's politics, science, technology, religion, and much of its philosophy are all founded on the notion that the world is a vale of tears and danger); it just as certainly manifests in an otherwise incomprehensible rage at (and fear of) the indigenous everywhere, as well as the natural world; and of course those of us who hate the destruction consistently fail to resist in anything approaching a meaningful fashion.[86]

But there's more to it than this. Judith Herman defined a new type of PTSD. She asked, what happens to people who have been traumatized not in one discreet incident—for example, an earthquake or a rape—but instead have suffered "subjection to totalitarian control over a prolonged period (months to years)"?[87] Or, I would add, for the six thousand years of civilization. She includes not only hostages, prisoners of war, and the like, but also those who have survived the captivity of long-term domestic violence. Concerning this latter, she asks what happens to those whose personalities are not only deformed by extended violence, having suffered it as adults, but to those whose personalities are *formed* as children in such a crucible of totalitarian violence. The answer is that they may suffer amnesia, forgetting the violence of their childhood (or, I would once again add in our larger case, the violence on which, to choose just one example, white title to land in North America is based). They may suffer a sense of helplessness. They may identify with their abuser. They may come to perceive mutually beneficial relationships as impossible, and to believe instead that all relationships are based on force, on power. They may come to believe that the strong dominate the weak, the weak dominate the weaker, and the weakest survive as they can.

The understanding that the entire culture could reasonably be said to be suffering from complex PTSD helps to make sense of many of the culture's otherwise absurd actions and philosophies. Our hatred of the body. The certainty that nature is red in tooth and claw. The long-standing movement toward centralized control. The neurotic insistence on repeatability (and control) in science, and the insane exclusion of emotion—which means the exclusion of life—from both science and economics. Using the lens of domestic violence to look at civilization's unwavering violence helps to make sense of all of these symptoms, but the important thing about using this lens as it pertains to the sixth premise of this book, that of civilization's unredeemability, is that perpetrators of domestic violence are among the most intractable of all who commit violence, so intractable, in fact, that in 2000, the United Kingdom removed all funding for therapy sessions designed to treat men guilty of domestic violence (putting the money instead into shelters and other means of keeping women safe from their attackers). Sandra Horley, chief executive of *Refuge*, that country's largest single provider of support to abused women and children, said: "I am not a hard-line feminist and I am not against men receiving help, but in many years of experience I have known only one man who has changed his behaviour." *The Guardian* put it simply: "There is no cure for men who beat their wives or partners, according to new Home Office research."[88]

If perpetrators of domestic violence cannot be cured, they must simply be stopped. If you believe, as I think I sufficiently showed in *A Language Older Than Words*, that familial violence within this culture is in many ways a microcosm of the violence the culture tricks out on the larger stages of history and the landscape, the implications for the culture—and its human and nonhuman victims—are, I think, sobering. As well as exploring the psychological irredeemability of this culture I discussed in that book many of the reasons for the culture's death urge—its urge to destroy all life, including our own—and the reasons for this urge's intractability.

In *The Culture of Make Believe*, I approached the question of the culture's essential destructiveness, and its death urge, from an entirely different direction, exploring the mutually reinforcing interplay of an economic and social system based on competition; the belief that humans are the apex of creation and our culture is the apex of this apex (it's always been pretty clear to me that all of evolution has taken place simply to bring me into existence, so that I can watch television); the valuing of material production over all else, including (most especially) life; the consistent preference for abstraction over the particular (manifesting, to provide three quick examples among many, as the pro-

mulgation of moral systems based on abstract principles rather than circumstances; as the flood of pornography (abstract images of naked women on the
internet account for $90 billion in revenue per year, making porn the number
one cash generator online, accounting for 13 percent of all revenue); and as the
ability, and proclivity, to kill at ever-greater psychic and physical distances); and
the increasing bureaucratization of this society. I showed how all of these vec-
tors come together to lead ineluctably to the attempted elimination of all diversity, to the attempted killing of the planet, and to the increasingly routine mass
murder of fellow humans (and of course nonhumans).

I'm taking a more fundamental approach here to understanding the reasons
for the implacability of this culture's violence, and I'm discovering that just as
all roads lead, as the saying goes, to Rome, all pathways here lead to the perception and articulation of civilization's basis in exploitation. In other words, it
doesn't really matter whether we're talking about the psychological, social/eco-
nomic, or physical/resource levels (none of which are separable anyway), we
come to the same conclusion. To put this yet another way, the micro manifests
the macro, which mirrors back the micro. Or to change terms once again, we're
in trouble, and we need to figure out what we're going to do about it.

Because every city-state (and now the entire globally interconnected industrial economy) relies on imported resources, our entire culture's basis in
exploitation must remain in place no matter how spiritual, enlightened, or
peaceful we may seem to ourselves, may claim to be, or may in fact personally
become. This basis in violence is in place whether or not we choose to acknowledge it. It is in place whether or not we call ourselves peaceloving, and whether
or not we tell ourselves (each time) that we are fighting to bring freedom,
democracy, and prosperity to people who, unaccountably, often do not seem
to want what we have to offer. Stripped of all lies, we are fighting, or rather
killing (remember premise four), to take their resources. More precisely, those
in power are doing so. More precisely yet, those in power are ordering their servants to do so, servants who have bought into the belief that those in power are
entitled to take these resources.[89]

This culture has killed a lot of people, and will continue to do so until it collapses, and probably long after. It must, because these killings inhere in the
structure and physical needs of the society, and so are not amenable to change.
Appeals to conscience, to humanity, to decency are thus doomed even before
they're made (and in fact can be harmful insofar as they allow all of us—from
presidents to CEOs to generals to soldiers to activists to people who don't much
think about it—to pretend those in power could maintain that power without

violence, and that the material production on which the entire culture is based could continue also without violence), not only because those in power have shown themselves—similarly to abusers in family violence, for similar reasons—eager to commit precisely as much violence as they can get away with, and not only because those in power have shown themselves psychologically impervious to such appeals (*Dear Adolf, Please don't hurt the Jews, nor take land from the Slavs or Russians. Be a pal, okay?*) but more importantly—and more implacably—the institutions these individuals serve are functionally just as impervious to the appeals as the individuals are psychologically. They need the resources, and will get them, come the hell of depleted-uranium induced malformations or the high water of melted ice caps. All of this means that movements for peace are damned before they start because unless they're willing to unmake the roots of this culture, and thus the roots of the violence, they can at best address superficial causes, and thus, at best, provide palliation.

There are many superficial causes of the culture's violence. There is the fact that those who make the political decisions that guide this culture are more interested in increasing their own personal power and the power of the state than they are in human and nonhuman well-being. Another way to say this is that gaining and maintaining access to resources, and facilitating production, are more important to them than life. Another way to say this is that power is more important to them than life. Another way to say this is that they are insane. If this were a root of the problem instead of a superficial manifestation, we could undermine the violence of this culture by simply replacing these decision-makers with those more reasonable, with those more sane, with those more humane, with those more human. But imagine if an American president decided tomorrow that the U.S. would no longer allow corporations to take oil from any region where the people themselves (*not* the government) did not want to relinquish it. The same would hold for metals, fish, meat, wood. Everything. What's more, no resources would be extracted if their removal would harm the natural world in any way. In other words, the president decided to put in place a truly non-exploitative, sustainable economy, the sort of economy all but psychopaths would *say* they want, the sort of economy that environmental and social justice activists say they're working toward. Presuming Congress and the Supreme Court went along—an extraordinarily dubious presumption—and presuming the president wasn't assassinated by CIA operatives or oil or other company hirelings—even more dubious—prices would skyrocket, the American way of life would implode, and riots would (probably) fill the streets. The economy would collapse. Soon, the president's head would be displayed atop the fence at 1600 Pennsylvania Avenue. The point is that the only people fit to be President are those who can institute policies that value economic production over life. A sane and humane person would not and could not last in that position.

Another superficial cause of the violence is that those who make the economic decisions (as opposed to political decisions, insofar as there is a difference) in this culture, too, are more interested in accumulating power—in this case monetary wealth—than they are in enriching the human and nonhuman communities that surround them. By itself, their interest in mining these communities would not be any more of a problem than any other compulsion, like excessive cleaning or obsessive hand-wringing. It really only becomes a problem because the power-hungry and the greedy work closely together as (somewhat) separate parts of the same corporate state, with the power-hungry wielding the military and police as muscle for the greedy, guaranteeing that the rich will get the resources required for them to increase their wealth—at gunpoint, if necessary—and guaranteeing also that those who effectively oppose these transfers of resources will get killed.

But even the conjoining of commerce and politics is, by itself, not a *source* of the violence, but a mechanism for it. If the lock-step march of government and industry were the essential cause of the culture's violence, we could solve it relatively easily by calling a constitutional convention and inserting new checks and balances to prevent this in the future. And if those in power were to oppose us, continuing their current policy of taxing us without representing us, well, we could simply follow the advice of Thomas Jefferson, Abraham Lincoln, and the Beatles and say we want a revolution (recognizing that the Beatles waffled a bit more than the other two, although listening carefully to the doo-wop version I think provides a clue to their beliefs). But we would find, after the dust settled and the blood stopped flowing in the streets, that our glorious new revolutionary government faced the same old problem of how to take resources from the country and give them to the city, to the producers. Our new bosses would of necessity be as violent as our old bosses.

We could easily assemble a long list of other mechanisms or superficial causes of violence. There is the fact that those in power have surrounded themselves with institutions such as the military and judicial systems (in fact the entire governmental structure) in order to protect and maintain their power. There is the fact that the social system rewards the insatiable accumulation of wealth and power. There is the fact that we are all immersed in a mythology that, far from causing us to see this accumulation as a great source of violence, causes us to see it as not only acceptable, reasonable, and desirable, but the only way to be, the way, in fact, that "the real world" works. There is the fact that this same mythology glorifies violence, so long as it is perpetrated only by those in power or their surrogates: top Hollywood executives recently met with the president's senior advisor to, in the words of *The New York Times*, find "common ground on how the entertainment industry can contribute to the war effort, replicating in spirit if not in scope the partnership formed between filmmakers and war planners in the 1940s"; simultaneously, Tom Cruise is said to be concerned about his role in his next movie as a garbage collector, oh, sorry, a CIA operative, wanting to show the "CIA in as positive a light as possible."[90] There is the arrogance of the civilized, who consider themselves morally and otherwise superior to all others, and who therefore may exploit or exterminate these others with moral impunity (and immunity). There is the arrogance of the humanists, who believe us separate from and superior to nonhumans, who may also then be exploited or exterminated at will. And there is the culture's death urge, pushing us all to end all life on the planet while simultaneously driving each and every one of us as much out of our bodies as we are out of our minds.

All of these are in place, and there is good reason to work on halting or slowing all of these. In no way am I suggesting we shouldn't work to reduce the harmfulness of these mechanisms or superficial causes, anymore than I would suggest people not work on rape crisis hot lines, or that people not attempt to stop individual rapists. But I would also not suggest that working on a rape crisis hotline will in any way halt the very real crisis of rape. No one I know who has ever worked on issues of men's violence against women has suggested that it will. Nor have they suggested that if only women will think nice enough thoughts, or practice the right sort of spiritual exercises, that men will stop raping women. Mitigation can be wonderful, and important, but we should not delude ourselves into thinking it is anything more than mitigation. Begging government and industry to stop destroying the planet and to stop killing people the world over is never going to work. It can't.

❨ ❨ ❨

This might be a good place to mention the primary stated goal of the United States military. No longer is it simply, as it was in the days of Manifest Destiny, the coast-to-coast conquest of the continental United States and the dispossession and/or extermination of the land's original inhabitants. Nor is it what it was at the turn of the twentieth century—the time of Theodore Roosevelt's ironically named Good Neighbor Policy—when the frontier was extended westward to the Philippines and beyond, where the U.S. killed one out of every ten Filipinos and did the same to residents of other countries in order to liberate them from themselves, and brought those they did not kill under their control so they could better use their land. Of course it was not only westward that they looked, but south and east as well, to bring as much of the globe as possible under U.S. control. Nor is the goal merely what it was fifty years ago, when National Security Council documents stated the obvious need for "a political and economic climate conducive to private investment,"[91] and when State Department Policy Planning staff head George Kennan said that if "we" are to maintain a "position of disparity" over those whose resources "we" must take, "We should cease to talk about vague and . . . unreal objectives such as human rights, the raising of the living standards, and democratization," and instead should "deal in straight power concepts," not hampered by idealistic slogans about "altruism and world-benefaction."[92] All of this is merely another way to say the same thing I've been hammering so far, that in order to move resources into cities—in order to steal resources—you have to use physical force. Nor does the present goal leave as much to the imagination as it did a mere decade ago, when a Defense Planning Guide (written when current Vice President Dick Cheney was Secretary of Defense) stated explicitly that the U.S. must hold "global power" and a monopoly of force,[93] and that it must make certain that no others are allowed even "to protect their legitimate interests."[94]

Instead, after all this time, those in power have finally gotten to the point. Or rather, their powers to surveil and kill have finally caught up with their lust for control. And they have articulated this clearly. The U.S. Joint Chiefs of Staff recently put out their *Joint Vision 2020*, which defines their goals for the next twenty years and beyond. The U.S. military, according to the first words of this document, consists of: "Dedicated individuals and innovative organizations transforming the joint force for the 21st Century to achieve full spectrum dominance." To make sure we get the point, the military bolded the phrase "full spectrum dominance." Just in case we still don't get it, the phrase is repeated thirteen more times in this brief, 8,700-word document, and is specified in U.S. military press releases and articles as the "key phrase" of the vision statement.[95]

I suppose we should at least thank them for their directness, although the question remains, as always: do we really get the point?

COUNTERVIOLENCE

The condemnation of liberation movements for resorting to violence or armed struggle is almost invariably superficial, hypocritical, judgmental, and unfair, and tends strongly to represent another example of the generalised phenomenon of "blaming the victim." The violence of the situation, the pre-existing oppression suffered by those who eventually strike back, is conveniently ignored. The violence of the oppressed is a form of defensive *counterviolence* to the violence of conquest and oppression. In no armed national liberation movement I know of in history has this not been the case.

Jeff Sluka [96]

THIS BOOK ORIGINALLY WAS GOING TO BE AN EXAMINATION OF THE circumstances in which violence is an appropriate response to the ubiquitous violence upon which this culture is based. More specifically, it was going to be an examination of when counterviolence, as termed by Franz Fanon, is an appropriate response to state or corporate violence. I wanted to write that book because whenever I give talks in which I mention violence—suggesting that there are some things, including a living planet (or more basically clean water and clean air, by which I mean our very lives), that are worth fighting for, dying for, and killing for when other means of stopping the abuses have been exhausted, and that there exist those people (often buttressed or seemingly constrained by organizations) who will not listen to reason, and who can be stopped no way other than through meeting their violence with your own—the response is always the same. Mainstream environmentalists and peace and justice activists put up what I've taken to calling a "Gandhi shield." Their voices get thin, and I can see them psychically shut down. Their faces turn to stone. Their bodies do not move, but the ghosts of their bodies form fingers into the shapes of crosses as they try to keep vampires and evil thoughts at bay, and they begin to chant "Gandhi, Dalai Lama, Martin Luther King, Jr., Gandhi, Dalai Lama, Martin Luther King, Jr." in an effort to keep themselves pure. Grassroots environmentalists generally do the same, except after the talk some will sidle up to me, make sure no one is watching, and whisper in my ear, "Thank you for raising this issue." Often, young anarchists get excited, because someone is articulating something they know in their bones but have not yet put words to, and because they've not yet bought into—and been consumed by—the culture. The most interesting response comes from some of the other people with whom I've spoken: survivors of domestic violence; radical environmentalists; Indians; many of the poor, especially people of color; family farmers; and prisoners (I used to teach creative writing at Pelican Bay State Prison, a supermaximum security facility here in Crescent City). Their response is generally to nod slowly, look me hard in the eye, then say, "Tell me something I don't already know." Some will say, "What are you waiting for, bro? Let's go."

A major reason for the difference in response, I realized a long time ago, was that for these latter groups violence is not a theoretical question to be explored abstractly, philosophically, or spiritually,[97] as it can often be for more main-stream activists, for those who may not have experienced violence in their own bodies, and who can then be more distant, even—and I've seen this a lot—act-ing as if these were political or philosophical games instead of matters of life and death. The direct experience of violence, on the other hand, often brings these questions closer to the people involved, so the people are not facing the ques-tions as "activists" or "feminists" or "farmers" or "prisoners," but rather as human beings—animals—struggling to survive. Having felt your father's weight upon you in your bed; having stood in clearcut-and-herbicided moon-scape after moonscape, tears streaming down your face; having had your chil-dren taken from you, land stolen that belonged to your ancestors since the land was formed, and your way of life destroyed; having sat at a kitchen table, fore-closure notice in front of you for land your parents, grandparents, and great-grandparents worked, shotgun across your knees as you try to decide whether or not to put the barrel in your mouth; feeling the sting of a guard's baton or the jolt of a stun gun ("I was tired," one of my students wrote of being tasered, "I was 50,000 volts of tired")—to suffer this sort of violence directly in your body—is often to undergo some sort of deeply physical transformation. It is often to perceive and *be in* the world differently.

Not always. We can all list political prisoners who have been tortured, nuns who have been raped, who have emerged from these horrors uttering forgive-ness for their tormentors. But this is not, for the most part, the experience of the people I have met—(funny, isn't it, how the ones who forgive are the ones whose stories we're most likely to hear: could there possibly be political reasons for this? Remember, all writers are propagandists)—and I'm not convinced that this forgiving response is necessarily and generically better, by which I mean more conducive to the survivor's future health and happiness, and by which I mean especially more conducive to the halting of future atrocities. Sometimes it may be, and sometimes, as we shall eventually see, it may not.

❨ ❨ ❨

A story. Seattle, late November, 1999. Massive protests against the World Trade Organization, and more broadly against the consumption of the world by the rich, turn violent, as police shoot tear gas, pepper spray, and rubber bul-lets against nonviolent, nonresisting protesters. Among the tens of thousands of protesters are several hundred members of what is called the Black Bloc, an anarchist group that doesn't play by the rules of civil disobedience. Civil disobedience is normally a fairly straightforward dance between police and protesters. There are certain rules, such as trespassing, that protesters and police generally agree protesters will break, after which it is just-as-generally agreed that protesters will be arrested, often roughed up a little bit, and then usually given nominal fines. Sometimes, as in the case of Plowshares activists, whose courage can never be questioned, the dance becomes surreal. The activists show up at military installations, beat on pieces of military technology with hammers (thus the name; beating weapons into plowshares), and pour their own blood onto the devices in symbolic protest of the blood these weapons shed. They then wait for the military police to show up—or call the police themselves to *make sure* they show up—get arrested, and sentenced to years and years in prison. Other times the dance becomes comical, as when protest organizers provide police with estimates of the numbers of people who have volunteered to be arrested (so police can schedule the right number of paddy wagons), and also provide police with potential arrestee's IDs so the process of arrest will be smooth and easy on everyone involved. It's a great system, guaranteed to make all parties feel good. The police get to feel good because they've kept the barbarians from the gates, the activists feel good because they've made a stand—*I got arrested for what I believe in*—and those in power feel good because nothing much has changed.

The Black Bloc doesn't play by these rules (not, as we'll eventually see, that their rules necessarily work better). In Seattle, they broke windows of targeted corporations in order to protest the primacy of private property rights, which they distinguish from personal property rights: "The latter," a subgroup of the Black Bloc stated, "is based upon use while the former is based upon trade. The premise of personal property is that each of us has what s/he needs. The premise of private property is that each of us has something that someone else needs or wants. In a society based on private property rights, those who are able to accrue more of what others need or want have greater power. By extension, they wield greater control over what others perceive as needs and desires, usually in the interest of increasing profit to themselves."[98]

Although the actions of the Black Bloc have been painted as violent by pacifists, members of the corporate media, and, ironically enough, by gun-wielding police, Black Bloc members themselves deny this: "We contend that property destruction is not a violent activity unless it destroys lives or causes pain in the process. By this definition, private property—especially corporate private property—is itself infinitely more violent than any action taken against it."[99] It seems pretty obvious that unless you're a hard-core animist, it's not really possible to perceive breaking a window—especially a store window, as opposed to a bedroom window at three in the morning—as violent. But because of Premise Five, when the window belongs to the rich and the rock to the poor, that act becomes something akin to blasphemy. The anarchists continued, "Private property—and capitalism, by extension—is intrinsically violent and repressive and cannot be reformed or mitigated."[100] Their stated reason for the property destruction was, "When we smash a window, we aim to destroy the thin veneer of legitimacy that surrounds private property rights."[101] Of course the Black Bloc did not target just any property—so don't worry, they're not going to break *your* windows next— but they instead targeted the property of egregiously violent corporations, such as, "Fidelity Investment (major investor in Occidental Petroleum, the bane of the U'wa tribe in Columbia); Bank of America, U.S. Bancorp, Key Bank and Washington Mutual Bank (financial institutions key in the expansion of corporate repression); Old Navy, Banana Republic and the GAP (as Fisher family businesses, rapers of Northwest forest lands and sweatshop laborers); NikeTown and Levi's (whose [sic] overpriced products are made in sweatshops); McDonald's (slave-wage fast-food peddlers responsible for destruction of tropical rainforests for grazing land and slaughter of animals); Starbucks (peddlers of an addictive substance whose [sic] products are harvested at below-poverty wages by farmers who are forced to destroy their own forests in the process); Warner Bros. (media monopolists); Planet Hollywood (for being Planet Hollywood)."

Now here's the interesting thing. As members of the Black Bloc broke windows, the police, who already had their hands full shooting at the civil disobedience crowd (many pacifists later claimed police fired in response to Black Bloc actions, but this is demonstrably untrue: police were shooting long before the first Starbucks window exploded into shards), were unable to protect this corporate property. That's a good thing, right? Well, according to some of the pacifists, evidently not. They stepped in to protect the corporations, going so far as to physically attack individuals targeting corporate property.[102]

These protectors of corporate property included many people who otherwise do a lot of good work. For example there were longtime liberal/Green politicians and activists associated with Global Exchange, a "fair trade organization" focusing on corporate accountability and on eradicating sweatshops around the world. One can go to Global Exchange's website, and learn that "Global Exchange and other human rights organizations have taken steps to eradicate sweatshops by organizing consumer campaigns to pressure corporations such as GAP Inc. (GAP, Old Navy, and Banana Republic) and Nike to pay workers a living wage and respect workers' basic rights."[103] One can also learn that "Sadly, there is not one major clothing company that has made a commitment to completely eradicate abusive labor practices from its garment factories. While we [Global Exchange] continue to pressure corporations to become socially responsible, we as consumers can support the following alternatives."[104] It's misleading for Global Exchange to use the plural on *alternatives*, since the only alternative that follows consists of variations on the theme of their next three words (bolded!): "**Buy Fair Trade!**"[105] Coincidentally enough, shoppers can Buy Fair Trade! right there at the website, as the good people at Global Exchange "offer consumers the opportunity to purchase beautiful, high quality gifts, housewares, jewelry, clothing, and decor from producers that [*sic*] were paid a fair price for their work."[106] Thus I could buy a Guatemalan Shopping Bag ("for her") for $43, or a "Traveler's Basket" ("for him") priced at a mere $59 ("Say the perfect Bon Voyage to a loved one on pursuit of the next adventure [or treat yourself before the journey begins]. The Traveler's Basket offers a warm collection of traveling essentials from around the world. Guatemalan Hemp Trifold Wallet from Hempmania, Zip Passport Holder from Guatemala, Handmade Natural Paper Journal from Nepal, Guatemalan Hacky Sack"). The Traveler's Basket would be really handy if you also have several thousand dollars you can pony up to go on one of Global Exchange's "Reality Tours" of third world nations (Sheesh, would you quit your worrying? *Of course* you'll stay in three-star hotels). Afterwards you'll be able to tell your friends that you "watched a performance of the band made up of young people (with tin cans for drums) and toured the favela." (And geez-Louise, will you get over the whining thing? *Of course* when you've finished the Reality Tour, you won't have to stay in the favela: *you* get to come home!)[107]

Perhaps I'm being too harsh. Global Exchange does offer people the opportunity to change the culture in more ways than merely buying things. For example, by following a link you can "Send a fax to [CEO] Philip Knight asking that Nike take immediate and concrete steps to ensure that the people making the company's products aren't facing abuse and intimidation."[108] I'm sure Phil will personally read your fax, and I'm sure yours will be the one that convinces him to give up the practices that have made him one of the richest men in the world.

If the fax doesn't work, you can always try a rock through his window. But be warned: folks at Global Exchange probably won't approve (see Premise Five).

Back to Seattle, where black-clad anarchists were throwing rocks through the windows of Nike and other stores, and police were nowhere to be seen. Who was going to protect the stores? Pacifists to the rescue. Many shouted "You're ruining our demonstration"[109] as they formed human chains in front of chain stores. Others began "physically assaulting window smashers while yelling 'This is a non-violent protest.'"[110] One shared her thoughts with a reporter for *The New York Times*, "Here we are protecting Nike, McDonald's, the Gap, and all the while I'm thinking, 'Where are the police? These anarchists should have been arrested.'"[111] Local kids—mainly people of color from the Seattle equivalent of the favelas (*favela* in Brazil, *poblacione* in Chile, *villa miseria* in Argentina, *cantegril* in Uraguay, *rancho* in Venezuela, *banlieue* in France, *ghetto* in the United States[112])—joined the anarchists, smashed some windows, and started liberating some of the goods (I believe the technical term for this is *looting*). The crowd of vandals—from the Latin *Vandalii*, of Germanic origin: a member of a Germanic people who lived in an area south of the Baltic between the Vistula and the Oder, overran Gaul, Spain, and northern Africa in the fourth and fifth centuries CE, and in 455 sacked Rome—was the most multicultural and multiracial group of the protest. As one anarchist later commented: "When [writer] Jeffrey St. Clair started to leave town on December 3rd, a black youth rushed up to him and excitedly asked if this WTO thing will come back next year. Sure, the labor march and enviro's were mostly white folks. But the action against corporate property was the one truly diverse, inclusive, festive action."[113] Pacifists were caught on videotape assaulting young black men—the whole time chanting "non-violent protest"—and attempting to hold them to turn over to police.[114] I'm sure that had these youths wanted to do some *real* damage to Nike, they could have gone to the library, logged onto computers, and sent Phil Knight a bunch of faxes. And when they'd finished at the library, they could have gone back to their ghetto and played tin can drums for tourists.

All of which is to say that pacifism makes strange bedfellows.

❨ ❨ ❨

To keep dogmatic pacifists from calling the cops and then holding me till they arrive, I need to say that I no more advocate violence than I advocate nonviolence. Further, I think that when our lifestyle is predicated on the violent theft of resources, to advocate nonviolence without advocating the immediate dismantling of the entire system is not, in fact, to advocate nonviolence at all, but to tacitly countenance the violence (unseen by us, of course: see Premise Four) on which the system is based. I advocate speaking honestly about violence (and other things), and I advocate paying attention to circumstances. I advocate not allowing dogma to predetermine my course of action. I advocate keeping an open mind. I advocate a rigorous examination of all possibilities, including fair trade, "Reality Tours," lawsuits, writing, civil disobedience, vandalism, sabotage, violence, and even voting. (Recently I was talking to a number of college students about the fix we're in and said, "We need to stop civilization from killing the planet by any means necessary." An instructor at the college, a longtime pacifist, corrected me, "You mean by any nonviolent means, of course." I replied that I meant precisely what I said.) I advocate listening to my body. I advocate clean water and clean air. I advocate a world with wild salmon in it, and grizzlies, and sharks, whales (just yesterday I read—not in the capitalist press, obviously—that the federal government recently refused to provide protection for the North Pacific right whale, the world's most imperiled large whale, because, in the words of an industry spokesperson—oh, sorry, a government spokesperson— "the essential biological requirements of the population . . . are not sufficiently understood"[115]), red-legged frogs, and Siskiyou Mountain salamanders (then tonight I read—also not in the capitalist press, silly: what do you think their purpose is, to provide useful information?— "The rare Siskiyou Mountain salamander may be facing extinction because the Bureau of Land Management will soon allow Boise Cascade to begin logging in the amphibian's [last remaining] habitat"[116]). I advocate a world in which human and nonhuman communities are allowed to live on their own landbases. I advocate not allowing those in power to take resources by force, by law, by convention, or any other real or imagined means. Beyond *not allowing*, I advocate actively stopping them from doing so.

❨ ❨ ❨

Most of our discourse surrounding counterviolence in this country runs from nonexistent all the way to superficial. So the course for this book seemed clear. One-by-one I would carefully examine the arguments that are commonly— and I have to say, I've learned through long and tedious experience, most often unthinkingly—thrown out against any use of violence in any (especially political) circumstances. *You can't use the master's tools to take down the master's house*, says the person still attempting to work within religious, philosophical, economic, and political systems—Can you say "green capitalism?"—devised explicitly to serve the rich (John Locke put it succinctly: "Government has no other end than the preservation of property"[117]). *You will become just like they are*, say people whose knowledge of violence is almost entirely theoretical (I asked some of my students, in for murder, if killing someone is a psychological or spiritual Rubicon, and some said *yes* while some said *no*; unfortunately, Sitting Bull, Crazy Horse, and Geronimo aren't available for comment, although I'd wager their wars against civilization did not civilize them). *Violence doesn't work,* say those who tell us to shop and fax our way to sustainability, and who have to ignore—as is true for most of us working on these issues, else we'd probably go mad—that nothing is working to stop or even significantly slow the destruction. Hell, as I mentioned before, we can't even slow the destruction's *acceleration*! I would say this is partly because those in power have on their side so many tanks and guns and airplanes, as well as writers, therapists, and teachers; partly because we're all crazy (and our sickness is very strong); partly because in the main neither our violent nor nonviolent responses are attempts to rid us of civilization itself—by allowing the framing conditions to remain we guarantee a continuation of the behaviors these framing conditions necessitate—and partly because we're all scared spitless about doing what we all know needs to be done.

But in the couple of years between the book's conception and the start of writing I realized that the question of whether or when to use violence is only a small yet integral part of the real question I'm after. I'm after much bigger game indeed.

LISTENING TO THE LAND

To be civilized is to hold oneself in opposition to nature, which is to hold oneself in opposition to oneself, to be ashamed of the animality of the self, which to the fully civilized means the "filth" of the self. All of this destroys any possibility of communication or entering into communion with anyone but other civilized humans. If we listen to the creatures and to the elements, and even to our bodies, we are then primitive, backwards. So we learn very early to put that away. We learn to despise ourselves and to feel ashamed of our bodies, to hate the dirt and to hate everything about us, because we're human, which means we're humus: they come from the same Latin root, earth and dirt. But self-loathing is a difficult thing to acknowledge—maybe the most difficult—so all those characteristics we must loathe if we are to be civilized, if we are to dominate, get dumped into others who bear the shame and who end up feeling dirty.

Jane Caputi[118]

I THINK FOR THE MOST PART IT'S NOT ONLY ABUSERS WHO DON'T change. Most of us don't. Sure, sometimes somebody or another may have an epiphany, like Saul of Tarsus did in the Bible. But let's be honest about that one, too: after he saw the light of God and got knocked off his ass, he may have changed his name to Paul, but he was still a domineering asshole. It's just that now instead of persecuting Christians he used Christianity as a vessel for his pre-existing rigidity, making certain the reasonably new religion mirrored his hierarchical perception of the world.

Most often, change, at least on a social level, occurs the way Max Planck described it: "A new scientific truth does not triumph by convincing its opponents and making them see the light, but rather because its opponents eventually die, and a new generation grows up that is familiar with it."[119] Years ago I read Oswald Spengler's *Decline of the West*. It's a long book, from which I really remember only one image. I think Spengler would be pleased at which one. A culture is like a plant growing in a particular soil. When the soil is exhausted— presuming a closed system (i.e., the soil isn't being replenished)—the plant dies. Cultures—or at least historical (as opposed to cyclical) cultures—are the same. The Roman Empire exhausted its possibilities (both physical, in terms of resources, and psychic or spiritual), then hung on decadent—I mean this in its deeper sense of decaying, although the meaning having to do with debauchery works, too—for a thousand years. Other empires are the same. The British Empire. The American Empire. Civilization itself has continued to grow by expanding the zone from which it takes resources. The plant has gotten pretty big, but at the cost of a lot of dead soil.

I think the exhausted soil metaphor works for individuals, too: they don't generally change until they've exhausted the possibilities of their previous way of being.

Last year I received an email from a woman who said that my work had saved her life. She had many times tried to kill herself, and was contemplating suicide again when she came across a passage in my work describing part of the reason this culture's death urge manifests the way it does, in the widespread killing of humans and nonhumans, and in the killing of the planet. This death urge is partly a simple desire to die to this way of living that does not serve us well, but because we in this culture have forgotten that the spiritual exists, and have deval-ued the metaphorical, we do not understand that this death does not have to be physical, but could be transformative. Dying to one way of being so you can be reborn transformed is the oldest metaphor in the world, one the world is built on. But we forget, and so we build daisy-cutters and depleted-uranium shells, and we kill without ceasing. The woman said her own death urge might not have to manifest in the taking of her own life. Maybe she just wanted to transform. We corresponded a bit, she asked if we could take a walk when she was passing through town, and I agreed. It was a good walk, through meadows of thick sharp-edged grasses perfect for ground-nesting birds, into a sandy-soiled scrub pine for-est near the ocean, and along the ocean beaches themselves. She was a good woman, smart, dedicated, knowledgeable about wild things. She was also in agony. Her agony derived partly from the aftereffects of the horrendous violence her father visited upon her as a child, and partly from her sensitivity to the sim-ilarly horrendous violence our culture perpetrates on the natural world. She said that instead of killing herself, she was going to spend three months alone in the desert, talking and listening to coyotes, clouds, ravens, rabbitbrush, and a cool, clear river. She hoped to return a new person.

She wrote me briefly when she returned, and then again a couple of months ago. She seemed to be doing well.

And then yesterday I received the letter. Evidently other people got the letter, too. It began, "Dear Friend, By the time you read this I will have done something that will come as no surprise to many of you. I will have committed suicide." The letter went on to describe her attempts to overcome her pain, and ended with her arrangements for what should come after her death. She expressed regret that the law would not allow her to become food for wild animals.

After I got over my shock and had begun to move through my sorrow over the death of a good person I did not really know, I began to feel a stirring that within a few hours became the understanding that people usually don't change. She may have thought she changed when she read my work, but she didn't. She continued to daily ask the question of whether she should live or die until finally the answer came up *die*.

I know that I, too, carry scars both physical and emotional from my child-hood that will never be healed. I know also that I will ask the same questions when I am old as when I was young. And I have to ask (the genesis of my question in no way negating its current relevance or importance): how much did my early experience of my father's violence lead me to ask the question I'm asking now, about when is counterviolence an appropriate response to the violence of a dominator?[120] Similarly, my mother is the same person she was twenty years ago, only wiser, and more tired. Most of my students at the prison love drugs—or at this point love writing about them—as much as they ever did. Often I only have to mention *blunt, dub, heroin, crack, crank,* and they'll reminisce and laugh as those possessed. And even though they hate being imprisoned with an intensity I've rarely seen matched, and even though in many cases it was drugs that got them there, when I ask if they will use again when they get out (or for the lifers, *would* and *if*), most say *yes.* Statistics on addicts remaining sober (much less free of craving) are fairly dismal, and run from a low of ten to a high of 40 percent, with one writer commenting, "Chronic relapse is part of the etiology of addiction."[121]

But of course I'm overstating when I say people don't change. They do. I did. I could have turned out like my father. I could have remained a scientist. I could have—god help me—remained a Republican, as I was in my teens, or just as bad, a Democrat, as I was in my early twenties. People do change. But change takes hard work, luck, and some treasured reward on the other side; even when these are all present, it still doesn't happen often. And that's only on the scale of one person, with only one lifetime of momentum built into that trajectory. How much more difficult is it to expect change when we have six thousand years of history, as well as space heaters, major league baseball, tomatoes in January, strawberry cheesecake, and the capacity at any time to bid on 1,527,463 products ("most with 'NO RESERVE PRICE'") at ubid.com? And how much more difficult than that when those in power have prisons, guns, and sophisticated surveillance technologies at their disposal? And how much more difficult than that when those in power have television, newspapers, and compulsory schooling to promulgate their perspective? And how much more difficult than that when we promulgate it ourselves?

⟨ ⟨ ⟨

Several years ago the environmentalist and physician John Osborn pointed out to me that many environmentalists begin by wanting to protect a piece of ground and end up questioning the foundations of Western civilization. I agree, obviously, but would emend his comment in two ways. The first is that it's not only environmentalists whose involvement in their particular struggle leads them to question the basis of this whole way of living. Feminists, conservation biologists, anthropologists, historians, economists, anti-imperialists, anti-colonialists, prison activists, American Indian activists (obviously), other people of color, those who simply hate the wage economy: I've spoken with people who are each of these, and they've reached the same conclusions. Why? Because once the questioning begins the search for root causes leads you back to the primary problem: the culture itself. And why is the problem the culture itself? Because this way of life is based on exploitation, domination, theft, and murder. And why is this culture based on exploitation, domination, theft, and murder? Because it's based on the perceived right of the powerful to take whatever resources they want. If you perceive yourself as entitled to some resource—and if you're unwilling or incapable of perceiving this other as a being with whom you can and should enter into a relationship—it doesn't much matter whether the resource is land, gold, oil, fur, labor, or a warm, wet place to put your penis, nor does it matter who this other is, you're going to take the resource.

The second way I would emend his comment is by adding the words *in private*. This questioning—and in fact rejection—of civilization happens almost exclusively in private, because a lot of these activists are afraid that if they spoke this in public, people would laugh at them, and they would lose whatever credibility they have—or feel they have. It's always a difficult question. Do I stop this clearcut now, even knowing that without a fundamental change in the culture (see Premise Six) I'm merely putting off the date of execution till the next corporate Congressman figures out the next way to make sure the timber companies get out the cut? Or do I tell the truth, stand by, and watch the trees fall? The environmentalists I know are hanging on by our fingernails, praying that salmon, grizzlies, lynx, bobcat, Port Orford cedars survive till civilization comes down. If they survive, they'll have a chance. If they don't, they're gone forever.

I'm sick of these options. I want to stop the destruction. I want to stop it now. I'm not satisfied to wait for civilization to exhaust its physical and metaphorical soil, then collapse. In the meantime it's killing too many humans, too many nonhumans; it's making too much of a shambles of the world.

The seventh premise of this book is: *The longer we wait for civilization to crash—or before we ourselves bring it down—the messier will be the crash, and the worse things will be for those humans and nonhumans who live during it, and for those who come after.*

Had somebody snuffed civilization in its multiple cradles, the Middle East would probably still be forested, as would Greece, Italy, and North Africa. Lions would probably still patrol southern Europe. The peoples of the region would quite possibly still live in traditional communal ways, and thus would be capable of feeding themselves in a still-fecund landscape.

Fast forward a few hundred years and we can say the same in Europe. Somehow stop the Greeks and Romans, and the indigenous people of Gaul, Spain, Germany probably still survive. Wolves might howl in England. Great auks might nest in France, providing year-round food for the humans who live there. Salmon might run in more than token numbers up the Seine. The Rhine would be almost undoubtedly clean. The continent would be forested. Many of the cultures would be matrifocal. Many would be peaceful.

Had someone brought down civilization before 1492, the Arawaks would probably still live peacefully in the Caribbean. Indians would live in ancient forests all along the Eastern seaboard, along with bison, marten, fisher. North, Central, and South America would be ecologically and culturally intact. The people would probably have, as always, plenty to eat.

Had someone brought down civilization before the slave trade took hold, 100 million Africans would not have been sacrificed on that particular altar of economic production. Native cultures might still live untraumatized on their own land all across that continent. There probably would be, as there always was, plenty to eat.

If someone had brought down civilization one hundred and fifty years ago, those who came after probably could still eat passenger pigeons and Eskimo curlews. They could surely eat bison and pronghorn antelope. They could undoubtedly eat salmon, cod, lobster. The people who came after would not have to worry about dioxin, radiation poisoning, organochloride carcinogens, or the extreme weather and ecological flux that characterize global warming. They would not have to worry about escaped genetically engineered plants and animals. There probably would have been, as almost always, plenty to eat.

If civilization lasts another one or two hundred years, will the people then say of us, "Why did they not take it down?" Will they be as furious with us as I am with those who came before and stood by? I could very well hear those people who come after saying, "If they had taken it down, we would still have earthworms to feed the soil. We would have redwoods, and we would have oaks in California. We would still have frogs. We would still have other amphibians. I am starving because there are no salmon in the river, and you allowed the salmon to be killed so rich people could have cheap electricity for aluminum smelters. God damn you. God damn you all."

❆ ❆ ❆

I know someone whose brother demolishes buildings. The trick, he says, is to position the charges precisely so the building collapses in place, and doesn't take out the surroundings. It seems to me that this is what we must do: position the charges so that civilization collapses in on itself, and takes out as little life as possible on its way down.

Part of the task of the rest of this exploration is to discover what form those charges will take, and where to put them.

(((

The past few weeks I've been in crisis. I'm scared. Scared of the implications of this work. Scared to articulate what I know in my heart is necessary, and even more scared to help bring it about. I mean, we're talking about taking down civilization here.

Last night I was at my mom's eating dinner and watching a little March Madness—the NCAA basketball tournament—and I kept thinking, as I watched UNC-Wilmington hold off USC in overtime after blowing a nineteen-point lead, a variant of the question my friend asked about what right I have to not let people live in cities. There are hundreds of thousands, if not millions, of people having fun watching these games. They're not trying to exploit anyone. They're not trying to kill the planet. What right do I have to so alter their lives? I'm not saying there would never be games again, because the lives of traditional indigenous peoples the world over are far more full of leisure and play than ours. I'm just saying that bringing down civilization would cause substantive changes in the way these people spend their time. And they may not—evidently they *do* not—want to change.

The answer came to me today. It's the same answer I gave my friend, which is that I think it's the wrong question. The question is: what right do all of these people have to destroy the lives of others by their very lifestyle?

It's hard. I would have no moral or existential problem destroying the lifestyles of those in power. The politicians, CEOs, generals, capitalist journalists. Those who, if faced with a Nuremberg-style tribunal, should and would find themselves at the end of a rope for their crimes against both the natural world and humanity. But what about Americans just trying to love their children and take them to the amusement park once a month, to buy them toys, to get them an education so they can get a job? If I were directing a movie instead of writing a book, it might be appropriate for me to add a montage of images of everyday life in civilization. Young children dancing to

"Y.M.C.A." at a minor-league baseball game. An audience watching Hamlet try-ing to decide whether he should kill the murderous king (You *do* regularly go to Shakespeare festivals, don't you?). People walking the aisles of independent bookstores, stopping to pick titles from the shelves. An ice cream truck. A pic-nic. But then to round out the montage I'd have to include children starving because the resources they need to live have been stolen; denuded hillsides, blasted streams, dammed and polluted rivers (I just heard that most of the rivers of southern England are so hormone-polluted that more than half of the male fish—in some cases all—are changing gender); prisons full of bored adults who've been convicted of crimes; factories full of bored adults who've not been convicted of crimes but are nonetheless sentenced to years of tedium; classrooms full of bored children being prepared for their boring lives in office or factory; factory farms full of bored (and tortured) chickens, pigs, cows, or turkeys; laboratories full of bored (and tortured) chimpanzees, rats, rhesus monkeys, mice.

The question quickly becomes: what rights do people have? More specifically, does anyone have the right to enslave another? More specifically yet, does any group of people have the right to enslave others—human or nonhuman—sim-ply because they have the power to do so, and because they perceive it as their right (and because they have created a propaganda system consisting of inter-twined religious, philosophical, scientific, educational, informational, eco-nomic, governmental, and legal systems all working to convince themselves and at least some of their human victims it is their right)? If not, what are you going to do about it? How much will it take? How far will you go in order to stop those in power from enslaving—and killing—the mass of humans, and in fact the planet?

❨ ❨ ❨

I often give talks, at universities and elsewhere. I gave one such talk last week. Just before I walked on stage, the person who brought me there whispered, "I forgot to tell you, but I publicized this as a speech about human rights. Can you make sure to talk about that?"

I nodded agreement, although I had no idea what to say. Everything that came to me was tepid, along the lines of "Human rights are good." I may as well say I'm for apple pie and the girl next door. Even though I didn't tell her this, I think she read my face. She smiled nervously. I smiled twice as nervously back. It's a good thing we weren't playing poker.

She went out to introduce me. I thought and thought, and wished there were a lot more upcoming events for her to talk about. I wished she would start announcing the day's major league baseball scores. I wished she would forecast the weather, and tell the fortunes of the people in the front row. But she didn't do any of that, and soon enough it was my turn. As I walked on stage, however, I suddenly knew what I had to say, and was reminded, as I often am, how quickly the mind can work under pressure, or at least how quickly it can work those times it doesn't seize up altogether. "Most people," I said, "who care about human rights and who talk about them in a meaningful fashion, as opposed to those who use them as a smokescreen to facilitate production and implement policies harmful to humans and nonhumans, usually spend a lot of energy demanding the realization of rights those in power give lip service to. Sometimes they expand their demands to include things—like a livable planet—people don't often associate with human rights. People have a right to clean air, we say, and clean water. We have a right to food. We have a right to bodily integrity. Women (and men) have the right to not be raped. Some even go so far as to say that nonhumans, too, have the right to clean air and water. They have the right to habitat. They have the right to continued existence."

People nodded. Who but a sociopath or a capitalist—insofar as there is a difference—could disagree with any of these?

"But," I continued, "I'm not sure that's the right approach. I think that instead of adding rights we need to subtract them."

Silence. Frowns. The narrowing of eyes.

"No one," I said, "has the right to toxify a river. No one has the right to pollute the air. No one has the right to drive a creature to extinction, nor destroy a species' habitat. No one has the right to profit from the labor or misery of another. No one has the right to steal resources from another."

They seemed to get it.

I continued, "The first thing to do is recognize in our own hearts and minds that no one has any of these rights, because clearly on some level we *do* perceive others as having them, or we wouldn't allow rivers to be toxified, oceans to be vacuumed, and so on. Having become clear ourselves, we then need to let those in power know we're taking back our permission, that they have *no right* to wield this power the way they do, because clearly on some level they, too, perceive themselves as having the right to kill the planet, or they wouldn't do it. Of course they have entire philosophical, theological, and judicial systems in place to buttress their perceptions. As well as, of course, bombs, guns, and prisons. And then, if our clear statement that they have no right fails to convince them—and I wouldn't hold my breath here—we'll be faced with a decision: how do we stop them?"

A lot of people seemed to agree. Then after the talk someone asked me, "Aren't these just different ways of saying the same thing?"

I wasn't sure what she meant.

"What's the difference between saying I have the right to not be raped, and saying to some man, 'You have no right to rape me'?"

I was stumped. Maybe, I thought, my mind actually *had* seized up, only so completely that I hadn't known it. The reason the words had come so quickly is because they were just a recapitulation of the obvious. I have a few male friends who routinely take something someone else says, change a word or two or invert the sentence structure, and then claim it as their own great idea. I've been known to do that myself. But then I realized there's an experiential difference between these two ways of putting it. A big one. Pretend you're in an abusive relationship. Picture yourself saying to this other person, "I have the right to be treated with respect." Now, that may developmentally be important for you to say, but there comes a point when it's no longer appropriate to keep the focus on you— *you're* not the problem. Contrast how that former statement *feels* with how it feels to say: "You have *no right* to treat me this way." The former is almost a supplication, the latter almost a command. And its focus is on the perpetrator.

For too long we've been supplicants. For too long the focus has been on us. It's time we simply set out to stop those who are doing wrong.

❨ ❨ ❨

Before I go any further, I need to be clear that it's not up to all of us to dismantle the system. Not all of us need to take down dams, factories, electrical infrastructures. Some of us need to file timber sale appeals, some need to file lawsuits. Some need to work on rape crisis hot lines, and some need to work at battered women's shelters. Some need to help family farmers or work on other sustainable agriculture issues. Some need to work on fair trade, and some need to work on stopping international trade altogether. Some need to work on decreasing birth rates among the industrialized, and some need to give all the love and support they can to children (I've heard it said that the most revolutionary thing any of us can do is raise a loving child[122]).

One of the good things about everything being so fucked up—about the culture being so ubiquitously destructive—is that no matter where you look— no matter what your gifts, no matter where your heart lies—there's good and desperately important work to be done. Know explosives? Take out a dam. Know how to love and accept children, how to teach them to love themselves, to think and feel for themselves? That's what you need to do.

If you agree with my premises and arguments, yet find yourself for whatever reason unable or unwilling to take the offensive, your talents are still needed. I think often of the military tactic called Hammer and Anvil, used most famously by Robert E. Lee at the battle of Chancellorsville. Lee kept Anderson's and McLaws's divisions in place while sending Stonewall Jackson's corps around the enemy's flank to crush that part of the opposing army between Jackson's hammer and Anderson's and McLaws's anvil. Both parts—offense and defense— were, and are, necessary.

(((

At another talk, this one last fall, a man asked a question I'd never heard before: "If ten thousand people lined up ready to do your bidding, what would you say to them?"

My answer was immediate: "I'd tell them sure as hell not to listen to me."

His was just as fast: "That's a copout. How many dams could ten thousand people take down? People know how bad things are, but they don't know what to do. They want to be told. That's your responsibility. What's the purpose of writing if you don't tell us what to do?"

I shot back: "Instead of telling me what hypothetical readers want, tell me what you want."

"Tell me—"

"Do you want me to tell you—"

"—Yes—"

"—what to do?"

He nodded, then said, "You've had more time. . . ."

"Okay," I said. "Tomorrow, go to Barton Springs"—Barton Springs are a set of defining, and critically imperiled, springs in Austin, huge and beautiful, dying before the eyes of those who live there and love them—"and sit."

"Then what?"

"Wait until the springs tell you what to do."

"Why won't you—"

"I just did. Barton Springs know this region much better than I. They know what this region needs, know what sustainability looks and feels like here. The springs are much smarter than I am. They'll tell you exactly what to do."

Somebody else asked, "Is it Barton Springs?"

"Yes," I said, "And no. It's everywhere. Just listen. Not to me. To yourself. And to the land."

CARRYING CAPACITY

It is axiomatic that we are in no way protected from the consequences of our actions by remaining confused about the
ecological meaning of our humanness, ignorant of ecological
processes, and unmindful of the ecological aspects of history.

William R. Catton, Jr.[123]

I'VE BEEN THINKING A LOT LATELY ABOUT CARRYING CAPACITY, AND WHAT that will mean for life through the crash. The best book I've read about carrying capacity—what it is and what it means—is *Overshoot: The Ecological Basis of Revolutionary Change*, by William R. Catton, Jr. Any environment's carrying capacity, he states, is the number of creatures living a certain way who can be supported permanently on a certain piece of land, for example how many deer could live on a certain island without overgrazing and damaging the capacity of that island to grow food for them. *Permanently* is the key word here, because it's possible to overshoot carrying capacity—to temporarily have more creatures than the land can support—but doing so damages the land, and permanently lowers future carrying capacity. This is true when we talk about nonhumans, and it's just as true when we talk about humans.

Consider the land where you live. How many people could it have permanently supported before the arrival of our extractive culture? How many people *did* it support? What did these people eat? What materials did those who came before use to make their homes?

And now? What will those who come after eat? If you were to rely only on local foods harvested sustainably—by which I mean entirely without the assistance of civilization or its technologies (e.g., *no* fossil fuels or mining)—what would you eat? Do the plants and animals eaten there before still call this their home? How many people could live in your place forever? How many people *will* live there after the crash?

There are a few ways one can temporarily exceed a place's carrying capacity (I first wrote, "There are a few ways one can temporarily exceed the carrying capacity of one's home" but realized that the sentence is absurd: given the obvious consequences, no sane and intelligent group of people would ever intentionally exceed the carrying capacity of their home). One is by degrading the landscape; for example, eating all of the local fish this year instead of eating few enough that the fish remain fecund as always. Another example would be killing off species you don't eat—salamanders, owls, bees, grasshoppers, and others—and in doing so almost undoubtedly impeding the eventual viability of your food sources.

Once you've undercut the carrying capacity where you live, you can continue to exceed your carrying capacity by degrading someplace else, for example, by eating all of *that* place's fish. This is just another way of saying that cities must import resources, a process also known as conquest, colonialism, and these days, the global economy. As we've seen, when the resources of that other place get depleted—when its carrying capacity has more or less been permanently reduced—those who are importing resources will attempt to find another place to exploit. Because the power of those at the center of empires always depends on this importation/exploitation, the powerful have become quite adept at it. It is, at this point, nearly ubiquitous. As long ago as 1965, more than half of Great Britain's foods were coming from what Catton and others call "ghost acreage," that is, from sources invisible to those at the center. Catton writes, "If food could not be obtained from the sea (6.5%) or from other nations (48%), more than half of Britain would have faced starvation, or all British people would have been less than half nourished. Likewise, if Japan could not have drawn upon fisheries all around the globe and upon trade with other nations, two-thirds of her people would have been starving, or every Japanese citizen would have been two-thirds undernourished."[124] This importation not only makes the lifestyles (and lives) of those who import dependent on the military and economic violence I've been talking about so far in this book, but also makes them strangely dependent on those from whom they steal.

The United States economy is dependent on oil from the Middle East, South America, and around the world. American lives are dependent on it: the agricultural infrastructure—from gasoline to pesticides—rests on the foundation of oil and natural gas. It's not too much to say that we eat refined and transformed oil. It's like Catton wrote, "Everything human beings do requires energy. At the barest minimum, animals human in form but with no technology would have been converting in their own bodies about 2,000 to 3,000 kilocalories of chemical energy (from food) into heat in the course of a day's activities."[125] That changed with domestication—more properly called enslavement—as some humans were able to harvest the energy—work—of those they enslaved, whether it was an ox pulling a plow or a bunch of humans pulling big blocks of stone to make mausoleums for the rich.

And it changed again with oil.

James Watt is one of the most important names in the history of enslavement, a first vote inductee into the Enslavers Hall of Fame, which is quartered neither in Cooperstown nor Cleveland, but in every city on the planet, and increasingly, in every head. He ranks up there with the first of the domesticators, who not only enslaved plants, animals, and land to agriculturalists but all of us to the process of agriculture. He ranks with those who first created a god in the sky, in so doing denying the divinity present in every rock, plant, animal, river, and raindrop, as well as every moment of every being's life, and in so doing also created a heaven beyond the earth where the wretched could receive a reward perhaps (they hope) commensurate with their enslavement here. He ranks with the founders of the first cities, whose kingship, we learn from the ancient King List of Sumer, "was lowered down from heaven,"[126] showing, if little else, that from the beginning, all writers have been propagandists, and mainly for the wrong side. He ranks with those who first used force to steal another's resources. He ranks with those who discovered—after agriculture had enslaved us all—that, as Lewis Mumford put it, "He who controlled the agricultural surplus exercised the powers of life and death over his neighbors. That artificial creation of scarcity in the midst of increasing natural abundance was one of the first characteristic triumphs of civilized exploitation: an economy profoundly contrary to the mores of the village."[127] Others in the Hall of Fame would include those who discovered, as Mumford also wrote, that any "crude system of control had inherent limitations. Mere physical power, even if backed by systematic terrorism, does not produce a smoothly flowing movement of goods to a collecting point, still less a maximum communal devotion to productive enterprise. Sooner or later, every totalitarian state, from Imperial Rome to Soviet Russia, finds this out. To achieve willing compliance without undue waste in constant police supervision, the governing body must create an appearance of beneficence and helpfulness, sufficient to awaken some degree of affection and trust and loyalty."[128] Entire histories could be filled with those who are in the Enslavers Hall of Fame (indeed, this is precisely what history consists of): the Benedictine Monks who developed clocks in order to regiment work, enslaving themselves and those who followed to time itself, and enslaving each moment also to this artificial creation, the clock, the second; Columbus, Cortés, Frobiscer, Cartier (and the kings and queens [and bankers] they served), who sought out new peoples and new lands to enslave; today's mineralogical and biological prospectors (and the CEOs they serve) who seek to enslave ever more of the planet; the engineers, scientists, and technicians from the earliest cities till now who conceptualize ever-more-efficient ways to bring everything we see (and things we do not see) under their control; and many many more.

James Watt invented an effective means to enslave the dead. The bodies of the dead are burned in a confined space, heating the air around them and causing it to expand. Because the space is confined, pressure goes up, pushing out a piston which is attached to, and turns, a crankshaft. This enslavement device is called the steam engine, and has evolved now into the internal combustion engine.

At first the burned dead were trees, and later the longer dead, in the form of coal and oil. The energy released in this burning originally struck the earth when these plants and animals were alive, and had been stored in their bodies. Of course using energy stored in the bodies of others is old news: everybody's been doing that since they learned how to metabolize. And everybody who has ever used fire to keep themselves warm has used energy stored in trees, or coal, for that matter. The big change was in the conversion of these dead into mechanical energy, into what Catton and others call "ghost slaves."[129]

A ghost slave would be the equivalent to how much energy one human would spend in one day (that 2,000 to 3,000 kilocalories Catton mentioned). Yesterday, for example, I went to a traditional Yurok (Indian) brush dance pit, where they hold their annual brush dances. The pit is perhaps four feet deep, and about ten by ten. A narrow ramp leads into it. The walls are lined with weathered wooden planks, and a pole stands one per side. There is effectively no roof. I was told that the design is similar to that of a traditional Yurok home, except, of course, that the houses have roofs. The point as it relates to ghost slaves is this: this home could be constructed by hand by a few people in a day with materials close by. I pictured how the Yurok traditionally lived, there on the banks of the Klamath River. Fishing for salmon. Hunting for elk and deer. Gathering greens and berries. Performing rituals. Building their homes. Playing. Sustainably. Using their own energy, energy gained from eating, metabolizing.

No more.

We have come to base our way of living on these ghost slaves, and our use of them has turned us into slavers on a degree unimaginable to the most megalomaniacal of our forebears. More energy was used in a few minutes to propel a Saturn V rocket toward the moon—and perhaps to an even less life-serving purpose—than was used by two decades of Egyptians stacking 2.3 million blocks of stone (each stone weighing 2.5 tons) to form the Great Pyramid of Cheops.[130] A little closer to the experience of most of us is the truth that, as Catton points out, "Within two eventful centuries of the time when James Watt started us substituting fossil energy for muscle power, per capita energy use in the United States reached a level equivalent to eighty or so ghost slaves for each citizen. The ratio remained much lower than that in many other parts of the world. But, dividing the energy content of total annual world fuel consumption by the annual rate of food-energy consumption in an active adult human body, the world average still worked out to the equivalent of about ten ghost slaves per person.... More than nine-tenths of the energy used by *Homo sapiens* was now derived from sources other than each year's crop of vegetation."[131]

Because the amount of energy that struck the earth a very long time ago and ended up stored in coal, oil, natural gas, and so on is merely tremendous, and not infinite, its use is not sustainable. To base one's way of life on this energy is to live unsustainably. "To become *completely* free from dependence on prehistoric energy (without reducing population or per capita energy consumption)," wrote Catton, and remember this was more than twenty years ago, meaning that things have become far more extreme, "modern man would require an increase in contemporary carrying capacity equivalent to ten earths—each of whose surfaces was forested, tilled, fished, and harvested to the current extent of our planet. Without ten new earths, it followed that man's exuberant way of life would be cut back drastically sometime in the future, or else that there would someday be *many fewer* people."[132] Or maybe both.

❨ ❨ ❨

I'm not the only one to speak of civilization as being based on slavery. Even those who defend civilization often acknowledge this. In *The Culture of Make Believe* I quoted philosopher William Harper's 1837 defense of slavery: "President Dew [another speaker at the conference where he first delivered this message] has shown that the institution of Slavery is a principal cause of civilization. Perhaps nothing can be more evident than that it is the sole cause. If any thing can be predicated as universally true of uncultivated man, it is that he will not labor beyond what is absolutely necessary to maintain his existence. Labour is pain to those who are unaccustomed to it, and the nature of man is averse to pain. Even with all the training, the helps and motives of civilization, we find that this aversion cannot be overcome in many individuals of the most cultivated societies. The coercion of Slavery alone is adequate to form man to habits of labour. Without it, there can be no accumulation of property, no providence for the future, no taste for comforts or elegancies, which are the characteristics and essentials of civilization. He who has obtained the command of another's labour, first begins to accumulate and provide for the future, and the foundations of civilization are laid. . . . Since the existence of man upon the earth, with no exception whatever, either of ancient or modern times, every society which has attained civilization has advanced to it through this process."[133]

I received additional acknowledgment of the necessary relationship between civilization and slavery today, when I received this note from a graduate student

in engineering at Georgia Tech: "Here in the mechanical engineering department, we have a 'distinguished lecturer' each semester who comes to give an
hour long talk. These lecturers are usually CEOs of successful global companies, and we students fill the largest lecture hall on campus (about 400 seats!) to
hear them speak. This semester it was Roger L. McCarthy, chairman of 'Expo-
nent Inc.,' giving a talk on the importance of innovation and engineering to society, with an emphasis on 'learning' from history's disasters. My heart pounded
during the lecture, as I wanted to stand up like a magician and reveal to the tran-
quilized audience the well-disguised and tremendously destructive mythology that
serves as the foundation for this culture. Of course, I couldn't do this with the
twenty to thirty seconds allotted to me in the Q&A session after the talk. So at
the risk of appearing combative in front of my professors, I settled for these simple questions: 'Has technology done more harm or good for human life, or more
to the point, for life in general? And what metrics will you use in formulating your
opinion?'

"I might as well have put the microphone to my ass and farted. He was baf-
fled at the question, and probably wondered how someone could even think of
asking it. His response was insulting, but typical: 'You must not know anything
about history! You must not know anything about what life was like two hundred years ago! Do you even realize what life would be like without technology? You have the equivalent of three hundred slaves working for you every day
due to the advances made in technology over the last two hundred years. You
have the benefit of three hundred slaves but without actually having slaves.' The
implication was that I was 'ungrateful' even to ask such a question.

"I was even more interested in the questions he didn't answer than the one he
did. First, he made no mention of whether technology is good for life itself. He
simply ignored the human and nonhuman slaves the world over, as well as the
fact that we're killing the planet. Such topics are beneath consideration. In fact,
they do not exist. And though he thought he didn't answer my final question,
about how we measure whether something is good or not, in fact he did: we can
measure the success of technology with 'an equivalent number of slaves'
approach. If next year, my life is such that I have an equivalent of six hundred slaves
as opposed to my meager three hundred this year, well then, I have something to
celebrate, don't I? Meanwhile, I've become fatter and more clinically depressed
while I strap on my jogging shoes and run in a circle for exercise (but not outside, of course, today is a red alert). What this means is that if we as a culture
have chosen to value 'enslavement' in the most general and inclusive way possible, then we have done a tremendously good job implementing our design."[134]

❦ ❦ ❦

Several years ago I interviewed Jan Lundberg, founder of the Alliance for a Paving Moratorium, "a diverse movement of grass-roots community groups, individuals, and businesses with the common goal of halting road-building," because "a paving moratorium would limit the spread of population, redirect investment from suburbs to inner cities, and free up funding for mass transportation and maintenance of existing roads." But there's more to it than just roads. Phasing out massive fossil-fuel use, Lundberg says, is crucial not only to saving the earth's climate, but to lessening the impact of the crisis that will occur when the world's oil supply begins to run out. "The challenge before us all," he writes, "is to survive an ecological correction unprecedented for our species. The correction will likely include an economic collapse and a conversion to subsistence activities and trading."

Lundberg grew up around the oil industry. His father ran Lundberg Survey, Inc., a company that collected statistics on gasoline prices and industry trends. In 1973, just before the oil crisis, father and son began publishing the *Lundberg Letter*, which became the number-one trade journal for the oil industry and went on to predict the second oil shock of 1979.

After his father's death in the mid-1980s, Lundberg quit the family business and directed his efforts toward energy conservation. By that point, Jan had realized that this culture's "waste economy," as he calls it, is unsustainable and the cause of massive environmental damage and species extinctions worldwide. We are laboring, he says, under the false impression that we can "have it all": the physical comfort of the current way of living and a livable planet.

I said to him, "When my friends and I talk about the end of civilization, we often search for some sort of marker: one of the things we've come up with is the end of car culture. How do you see the end of car culture playing out? Even before that, do you agree the car culture is in its endgame? What will cause it to end? And if you don't think it's in its endgame, is that because they'll find new oil, or failing that, figure out a new fuel system?"

He responded, "A lot of these questions have to be gone over in basics because the mass media and the educational systems provide *zero* insight into them. They act as though how much oil there is and what it can be used for are of no concern to the public.

"Probably the best place to start is by talking about Marion King Hubbert, a geologist who died several years ago who became famous for charting the life of an oil field. Extraction follows a bell curve—called the Hubbert Curve—in which production rises as new wells are put in, reaches a maximum when about half of the 'Estimated Ultimately Recoverable' (EUR) oil has been extracted, and then tails off as wells begin to run dry. During the decline, technologies such as water flooding and gas injection may be introduced to slow the rate of depletion, but all they do is stave off the inevitable. The same pattern that is true for individual oil fields holds for geological basins as well: production rises as new fields are found and then tails off as the larger and more accessible fields are depleted. This pattern can be extended also to entire nations, and ultimately to the planet. The bottom line of all this—and this is so obvious we shouldn't need to say it, but we have to because there is so much ignorance and intentional deceit surrounding this subject—is that the production of any field starts at zero, rises to a peak, and then falls to zero.

"For the United States, production in the lower-forty-eight peaked about 1970—as predicted by Hubbert some forty years ago—and has been on the decline ever since."

I asked, "When will world oil production peak?"

He responded, "Before we can ask that—and that *is* the question, isn't it?— we need to ask another, which is, what is the world's volume of EUR oil? Once again, production will peak when half of this volume has been extracted.

"One of the best figures I've seen for EUR is about 1,800 billion barrels, which would mean that global production would peak by the year 2007. Even if EUR oil is as high as 2,600 billion barrels, that would move the peak back to only 2019. To be honest, both of these figures seem too far away, because I don't think they fully take into account that oil consumption continues to rise very quickly. I have seen other credible figures—and these seem far more feasible to me—suggesting that global oil production has already peaked.

"Now, when United States production peaked, that didn't mean the end of the oil age, since the U.S. could still import oil. But when global production peaks, as it either already has or will shortly, it means the beginning of the end of the economy as we know it. Five Middle East countries will regain control of world supply. This will make the oil shocks of the 1970s seem like nothing, because then there were plenty of new oil and gas finds to bring onstream. This time there are virtually no new prolific basins to yield a crop of giant fields sufficient to have a global impact. The growing Middle East control of the market is likely to lead to a radical and permanent increase in the price of oil long before physical shortages begin to appear, and they *will* appear within the next decade."

Of course the most recent U.S. invasion of Iraq took place in great measure to secure U.S. access to Iraqi oil.

He continued, "This will, of course, demolish the economy, which has been driven by an abundant supply of cheap energy for a century. We're going to live through an 'economic and political discontinuity of historic proportions,' as one analyst puts it, or the crash, as we more often refer to it. I like the language of oil industry geologist Dr. Walter Youngquist: 'My observations in some seventy countries over about fifty years of travel and work tell me that we are clearly already over the cliff. The momentum of population growth and resource consumption is so great that a collision course with disaster is inevitable. Large problems lie not very far ahead.'"

I responded, "Wait a minute. I've seen industry and government figures showing that "proven reserves of oil are enough to supply the world for forty-three years at current rates of production.""

He said, "I see two immediate problems with this. The first is that these figures come from government and industry. You don't think that either of those groups would lie to the American public, do you? For political reasons, proved oil reserves are consistently substantially overstated. It is in the interest of both oil-producing nations and companies to overstate their remaining oil, because their business agreements limit them to pumping and selling a proportion of their remaining resources. For example, if contracts limit you to pumping 10 percent of your proven reserves per year, you'll make a lot more money, and you'll make it a lot more quickly, if you simply lie about your proven reserves. But in fact the rate of oil discovery is falling sharply. Discovery of oil and gas peaked in the 1960s, and the situation has deteriorated enough that by now the world consumes more than three times as much oil each year as is discovered. Do you think the oil industry is aware of oil field depletion? Of course. It's their business. Why do you think no new supertankers have been built for twenty years? A report written for oil industry insiders and priced at $32,000 per copy concludes that world oil production and supply peaked in 2000, and will decline to half by 2025. The report predicts large and permanent increases in oil prices for the very near future.

"The second problem with that argument—that oil reserves will last forty-three years—is that it is based on 'current rates of production.' Their use of that language should clue us to the fact that they are dissembling, because the truth is that production is skyrocketing. At one time I thought that the downslope of the Hubbert Curve might be at least slightly gradual, but because in recent years production has accelerated to unanticipatedly high levels, I've come to believe that the downslope of the curve will be extremely steep."

I told him I didn't understand.

He said, "It means we're using up the oil far faster than anyone anticipated, so the crash will be sooner and harder than even environmentalists predict."

"But as oil becomes increasingly rare, it will become increasingly expensive, which will provide financial incentives to develop other forms of energy. Tar sands for example, or oil shale."

"Economists say this all the time. They like to argue that scarcity results in price increases, making it more profitable to access poorer deposits. It's too bad that economics and the real world so rarely intersect."

"True. I took a year of graduate study in Mineral Economics back in the 1980s, and I remember informally renaming one of my classes 'ME 514: Guessing at Things,' and another 'ME 525: Pretending to Have Answers.'"

"In this case the economists are confusing dollars with calories. The fact is that as an oil field ages, it takes increasing amounts of energy to pump out the remaining oil. You need to subtract that energy cost from the total value of the energy extracted. Even now, the average energy profit ratio for newly discovered oil in the United States has fallen to 1:1, meaning the energy required to find and extract a barrel of oil increasingly exceeds the energy contained in the barrel. At some point it will no longer make sense to use oil for energy, no matter how much you can sell it for. Too often, both economists and engineers forget that they cannot repeal the laws of thermodynamics. They forget, to switch ways of speaking here, something known to every child: that an orange only has so much orange juice in it."

"Energy profit ratio?"

"That's a measure of how much energy must go into a process to get a certain amount of energy out. The early oil wells in Pennsylvania had a ridiculously high ratio because you had almost zero energy input. You just had to go scoop it up and burn it. But the ratios for all these other forms of energy are much lower. Ethanol, for example, has an energy profit ratio of less than 1:1, meaning it takes more energy to make it than you get out of it."

"You make a great point," I said, "but I still have another concern. The government already subsidizes the oil industry, and subsidizes many other industries that make no fiscal, ecological, or economic sense. Why would we think that the same government wouldn't just continue with these subsidies, even when they make no sense from an energy perspective? Why wouldn't the government just use the full force and power of the state to hand over money, and energy, so that from the perspective of the corporation the tar sands are profitable?"

"That's a good question, especially because that's already happening. Our entire economic system is based on these subsidies, from agriculture to manufacturing to energy. Especially energy. That's why oil is so cheap right now. Just including the cost of the Persian Gulf military presence—for which we as tax-payers foot the bill—would at least double the price of oil.

"The thing that scares me even more than monetary subsidies, however, are the hidden subsidies that can never be accounted for. Can you put a price on global warming? Can you put a price on a pristine lake or river? The so-called economic view of our planet and of life is anti-life.

"So long as we cling to this economic view, we will be able to maintain the illusion of cheap oil for just a little bit longer, paying for the oil in ways that we don't know and don't necessarily feel.

"But I'll tell you what scares me the most about all this: everything in this economy is based on petroleum. It's not just cars. It's the food we eat, fertilized with petroleum products, transported by petroleum. It's the plastics we surround ourselves with. It's everywhere. Everything is oil. People don't even know. They don't even think. And it wasn't very long ago that we supported ourselves on a plant-based economy. Canvas, for example, was from cannabis, and now it's from DuPont. One reason they outlawed hemp was that DuPont was able to make substitutes. Medicines, clothes, it's all there."

"What about natural gas? Can the system keep going another couple of generations on natural gas and coal? Maybe coal gasification."

"There's not a lot of natural gas out there. And coal gasification is another one of those inefficient processes in which you have to put in a lot of energy but you don't get that much out. Now, there's a hell of a lot of coal, if you're willing to destroy the surface of the planet to get it out, and pump all the mine wastes into your rivers, and the soot into the air. I'm not certain that even *this* culture is crazy enough to do that."

"Let's cut to the chase. Do you think we'll see the end of car culture in our lives?"

"Yes. It may be because we run out of oil, or it may be because of economic collapse from which we do not get up, based on the demand for oil so greatly outstripping the supply that the price goes through the roof. And the end of car culture may ultimately be a liberating event, for those who survive, as we try to remember how to live with what the land will give us. But if the collapse is so pervasive that too many nuclear events occur, even the collapse may simply further the destruction that is the hallmark of this culture."

"Let's take this step-by-step," I said, "When we talk about the end of car culture, we're not talking just about the end of traffic jams and commuting..."

"Because the agricultural system is also petrochemically based, we're essentially eating oil. So we're really talking about the collapse of the agricultural infrastructure, and the associated transportation and distribution network, which goes beyond agriculture. It's the products, it's commuting, it's food.

"We're essentially fucked, and we don't know it. It's like Youngquist said, we're already over the cliff, but we aren't paying attention."

❨ ❨ ❨

Our discourse surrounding carrying capacity is generally as absurd as the rest of our discourse. Most often we simply ignore it. Failing that, talk of carrying capacity quite often falls into one of three camps, none of which are particularly helpful, all of which support the status quo.

The first begins and ends with population. There are simply too many people. You've seen the pictures. Crowded streets in Calcutta, impoverished babies with huge hungry eyes and bloated bellies in Mexico, refugee camps in Africa, masses of Chinese crammed into filthy cities. The earth can't support these numbers. Something's got to give.

And you've heard the arguments. The United States needs to close its borders to immigration from poor countries. Having finally gotten our own birthrate down sufficiently to more or less stabilize our population, the last thing we need is a bunch of poor (brown) people moving in to crowd us out (we know, also, that once they're here they'll breed faster than we do, and soon enough will outnumber us).

I often respond to this argument by saying I'm all for closing the border to Mexico (and everywhere else, for that matter, all the way down to closing bioregional borders), so long as we close it not only to people but to resources as well. No bananas from Mexico. No coffee. No oil. No tomatoes in January. Many of the people who leave their families in Mexico (or any other impoverished nation) to come to the United States to work do so not because they hate their husbands or wives yet have not gotten to the point in their therapy where they feel comfortable expressing (much less acting on) this. Nor is it generally because they're bored with Cancun, Acapulco, and their other normal vacation spots and have decided this tourist season to take a Reality Tour™ of the bean fields of the San Joachin Valley. They come, one way or another, because the integrity of their resource base and their community (insofar as there can meaningfully be said to be a difference) have already been compromised: the resources have been stolen, and the community is unraveling. Of course this migration, too, is part of the unraveling. From the beginning of history, this is why people have moved from country to city.

To want, on the other hand, to close the border to people yet leave it open to the theft of their resources (*importation* is the preferred term in polite society), is to show that your alleged concern over population is nothing but a cover for continuing the same old bigotry and exploitation. *I don't want you, but I do want the coffee grown on land that used to be yours.* Even those who don't specifically want to close borders, but merely want to talk about population while conveniently forgetting to talk about resource consumption are, too, pushing us ever closer to the abyss. For the real bottom line of overshooting carrying capacity is resource consumption and other damage. It wouldn't matter if there were a hundred billion deer on a tiny island if they didn't consume, trample, or otherwise destroy anything, and didn't pollute the place with their feces or anything else. Numbers by themselves are meaningless. It's the damage that counts.

Another way to talk about this is to notice the language: over*population,* zero *population* growth. How different would our discourse be if we spoke instead of overconsumption and zero consumption growth? This shift in discourse won't happen, of course, because zero consumption growth would destroy the capitalist economy.

The United States constitutes less than 5 percent of the world's population yet uses more than one-fourth of the world's resources and produces one-fourth of the world's pollution and waste. If you compare the average U.S. citizen to the average citizen of India, you find that the American uses fifty times more steel, fifty-six times more energy, one hundred and seventy times more synthetic rubber, two hundred and fifty times more motor fuel, and three hundred times more plastic.[135] Yet our images of overpopulation generally consist not of those who do the most damage, the primary perpetrators (there can't be too many [middle-class] Americans, can there?), but instead their primary (human) victims.

At least partially in response to the obvious arrogance and absurdity of those who want the poor to stop having babies but don't mind the rich having SUVs (and nuclear weapons), there are those who claim—equally absurdly, and equally arrogantly—that all talk of carrying capacity is racist and classist. To even use the phrase *carrying capacity* in this crowd is to invite hisses and catcalls, as well as spat epithets of *Neo-Malthusian.* I suppose the argument is that because some of those who want to protect this exploitative way of living use carrying capacity as a means of social control against the poor—as an American Indian activist friend said to me, "The only problem I have with population control is that you and I both know who is going to do the controlling"—then the notion of carrying capacity itself must be racist and classist. This seems similar to me to suggesting that because Hitler claimed (falsely) that Germany was being attacked by Poland, and that therefore the Germans needed to attack, and that because this same argument has routinely been used (just as falsely) by the United States as well as other imperial powers, that anyone who claims self-defense is lying.[136] These people seem to forget that the misuse of an argument does not invalidate the argument itself.

Worse, this argument, that the very concept of carrying capacity is a fabrication designed for social control, as opposed to a simple statement of limits, serves those in power as effectively as does ignoring or de-emphasizing resource consumption when speaking of overshooting carrying capacity, because it goes along with the refusal to acknowledge physical limits (and limits to exploitation) that characterize this culture. What would it take, I've heard peace and social justice activists ask, to bring the poor of the world to the fiscal standard of living of the rich? Well, another thirty planets, for one thing. It's a dangerous—and stupid—question. Within this culture wealth is measured by one's ability to consume and destroy. This means that attempts to industrialize the poor will further harm the planet. Because industrial production requires the exploitation of resources, the wealth of one group is always based on the impoverishment of another's landbase, meaning that on a finite planet, the creation of one person's (fiscal) wealth always comes at the cost of many others' poverty. Those reasons are why the question is stupid. It's dangerous because it serves as propaganda to keep both activists and the poor playing a game that doesn't serve them well, and which they can never win,[137] instead of quitting this game and working to take down the system.

For at least the past ten years, there has been a lot of talk, primarily among those whose alleged concern for sustainability is a cover for exploitation but also among those who should know better, of something called *sustainable development*. In this phrase, development is essentially a synonym for industrialization, for destruction, as in the *development of natural resources*. Under this rubric, *sustainable development* is an obvious oxymoron. Industrialized people consume more resources and cause more damage, than nonindustrialized people. The "development" of the industrialized nations has been and continues to be unsustainable for the industrialized nations and for the world at large, and the further "development" of the world will only make things worse.

Sometimes activists complain—sometimes *I* complain—that the United States spends boatloads of money on weapons, but gives comparatively little to the poor. I've grown to understand, however, that the best thing Americans could do for the poor is not to hand them crumbs, nor to give (or worse, loan) their government money for dams, factories, roads, and (of course) weapons, but instead to stop stealing their resources. I recently asked Anuradha Mittal, former co-director of Food First, if she thought the poor of her native India would be better off if the United States economy disappeared tomorrow. She laughed and said, "Of course. All the poor would be." She told me that former granaries in India now export dog food and tulips to Europe.

There's a third way to look at population, which is, I think, as useless and harmful as the others. Even when people *do* accept the existence of carrying capacity and aren't trying to use their talk of overshoot to maintain the rich's current stranglehold over the lives of the poor—and to extend this stranglehold into the most intimate aspects and decisions (sexuality and childrearing) of their lives—they more often than not talk of population in terms of mathematics, in terms of exponential increase, in terms of some "natural rate of population growth." It's very simple: turn on your computer, plug the appropriate numbers into your handy-dandy formula—X number of people on Y amount of land containing Z amount of resources, where W represents the industrial educational level of women—and watch the little black and brown dots representing people fill your screen. But this formulation carries with it many dangerous premises, including the essential premise of mathematics itself: those to be studied and described are not individuals who make choices, but instead are objects who—or rather which—act with no great measure of volition. It presumes people do not make rational short-, mid-, and long-term family-planning decisions based on their circumstances, experiences, and the social values into which they've been acculturated. Nor do they give any thought to the personal, social, or environmental consequences of their decisions. Heck, it presumes people—especially poor, brown, uneducated people—breed with no thought whatsoever: where does thought, or choice, fit into these or any equations? It presumes they breed like rabbits. But that's nonsense. I'm not even sure *rabbits* breed like rabbits.

Sure, we can make probabilistic predictions of what certain percentages of people (or rabbits) will do under certain social and ecological conditions, but to talk of any "natural rate of population growth" without talking about the culture that causes—acculturates, inculcates, coerces, rewards—people to not only ignore environmental limits but to perceive, accurately, that their larger social fabric would collapse without incessant growth is to naturalize—make normal, make invisible, make seem as inevitable as gravity—something that is not natural but cultural.

Non-linear—cyclical—cultures, those not predicated on growth but on dynamic equilibrium, maintain stable populations. Having reached the limits of what their landbase willingly supports, indeed—and this is well-nigh inconceivable to those of us raised in a culture where we are taught to perceive all life as horrific competition and humans as the bloody victors—having reached a population level that best serves the needs not only of their human community but of their nonhuman neighbors, they, believe it or not, reduce the number of children. They do this by breastfeeding their existing children for many years, by abstinence, by taboos, by the use of herbal contraceptives and abortions. Prior to conquest, American Indian women, for example, used more than two hundred plants, roots, and other medicines as means of birth control, making the decisions themselves as to whether to use them.[138] When all else fails, some cultures, and I'm not promoting this, practice infanticide. This infanticide is often not gender-based.

Beneath these techniques is the real point, which is an intimate and mutually beneficial relationship with their landbase.

☾ ☾ ☾

"What nonsense!" I can hear you say, "Humans *exploit* their surroundings! Human needs are in opposition to the natural world, otherwise why would politicians say we need to balance the economy versus the environment? Balance implies opposition. Whether it's a God-given right or an evolutionarily ordained mandate, humans chop down trees, deprive all others of their habitat. It's what we *do*." But to believe this is to mistake civilization for humanity, an unforgivable and fatal, if flattering, error.

One of the central myths of this culture concerns the desirability of growth, a parasitic expansion to fill and consume its host. This was manifest from the beginning, as we were told in Genesis, "And God blessed them, and God said unto them, Be fruitful, and multiply, and replenish the Earth, and subdue it: and have dominion over the fish of the sea, and over the fowl of the air, and over every living thing that moveth upon the Earth."[139] Of course we see the same absurd mythology of growth and exploitation today. Just last night I read, in language less theological yet expressing the same damn thing, a sentence by Joseph Chilton Pearce, an author well-respected for his attempts to change this culture's destructive path: "The amount [of gray matter] we have is just what we need for certain goals nature has in mind, such as our dominion over the earth."[140] From its opening to its endgame, civilization has been nothing if not consistently narcissistic, domineering, and exploitative. And it is consistent in its attempts to make these attributes seem natural, to make them seem as though nature itself is to blame for our exploitation of it ("She was asking for it," we can say with clean conscience as we pull up our pants and leave the darkened alley).

We can see the myth of growth at work in the Catholic church's continued hostility toward birth control, attempting to get us to believe, as the ironic bumper sticker so eloquently puts it, that "every ejaculation deserves a name." We can see it in the concern over falling birthrates in industrialized nations such as Greece and Russia. And we can see it in the commonplace acceptance of the very real fact that without constant economic expansion capitalism will collapse almost immediately.

This mythology is grounded in reality—cultural reality, that is—because from the beginning the very existence of city-states has required the importation of resources from ever-expanding regions of increasingly exploited countryside. It has required growth.

Well, that's going to stop someday. At some point, probably in the not-too-distant future, there will be far fewer people on this planet. There will be far fewer than the planet could have supported—and did support—prior to us overshooting carrying capacity, because the great stocks of wild foods are gone (or poisoned), the top soil lost in the wind.

My saying this doesn't mean I hate people. Far from it. A few weeks ago I received an email in response to my statement that the only sustainable level of technology is the Stone Age. The person said, "I don't think the stone-age will support anything near the current world population. [Of course I agree.] So to return to this level implies either killing a lot of people or not having many children and waiting for the population to diminish. Or do we allow war or other pestilence to do the job? Is this what you are proposing?"

I responded that what I'm proposing, startlingly enough, is that we look honestly at our situation. And our situation is that we have overshot carrying capacity. The question becomes: What are we going to do about it?

THE NEEDS OF THE NATURAL WORLD

Industrial technology is by nature exploitative and destructive of the materials that are necessary to maintain it.

Richard T. LaPiere[141]

A COUPLE OF MONTHS AGO I GAVE A TALK EXPLORING SOME OF THE THINGS in this book, and afterwards someone said, "I think what you're saying is pretty heartless. What are you going to say to people with diabetes, cancer, or leukemia who need medicines made by the pharmaceutical industry?"

I said, "I'd tell them the same thing I'd tell myself— I have Crohn's disease—which is, 'Stock up.'"

I could tell he didn't like my answer.

I didn't like it either. I continued, "There's a deeper point here, which has to do with our attempts to separate ourselves from the rest of the world, to pretend we're not natural, to consider ourselves exempt from the ways the world works. Consider our utter disregard for overshooting carrying capacity—our belief that somehow these ecological principles don't apply to us. Consider also our denial of death and our deification of humans, especially civilized humans, most especially rich white civilized humans. All of this has to stop. The truth is that I'm going to die someday, whether or not I stock up on pills. That's life. And if I die in the population reduction that takes place as a corrective to our having overshot carrying capacity, well, that's life, too. Finally, if my death comes as part of something that serves the larger community, that helps stabilize and enrich the landbase of which I'm a part, so much the better."

"By what right," asked someone in the audience, "can you make that decision for others? Don't they have the right to extend their life by any means possible?"

A third person raised her hand, then said in response to the original question, "Every disease mentioned here is a disease of civilization. Civilization *causes* those diseases. The questioner seems to be implying that to talk about taking down civilization is to somehow not care about sick people. But to want to get rid of the thing that's making them sick—civilization—seems far more compassionate than to allow civilization to continue, and then to try to palliate."

That reminded me of a question my friend Carolyn Raffensperger, cofounder of the Science and Environmental Health Network, likes to pose, not necessarily about civilization but more specifically about the medical industry: "What are we going to do with the irony that industrial health care is one of the most toxic industries on earth? We produce PVC medical devices to treat someone's cancer, then put them in the hospital incinerator to send back out and give someone else cancer. Or we use mercury in our thermometers in the hospital, and then send *that* up the incinerator to be deposited in fish and to eventually give more children—human and nonhuman—brain damage. Where does any of this make sense?"

Yet another person pointed out that when we talk about the wonders of modern medicine, we need to remember that on the main it is the rich who receive these ecologically and economically expensive treatments: "Modern industrial medicine cures the cancer of some rich American who became sick because of the toxification of the total environment, and these processes lead to even more toxification, causing yet more poor people—and nonhumans—to die. The real wonder of modern medicine is that the poor buy into this at all."

The room was abuzz. Someone else stated, "You've talked a lot about the power of unstated premises, and this is a great example. No one in here has mentioned two of the most important premises behind the belief that taking down civilization will harm the sick. The first is that the western industrial model of medicine does in fact save people. Sure, industrial medicine saved my life, but only after nearly killing me several times through misdiagnoses and the toxicity of the whole process. And industrial medicine never made me well: what accomplished that were so-called alternative medical treatments such as herbs, energetic work, and changing the emotional, relational, and physical circumstances of my life. That leads to the second premise, which is that if we don't have industrial medicine we don't have anything. People talk about how advances in western medicine have decreased morbidity, and on some levels that's clearly true, but they're only comparing a more refined version of the same model to a less refined version of it. There have been plenty of studies showing that traditional hunter-gatherers were extremely healthy, with long life spans. There were often high rates of infant mortality, as is true for many creatures, but once you got past that, you could plan on living a long healthy life. And that wasn't just because they hadn't fucked up their world. They knew the healing properties of plants that lived in their neighborhoods. And they understood the spiritual bases for many of their illnesses. Although much of this knowledge is fast disappearing, and although many of the plants on which these medicines are based are being extirpated, these other models still exist. Getting rid of industrial civilization means getting rid of industrial medicine. It doesn't mean getting rid of medicine, and the possibility of healing the sick."

A couple of days after the talk I got an email from someone else weighing in on this whole question: "I didn't have the wherewithal to speak up at the time, but I'd offer the following counter-question: 'What would a diabetic or heart patient do if the drugs she needed to stay alive were integral to an economic system that exploited workers, degraded the environment, and increased the suffering of indigenous peoples?' To answer that she still wanted the drugs would expose the narcissism—the extreme emphasis on the individual, even at the expense of the larger community—that so dominates Western Culture. That's the root of much of our trouble."[142]

❨ ❨ ❨

What do we do with this information: Phoenix, Arizona, could sustain a human population of maybe one hundred and fifty. What about the rest of them, living right now on stolen resources? The land under New York City could probably sustain several thousand, or at least it could have if there were still passenger pigeons, bison, salmon, eel, and Eskimo curlews. What happens to the rest? I'm a bit luckier here in Tu'nes. The population might be remotely sustainable at a hunter-gatherer level, if salmon, steelhead, elk, and lamprey were still here in significant numbers.

To reverse the effects of civilization would destroy the dreams of a lot of people. There's no way around it. We can talk all we want about sustainability, but there's a sense in which it doesn't matter that these people's dreams are based on, embedded in, intertwined with, and formed by an inherently destructive economic and social system. Their dreams are still their dreams. What right do I—or does anyone else—have to destroy them?

At the same time, what right do they have to destroy the world?

❨ ❨ ❨

I've been thinking more about rights, and I've come to the conclusion that defensive rights always take precedence over offensive rights. To take an example especially close to the heart of many women, given the high regard sexual coercion is evidently given within this culture, one person's defensive right to bodily integrity always trumps—or rather would always trump, within a workable morality—another's perceived right to sexual access.

In my life I've been in a couple of romantic relationships I would define as emotionally abusive. The women would call me names, and harangue me for days about this or that characteristic of mine they didn't like. When I asked (begged, pleaded with) them to stop, they became all the more angry, and of course refused. When I told them to stop, they exploded, and let me know in no uncertain terms that I had no right to censor them. "Don't you think it's ironic," they'd say, "that you, a writer who rails at the emptiness of this culture's discourse, are trying to limit mine?"

This clash of my defensive right to not be mistreated with the other's perceived right to shower displaced rage upon me has never been a problem in normal relationships, where the right to say *no*—to whatever action, for whatever reason—trumps all others. This doesn't mean there are no consequences for saying *no*. To go back to the sexual example, if one person wants to be sexual and another person does not, there is no sex. That cannot be in question. But within the context of a sexual relationship, if one person consistently doesn't want to be sexual, the two involved may wish to re-examine the form of their relationship. Similarly, I'm certainly not going to force anyone to talk about what's wrong with civilization and what we're going to do about it, but a consistent refusal will probably limit our friendship: I'm not going to fight six thousand years of history, the full might of the state, and my friends as well (another way to say this is that I'm not going to revisit *Civilization is Destructive 101* every time I open my mouth).

To hear and respect another's *no* is to accept that the other has an existence independent of you. People generally refuse to hear another's *no*—and this is certainly true of the entire culture's refusal to enter into relationship with the natural world—when the possibility of intimate and genuine interactions with the other is too frightening to allow. Or when acculturation and personal history combine to make someone believe the other doesn't even exist for its own sake.

❨ ❨ ❨

Two weeks ago, the Klamath River, just south of here, was full of the biggest runs of salmon and steelhead (ocean-going rainbow trout) in years. "You could have walked across on their backs," someone said to me. I talked to a Yurok Indian, whose culture is based on the salmon, who said the runs made him imagine what it must have been like to see the *real* runs before the white men arrived. It made me happy. I was going to go see them.

But I got another call. The fish were dying, piling up in mounds on the shore or floating bloated and bleeding from their vents. "Don't come," the caller said. "You don't want to see this."

Walt Lara, the Requa representative to the Yurok Tribal Council, said in a local newspaper interview, "The whole chinook run will be impacted, probably by 85 to 95 percent. And the fish are dying as we speak. They're swimming around in circles. They bump up against your legs when you're standing in the water. These are beautiful, chrome-bright fish that are dying, not fish that are already spawned out." There are probably, he said, a thousand dead fish per mile of river.[143]

Last summer the federal government decided there was no evidence that fish need water, and instead redirected the water to (a few heavily subsidized) farmers in the Klamath Basin in southern Oregon. The water in the Klamath is now too warm for the salmon.

This is the story of civilization. This culture is killing the planet.

❰ ❰ ❰

The defensive right of the salmon community to live, and the defensive right of the river to exist for its own sake, would in any meaningful morality trump the perceived rights of the farmers to take the water, and the perceived right of the government to give it to them.

But, you could ask, what about the right of the farmers to continue their traditional (and in this case, taxpayer- and environmentally subsidized) lifestyle?

This brings us to the eighth premise of this book: *The needs of the natural world are more important than the needs of any economic system.* This seems so self-evident I'm embarrassed to have to defend it, but it is a notion that entirely escapes our public (and private) discourse. Just yesterday I saw a tiny article on page seven of the *San Francisco Chronicle* stating that every single stream—*every single stream*—in the United States is contaminated with toxic chemicals (the completeness of this toxification should surprise me less than it does: surely if every mother's breast milk is contaminated with toxic chemicals, why should we expect streams to be any more immune?), and that one-fifth of all animals and one-sixth of all plants are at risk for extinction within the next thirty years. Page one carried a huge article about Elvis memorabilia, and another that began, "Congress took its first tentative step Wednesday toward mandating that all television sets by 2006 include technology to foil piracy of digitized movies and television shows."[144] Don't forget, once again, the entire sections of the paper devoted to sports, business, and comics/gossip.

Think about it for a second: what is the real source of our life? Of our food, our air, our water? Is it the economic system? Of course not: it is our landbase.

Just last week I learned that the air in Los Angeles is so toxic that children born there inhale more carcinogenic pollutants in the first two weeks of their lives than the EPA (which routinely understates risks so as not to impede economic production) considers safe for a lifetime. In San Francisco it takes about three weeks.[145] We're poisoning ourselves. Or rather, we are being poisoned. Another way to put the eighth premise is: *Any economic or social system that does not ben-efit the natural communities on which it is based is unsustainable, immoral, and really stupid. Sustainability, morality, and intelligence (as well as justice) require the dismantling of any such economic or social system, or at the very least disallow-ing it from damaging your landbase.*

❨ ❨ ❨

If someone put a plastic bag over your head, or over the head of someone you love, and said he would give you money if you leave it there, would you take the money?

And if you said *no*, what would you do if he insisted, even to the point of a gun?

Would you take the money?

Or would you fight back?

❨ ❨ ❨

When they don't have anything better to do—which frankly seems like most of the time—anti-environmentalists are fond of pointing out the hypocrisy of environmentalists. *You live in a house, don't you? You wipe your ass with toilet paper. Your books are made of paper. Every one of these activities is environmen-tally destructive. You are not pure. Therefore what you say is meaningless.*

It's an interesting argument on several levels. The first is that it reveals the weakness of their own position: they cannot rebut the substance of our mes-sage, so they simply attack the messenger. It's one of the most overused rhetor-ical tricks going. But there's something even more interesting about their arguments—fundamentally stupid as they are—which is that they're right, and in being right they make one of my central points better than I do. Build-ing houses is destructive. Manufacturing toilet paper is destructive. Printing books is destructive. But there's no reason to stop there. The industrial econ-omy itself is inherently destructive, and every act that contributes to the industrial economy is inherently destructive. This includes buying my books. This includes buying something from Global Exchange. If we care about the planet, we then have a couple of options. The first—and this one is often sug-gested by anti-environmentalists—is that we simply off ourselves. I prefer the second one, which is that we dismantle the industrial economy.

❨ ❨ ❨

I want to be clear. When people tell me population is the number one environmental problem we face today, I always respond that population is by no means primary. It's not even secondary or tertiary. First, there's the question of resource consumption I mentioned earlier. Second is the failure to accept limits, of which overpopulation and overconsumption are merely two linked symptoms. Beneath that is our belief we're not animals, that we're separate from the rest of the world, that we're exempt from the negative consequences of our actions, and that we're exempt from death. Beneath these beliefs is a fear and loathing of the body, of the wild and uncontrollable nature of existence itself, and ultimately of death. These fears cause us to convince ourselves not only of the possibility but the desirability of not being animals, of separating ourselves from the world. These fears drive us crazy, and lead us to create and implement insane and destructive economic and social systems.

❨ ❨ ❨

All of this is a roundabout way of getting to my ninth premise, which is: *Although there will clearly some day be far fewer humans than there are at present, there are many ways this reduction in population may occur (or be achieved, depending on the passivity or activity with which we choose to approach this transformation). Some will be characterized by extreme violence and privation: nuclear Armageddon, for example, would reduce both population and consumption, yet do so horrifically; the same would be true for a continuation of overshoot, followed by a crash. Other ways could be characterized by less violence. Given the current levels of violence by this culture against both humans and the natural world, however, it's not possible to speak of reductions in population and consumption that do not involve violence and privation, not because the reductions themselves would necessarily involve violence, but because violence and privation have become the default of our culture. Yet some ways of reducing population and consumption, while still violent, would consist of decreasing the current levels of violence—required and caused by the (often forced) movement of resources from the poor to the rich—and would of course be marked by a reduction in current violence against the natural world. Personally and collectively we may be able to both reduce the amount and soften the character of violence that occurs during this ongoing and perhaps long-term shift. Or we may not. But this much is certain: if we do not approach it actively—if we do not talk about our predicament, and what we are going to do about it—the violence will almost undoubtedly be far more severe, the privation more extreme.*

PREDATOR AND PREY

Kill every buffalo you can, for every buffalo dead is an Indian gone.

Colonel R.I. Dodge, Fort McPherson, 1867[146]

I'VE LONG UNDERSTOOD THAT CIVILIZATION REQUIRES LAND OWNERSHIP to be concentrated in the hands of the rulers, by force when necessary and tradition when possible. More basically, it requires that people be inculcated to believe land can be bought and sold. Ultimately, of course, it requires that people be inculcated to believe *everything* can be bought and sold, and requires also that ownership of everything be concentrated as completely as possible in the hands of the rulers.

Those in charge have always understood—and have often been explicit about it—that it's difficult to control people who have access to land. Depriving them of this access puts them at your mercy. Without access to land there can be no self-sufficiency: land provides food, shelter, clothing. Without access to land people obviously have no place to stay. If you can force people to pay just so they can be alive on this earth—nowadays these payments are usually called *rent* or *mortgage*—you've forced them into the wage economy. The same holds true for forcing them to pay for materials the earth gives freely: the salmon, bison, huckleberries, willows, and so on that are central to the lives, cultures, and communities not only of indigenous peoples but to all of us, even if we make believe this isn't the case. To force people to pay for things they need to survive is an atrocity: a community- and nature-destroying atrocity. To convince us to pay willingly is a scam. It also, as we see around us—or would see had we not been so thoroughly convinced—causes us to forget that communities are even possible.

Those in power have rarely hidden their intentions. Indeed, as I've written elsewhere, the need to separate the majority of people from their food supplies—thus separating them also from their freedom—was central to the design of civilization's early cities.[147] I've written, too, how slave owners described the land-ownership conditions under which chattel slavery was the optimal means to control a workforce, and described also the conditions under which not chattel but wage slavery was the owners'/capitalists' best option. If there's a lot of land and not many people, you'll need to use force in order to convert free human beings into laborers. If, on the other hand, there's a lot of people and not much land, or if those in power otherwise control access to land, those who do not own the land have no choice but to work for those in power. Under these conditions there's no reason for owners to go to the expense of buying or enslaving people, then paying for their slaves' food, clothing, and shelter: it's much cheaper to simply hire them.[148] As one pro-slavery philosopher put it: "In all countries where the denseness of the population has reduced it to a matter of perfect certainty, that labor can be obtained, whenever wanted, and that the laborer can be forced, by sheer necessity, to hire for the smallest pittance that will keep soul and body together, and rags upon his back while in actual employment—dependent at all other times on alms or poor rates—in all such countries it is found cheaper to pay this pittance, than to clothe, feed, nurse, support through childhood, and pension in old age, a race of slaves."[149]

Today, of course, we have so internalized the ideology of centralized control, of civilization, that most of us do not consider it absurd that people have to pay someone simply so they may exist on the planet, except perhaps to grumble that without rent or mortgage (or second mortgage) payments we wouldn't have to work so hard at jobs we don't like, and could spend more time with people we love, doing things we enjoy.

Although I have understood all of the above for a long time, it was only last week that I realized—and my indigenous friends are wondering where I've been these last six thousand years—that just as those in power must control access to land, the same logic dictates they must destroy all stocks of wild foodstuffs. Wild salmon, for example, cannot be allowed to live. Why would I go to Safeway if I could catch coho in the stream outside my door? I wouldn't. So how do those in power make certain I lack food self-sufficiency? Simple. Eliminate free food sources. Eliminate wild nature. For the same is true, obviously, for everything that is wild and free, for everything else that can meet our needs without us having to pay those in power. The push to privatize the world's water helps make sense of official apathy surrounding the pollution of (free) water sources. You just watch: air will soon be privatized: I don't know how they'll do it, but they'll certainly find a way.

This destruction of wild foodstuffs has sometimes been accomplished explicitly to enslave a people, as when great herds of buffalo were destroyed to bring the Lakota and other Plains Indians to terms, or as when one stated reason for building dams on the Columbia River was that dams kill salmon. The hope was that this extirpation would break the cultural backs of the region's Indians. But the destruction of wild foodstocks doesn't require some fiendishly clever plot on the part of those in power. Far worse, it merely requires the reward and logic systems of civilization to remain in place. Eliminating wild foodstocks is just one of many ways those in power increase control. And so long as the rest of us continue to buy into the system that values the centralization of control over life, that values the production of things over life, that values cities and all they represent over life, that values civilization over life, so long will the world that is our real and only home continue to be destroyed, and so long will the noose that is civilization continue to tighten around our throats.

❨ ❨ ❨

Once again I had dinner with my friend who used to date the philosopher. We sat down. She jumped right in. "What is the relationship between drinkable quantities of clean water being good, and rape being bad?" In the days after our dinner conversation, her enthusiasm had run up against the clear leap of logic—her ex-boyfriend would have said faith—in my argument.

"We're animals," I said.

"I know that. So?"

"So we have needs."

"I've heard some people—men, mainly—say that's one reason for rape."

"No. Needs to survive, to develop into who we really are."

"Who are we?"

"That's the question, isn't it?"

"I've read science-based analyses suggesting rape is a demonstration of power—"

"No arguments from me there."

"—and serves the evolutionary purpose of getting women to bond with powerful men," she said.

"Lemme guess," I responded, "the scientists were males, right?"

"They also say rape serves to pass on the genes of more aggressive men—"

"Which might seem to make superficial sense if you presume life is based on competition, not cooperation."

"Right, and if you presume relationships don't exist, and presume also that sperm is way, way more important than love, joy, or peace."

"Very odd presumptions, aren't they? Makes you wonder about the sanity—and social lives—of those who make them," I said, then continued, "Scientists and economists can't measure or control love, joy, or peace . . ."

"So love, joy, and peace must not exist," she said. "It's all pretty fucked up."

"It also projects the presumptions of industrial production onto women, and to a lesser degree, men."

"That women are here to make babies . . ."

"To manufacture them, as it were."

"Pop them out like Model-Ts on an assembly line."

"Or buns in a factory oven."

"So why are we here?" she asked.

"It presumes the same for sex. That the purpose is reproduction."

"Is it?"

"Maybe the purpose of both—sex and life—is to have fun, and to enter into relationships with those around us, and to become who we are."

"So who are we?" she pressed.

"Humans, and this is just as true for rocks and trees and stars and catfish, have a natural mode of development, or many natural modes. And there are commonalities across all humans, just as there are commonalities across all mammals, all animals, all 'living beings,' all rocks, what have you. Humans start out physically small, we grow, we stop growing, eventually our bodies wear out, and we die. Emotionally we follow certain patterns as well: we live for a long time with those who nurture us, we learn from them what it means to be human, and what it means to be human within our communities (or in the case of the civilized, how to be inhuman, and how to live in cities). There exist normal patterns for how humans grow. Joseph Chilton Pearce, for example, has done as fine a job as anyone describing patterns of human cognitive and emotional development."

"What does this have to do with rape?"

"I think we can say, or at least those of us with any sense at all can say," and she knew I was taking a dig at her philosopher ex-boyfriend, "that just as we have physical needs that, if they're not met, cause us to end up malnourished or our bodies to not develop to their full potential, to not work very well, so, too, we have emotional needs. Failure to meet these needs can stunt us emotionally, leave us emotionally undeveloped, leave us incapable of experiencing, expressing—*participating in*—the full range of human emotions. I think it's safe to say that all other things being equal, it's better to not be emotionally stunted than to be so."

"And rape?"

"It can stunt you. Impede your emotional development. Let's take this even on a fairly basic level. It's one thing to be abstinent by choice. That's a fine choice. But what of those people—women, mainly, but some men—who've been deprived of their capacity to take pleasure in sexuality because they've been raped? Their choice to participate in sexuality was taken away from them. Their ability to fully express and experience the emotions associated with that has been stunted."

She thought a moment, then said, "Not only that, but they've been deprived of their capacity to simply *be* in the world without being terrified. If any woman, anywhere in the world, hears footfalls behind her on a darkened street, she has reason to be afraid. Robin Morgan called that the democracy of fear under patriarchy."

I responded, "It doesn't matter what stories anybody tells anyone else: these are all bad things. And of course I'm not just talking about rape, nor am I just talking about sex. I'm saying that just as we can say that drinkable quantities of clean water are good—once again, no matter the stories we tell ourselves—we can similarly say that actions causing us to move away from the development of the full range of human emotions are not good. Certainly an action that causes an entire gender to live their lives in fear is a very bad thing."

"But isn't it possible for trauma to open people out? You wouldn't be the person you are had your father not abused you."

"I've heard people say that. There were even a few who suggested I should have put him in the acknowledgments of *A Language Older Than Words*." That's the book where I describe his abuse, and my response. "But what do I have to thank him for? Insomnia? Nightmares and feelings of terror that lasted through my late thirties, until I exorcised them through writing that book? Fractured relationships with my siblings? Messed up relationships with other people?"

"But you've also gained wisdom and insight you might not otherwise have gained."

"Yes. *I've* gained it. And this is true for anyone who survives abuse. The perpetrator isn't responsible if the survivor is able to metabolize the horrors into gifts for the community. The survivor, and the humans and nonhumans who've supported the survivor, are responsible. I've not accomplished *any*thing because I was raped. I've accomplished it *through* and despite the rapes. The rapes did not help me develop. They were not and could never be good. My response can be and has been good. But the rapes? No."

《 《 《

"Does all of this mean predation is bad?" she asked.

"How so?"

"If a heron eats a tadpole, we can say for certain the tadpole will never develop into an emotionally healthy frog. It will never develop into a frog at all."

"I don't think we have any concept of what it means to participate in a larger-than-human community. A while ago I did a radio interview in Spokane. The interviewer said pre-conquest Indians exploited salmon as surely as do the civilized. I had two responses. The first: if that were the case, why were there so many salmon before, and so few now? Something clearly has changed. The second: Indians *ate* salmon, not exploited them. He asked what's the difference. I said Indians entered into a relationship with the salmon whereby they gave respect to the salmon in exchange for the flesh."

"I've read about that."

"I wasn't happy with that answer. It was true so far as it went, but also left out so much as to be effectively false.[150] There was another necessary condition to the agreement between predator and prey, but I didn't know what it was. Then that afternoon I took a walk to the coyote tree."

The coyote tree was a pine under which I'd fed coyotes when I lived in Spokane. I loved the tree, and part of the reason I moved from Spokane was that the forest of which the tree was a part was being destroyed to put in a sub-division. Each day I'd heard the clank and roar of heavy machinery, and I'd had no idea what I could do to stop the destruction. So, and I'm not proud of this, rather than watch the destruction of this place I loved, I fled, moved far away. But I was back in town, and I went to sit by the tree.

"I kept asking the questions: what are the bonds between predator and prey? What are the conditions on which their relationships are based? How is respect for the spirit of the eaten manifested by the one who eats?"

"And?"

"The coyote tree told me the answer." My friend knew me well enough to not be surprised, and to know I wasn't speaking metaphorically. "When you take the life of someone to eat or otherwise use so you can survive, you become responsible for the survival—and dignity—of that other's community. If I eat a salmon—or rather, *when* I eat a salmon—I pledge myself to ensuring that this particular run of salmon continues, and that this particular river of which the salmon are a part thrives. If I cut a tree, I make the same pledge to the larger community of which it's a part. When I eat beef—or for that matter carrots—I pledge to eradicate factory farming."

"Did Indians have this deal?"

"On one level I have no idea. I can't speak for them.[151] But on the other, it's clear to me that *every*one makes this deal. It's the only way to survive."

"In the case of nonhumans, do you think the exchange is conscious?"

"Once again, I have no idea. But I can't see any reason why not." I paused, then said, "And I have to say that none of this is woo-woo or particularly cosmic. It's very physical."

"How so?"

"Not only is this crucial on moral and relational levels, but if I eat salmon without devoting myself to their continued survival, I'll soon find myself hungry. The same is true for bears or anyone else eating them, or, to take your example, herons eating tadpoles."

We sat a long time without speaking before I said, "I don't think what the coyote tree said was precisely a rebuke for me having abandoned that forest to the machinery after having gained and shared so much there. But I did become extremely aware I'd abandoned my responsibility. I had abandoned a place I love. And that doesn't feel good. That doesn't feel right."

Far more silence. Finally she nodded, and said, "It's like poisoning that glass of water we talked about. Or letting it be poisoned."

"Yes," I said, "I acted immorally."

CHOICES

Whether it takes me four weeks or 14 hours to get to Hamburg from Munich is less important to my happiness and to
my humanity than the question: How many men who yearn
for sunlight just as I do must be imprisoned in factories, their
healthy limbs and lungs sacrificed in order to build a locomo-
tive? For me the only important thing is: The more swiftly our
thriving economy is completely brought to ruin, the more
pitilessly the last remnant of industry is wiped out, the sooner
will people have enough to eat and have a small measure of that
happiness to which every man has a right.

B. Traven[152]

RIGHT NOW BEAKED WHALES ARE BEING KILLED BY SCIENTISTS IN THE Gulf of California. The scientists, from the National Science Foundation and Columbia University, are on a ship (the *Maurice Ewing*) that has on board an impressive array of airguns that fire sonic blasts of up to 260 db. The scientists use these airguns at least ostensibly to map the ocean floor. They say they're exploring how continents rift apart, but honesty (on my part, not theirs) requires mention that data generated this way is crucial to underwater oil exploitation.

A 260 db sound is very intense. For comparison, damage to human hearing begins at 85 db. A police siren at thirty meters is about 100 db. And decibels are logarithmic, meaning every 10 db increase translates to ten times more intensity, and sounds (because human perception is also logarithmic) twice as loud. In this case, that means the blasts from the research vessel are approximately ten quadrillion times more intense than a siren at thirty meters, and would sound to humans about 16,384 times as loud (we could easily round this off to 16,000, since in either case the sound would have killed you). The sound of a jet taking off at 600 meters is about 110 db. The Ewing's blasts are a quadrillion times more intense, and sound 8,192 times louder. A loud indoor rock concert weighs in at 120 db (the threshold of human pain, by the way): whales and other creatures in the Gulf of California are subjected to sounds 100 trillion times more intense than that. The threshold at which humans die from sound alone is 160 db. People—including nonhuman people—die because sound is a pressure wave (which is why you can feel your body vibrate during loud, low sounds: one of the attractions of rock concerts for me was the feeling of the bass notes massaging and piercing my body). Too-intense waves rip ear, lung, and other vibrating tissues. They cause internal bleeding. Two hundred and sixty decibels: that's 10,000 times more intense than the sound of a nuclear explosion at a range of five hundred meters.

This is the intensity with which whales and other creatures in the Gulf of California are assaulted.

Whales live by their ears. They communicate with them, singing complex songs we will probably never understand. Babies find their mothers by them. Adults navigate by them. They find food by them. Whales subjected to loud

143

noises stop singing, sometimes for days: which means they do not eat, do not court, probably do not sleep. Whales subjected to loud enough noises lose their hearing. Eardrums rupture. Brains hemorrhage. They die.

Since the experiment began, dead beaked whales have been discovered stranded on beaches of the Gulf of California by senior marine biologists at the National Marine Fisheries Services, including several experts in beaked whales, the impacts of noise on marine mammals, and the stranding of marine mammals. These scientists, and others who care about whales, wrote letters to the expedition's sponsors. Columbia University failed to meaningfully respond. The National Science Foundation's response was to write a letter stating, "There is no evidence that there is any connection between the operations of the Ewing and the reported [sic] beached whales."[153]

I must be honest with you, even at the risk of offending or alienating you. When I read of the torture and murder of these whales by scientists from the National Science Foundation, and of their and their attorneys' response to concerns about the whales, my first impulse was to wish someone would put a gun to the heads of the scientists and pull the trigger. If somehow (unfortunately) caught by police, the person could respond, "There is no evidence that there is any connection between the operation of this gun and the reported holes in these men's heads."

Even if I *do* admit this fantasy, decorum requires my next paragraph to be a denial of it, a statement of its unthinkability, its immorality, a plea for forgiveness for my fall from grace. I should disown it as dishonorable, disgusting, and having no place in any discussion of social change. But I won't do that. I can't. There is no accountability in this culture, at least for those who work for the centralization of power. And that lack of accountability is not sustainable. It is killing the planet. It is killing those I love. This lack of accountability is itself obscenely immoral.

I have a student at the prison who, when he was seventeen, killed someone. He did so in what he has since described as drug-induced psychosis. He is now spending the rest of his life in prison. Never again will he put his feet into a stream. Never again will he even *see* a stream. Never again will he pull an apple from a tree. Never again will he feel the long, slow kiss of a woman, nor feel her breasts against his chest, feel the muscles of her vagina contract around his penis. Never again, unless and until civilization comes down, will he walk free. He is paying for his decision, his action, with every moment of his life.

Yet scientists decide fish don't need water, and a judge goes along with them. Activists, including me, wring our hands and cry. Salmon die. There's no accountability anywhere in this web of non-relationships except for the salmon. They pay with their lives. Engineers design oil processing facilities, CEOs and shareholders gain profits from them, politicians pass laws protecting the profits of the corporations against all environmental and human costs, police protect the property against all trespassers, and from this volatile stew of immorality emerges a cancer cluster. The ones who pay are the children who receive the gifts of asthma, leukemia, and other illnesses. And of course the land itself pays. The land always pays. And when that facility is no longer profitable? Those in charge move on to destroy some other place. But the children—those in graves, and those not yet there—remain. As does the land. There is no accountability anywhere. I will not back away from my fantasy. Accountability needs to be brought into this web of non-relationships. And it needs to be brought in quickly.

I've little doubt the whales would agree.

I need to say, furthermore, that scientists from the National Science Foundation and Columbia University aren't the only ones deafening, torturing, and killing whales, dolphins, and other sea life. In fact they're rank amateurs. The U.S. Navy has begun to deploy a system that will soon blanket 80 percent of the world's oceans with pulse blasts of at least 200 db. And oil companies routinely run ships around the ocean exploring for oil by blasting at 260 db. It's happening right now. It's got to stop.

We've got to stop it.

❲ ❲ ❲

In the particular case of whales being killed in the Gulf of California, the "accountability model" probably wouldn't have been the best choice. The good folks at the Biodiversity Legal Foundation were able to get a temporary restraining order against the organizations involved, and halt the experiment.

It ends up, also, that the judge in the case had, probably accidentally, an idea that would have an effect similar to my fantasy. Initially he was going to ask representatives of the National Science Foundation to bring soundmakers into the courtroom and match the volume of the ship's airguns, presumably to give him a tangible idea of what they're talking about. Someone who helped bring the case to the judge's attention told me, "One can only imagine how great it would have been if that had come to pass. I imagine the judge asking the NSF lawyer to test out the air guns in the courtroom. The NSF apologists would have been forced to explain that doing so would have blown out the windows in the federal building, not to mention what it would have done to everyone in the courtroom (all the better if NSF researchers had been there). I think that would have been a delectable example of accountability."

❪ ❪ ❪

Of course there's no accountability for those who work for the centralization of power. Their violence is not really violence (see Premise Four), or at least can never be seen as such. Thus the murder of whales is not *really* violence, nor indeed is the murder of entire oceans. The same is true for trees, forests, mountains, entire continents. The same is true for entire peoples. None of this violence can be considered violence, which means all talk of accountability makes no sense: there's nothing to be held accountable *for*.

❪ ❪ ❪

Today I'm driving through northern California. I started at the coast, passing through patches of afternoon fog that slid between the tops of tall redwoods like so many ghosts. Traffic was light, and in the two-lane stretches the slow cars inevitably (and unaccountably) used the turnouts, as they're supposed to.

Sometimes insects arced into my windshield, white or black dots that came upon me too quickly for me to swerve, then splattered yellow, orange, white, or transparent against the glass. I thought often, too, of the insects I did not even see, but killed nonetheless. Roads are free-kill zones for anything that enters.

I crossed the Klamath River, running much fuller now: after the salmon had been safely killed, the feds released water into the river. Federal biologists (political scientists, I guess you'd call them, although one friend prefers the term *biostitutes*) continue to claim—no surprises here—there's been no causal connection shown between the lack of water and the death of fish, and I continue to fantasize about accountability.

The road moved away from the coast, and the day warmed. Traffic remained light. I crossed the Eel and Russian Rivers, which are little more than braided streams good for warm foot baths and for little children wading. The Eel once had runs of lampreys great as the runs of salmon it also now no longer has. I don't know if the Russian ever had runs of Muscovites.

Then I entered wine country—Mendocino and Sonoma counties—and saw the reason for the rivers' deaths: great seas of grapes extending as far as I could see. Even though everyone—including teetotalers like myself—know that non-irrigated grapes make better wines, the rivers have been effectively dewatered to grow these grapes, and more importantly to grow the bank accounts of those wealthy enough to own wineries: a few huge corporations control production, as always, which means they also control politics, as always, which means they also control land use policies, as always.[154]

Last year an insect called the glassy-winged sharpshooter made news through the region, because it was helping to spread Pierce's Disease, an illness that threatened (or promised) to decimate grape plants. Federal, state, and local governments went all out to eradicate this threat (or promise), shoveling fistfuls of public moneys toward protecting these private (and especially corporate) investments.

But I must confess something else. Every time I see these dewatered rivers, and every time I see these miles upon miles of grapes (which are not used for food, nor for anything but an absolutely nonessential item commonly used for conspicuous consumption [note that I've nothing against luxuries; I do have something against luxuries that come at the expense of the landbase]), I think the same thing, that I'm in the wrong line of work. I need to quit writing, I think, and start raising glassy-winged sharpshooters to release in these fields.

❆ ❆ ❆

A few years ago I was watching television with two indigenous people. One was a Maori woman, the other an American Indian man. A newscaster was speaking, which is to say he was lying, spinning events to promote the interests of his bosses, more broadly of capital, more broadly still of the culture, of civilization, more broadly yet of destruction. The particular story he was spinning had to do with the environment and indigenous rights. No causal connection could be shown, he was saying, between deforestation and species extinction. In fact, he said, the worst enemies of these creatures were environmental extremists keeping timber companies from going in and cleaning up forests, and indigenous peoples insisting on archaic "treaty rights" allowing them to hunt and fish where white people couldn't. He made clear that decent people shouldn't stand for such blatant obstructionism on the part of environmentalists and racism on the part of the indigenous.

My two friends suddenly spoke at the same time. The Maori woman: "I want to hit him in the head with a taiaha," a Maori club. The Indian man: "I want to shoot an arrow through his throat."

I burst out laughing. They looked at me. I could tell they were hurt by my

laughter. I said, "No, it's not that. It's just this is such a wonderful example of parallel cultural evolution: different tools to accomplish the same important task."

They laughed now, also.

(((

We all face choices. We can have ice caps and polar bears, or we can have automobiles. We can have dams or we can have salmon. We can have irrigated wine
from Mendocino and Sonoma counties, or we can have the Russian and Eel
Rivers. We can have oil from beneath the oceans, or we can have whales. We can
have cardboard boxes or we can have living forests. We can have computers and
cancer clusters from the manufacture of those computers, or we can have neither. We can have electricity and a world devastated by mining, or we can have
neither (and don't give me any nonsense about solar: you'll need copper for
wiring, silicon for photovoltaics, metals and plastics for appliances, which need
to be manufactured and then transported to your home, and so on. Even solar
electrical energy can never be sustainable because electricity and all its accou-
trements require an industrial infrastructure). We can have fruits, vegetables, and
coffee brought to the U.S. from Latin America, or we can have at least somewhat
intact human and nonhuman communities throughout that region. (I don't
think I need to remind readers that, to take one not atypical example among far
too many, the democratically elected Arbenz government in Guatemala was
overthrown by the United States to support the United Fruit Company, now
Chiquita, leading to thirty years of U.S.-backed dictatorships and death squads.
Also, a few years ago I asked a member of the revolutionary tupacamaristas
what they wanted for the people of Peru, and he said something that cuts to the
heart of the current discussion [and to the heart of every struggle that has ever
taken place against civilization]: "We need to produce and distribute our own
food. We already know how to do that. We merely need to be allowed to do so.")
We can have international trade, inevitably and by definition as well as by function dominated by distant and huge economic/governmental entities which do
not (and cannot) act in the best interest of communities, or we can have local
control of local economies, which cannot happen so long as cities require the
importation (read: theft) of resources from ever-greater distances. We can have
civilization—too often called the highest form of social organization—that
spreads (I would say metastasizes) to all parts of the globe, or we can have a
multiplicity of autonomous cultures each uniquely adapted to the land from
which it springs. We can have cities and all they imply, or we can have a livable

planet. We can have "progress" and history, or we can have sustainability. We can have civilization, or we can have at least the possibility of a way of life not based on the violent theft of resources.

This is in no way abstract. It is physical. On a finite world, the forced and routine importation of resources is unsustainable. Duh.

Show me how car culture can coexist with wild nature, and more specifically, show me how anthropogenic global warming can coexist with ice caps and polar bears. And any fixes such as solar electric cars would present problems at least equally severe. For example, the electricity still needs to be generated, batteries are extraordinarily toxic, and in any case, driving is not the main way a car pollutes: far more pollution is emitted through its manufacture than through its exhaust pipe. We can perform the same exercise for any product of industrial civilization.

We can't have it all. The belief that we can is one of the things that has driven us to this awful place. If insanity could be defined as having lost functional connection with physical reality, to believe we can have it all—to believe we can simultaneously dismantle a world and live on it; to believe we can perpetually use more energy than arrives from the sun; to believe we can take more than the world gives willingly; to believe a finite world can support infinite growth, much less infinite economic growth, where economic growth consists of converting ever larger numbers of living beings to dead objects (industrial production, at its core, is the conversion of the living—trees or mountains—into the dead—two-by-fours and beer cans)— is grotesquely insane. This insanity manifests partly as a potent disrespect for limits and for justice. It manifests in the pretension that neither limits nor justice exist. To pretend that civilization can exist without destroying its own landbase and the landbases and cultures of others is to be entirely ignorant of history, biology, thermodynamics, morality, and self-preservation. And it is to have paid absolutely no attention to the past six thousand years.

(((

One of the reasons we fail to perceive all of this is that we—the civilized—have been inculcated to believe that belongings are more important than belonging, and that relationships are based on dominance—violence and exploitation. Having come to believe that, and having come to believe the acquisition of material possessions is good (or even more abstractly, that the accumulation of money is good) and in fact the primary goal of life, we then have come to perceive ourselves as the primary beneficiaries of all of this insanity and injustice.

Right now I'm sitting in front of a space heater, and all other things being equal, I'd rather my toes were toasty than otherwise. But all other things aren't equal, and destroying runs of salmon by constructing dams for hydropower is a really stupid (and immoral) way to warm my feet. It's an extraordinarily bad trade.

And it's not just space heaters. No amount of comforts or elegancies, what that nineteenth-century slave owner called the characteristics of civilization, are worth killing the planet. What's more, even if we do perceive it in our best interest to take these comforts or elegancies at the expense of the enslavement, impoverishment, or murder of others and their landbases, we *have no right to do so*. And no amount of rationalization nor overwhelming force—not even "full-spectrum domination"—will suffice to give us that right.

Yet we have been systematically taught to ignore these trade-offs, to pretend that we don't see them (even when they're right in front of our faces) they do not exist.

Yesterday, I received this email: "We all face the future unsure if our own grand-children will know what a tree is or ever taste salmon or even know what a clean glass of water tastes like. It is crucial, especially for those of us who see the world as a living being, to remember. I've realized that outside of radical activist circles and certain indigenous peoples, the majority has completely forgotten about the passenger pigeon, completely forgotten about salmon so abundant you could fish with baskets. I've met many people who think if we could just stop destroying the planet right now, that we'll be left with a beautiful world. It makes me won-der if the same type of people would say the same thing in the future even if they had to put on a protective suit in order to go outside and see the one tree left standing in their town. Would they also have forgotten? Would it still be a part of mainstream consciousness that there used to be whole forests teeming with life? I think you and I agree that as long as this culture continues with its preferred methods of perception, then it would not be widely known to the majority. I used to think environmental activists would at least get to say, 'I told you so' to every-one else once civilization finally succeeded in creating a wasteland, but now I'm not convinced that anyone will even remember. Perhaps the worst nightmare visions of activists a few hundred years ago match exactly the world we have out-side our windows today, yet nobody is saying, 'I told you so.'"[155]

I think he's right. I've long had a nightmare/fantasy of standing on a deso-late plain with a CEO or politician or capitalist journalist, shaking him by the shoulders and shouting, "Don't you see? Don't you see it was all a waste?" But after ruminating on this fellow's email, the nightmare has gotten even worse. Now I no longer have even the extraordinarily hollow satisfaction of seeing recogni-tion of a massive mistake on this other's face. Now he merely looks at me, his eyes flashing a combination of arrogance, hatred, and willful incomprehension, and says, "I have no idea what you're talking about."

And he isn't even entirely lying.

Except of course to himself.

❲ ❲ ❲

Sometimes lying awake at night in bed, I fantasize. I imagine how fun it would be to wrestle with the problems we face if only we weren't insane, if only the problems really were just technical, if only we could cling even to any remotely feasible, remotely forgivable hope of a soft landing rather than a hard crash, if only our culture were not driven to destroy all life on the planet, if only there were even the slightest chance our culture would undergo a voluntary transformation to a sane and sustainable way of living.

By now there can be few who do not understand that without massive public subsidies (far larger than total profits) the entire corporate economy would collapse overnight. People pay to deforest the planet, decapitate mountains, decimate oceans, destroy rivers.

Were we to suddenly find ourselves sane in this insane situation, we could easily and immediately shift subsidies. So long as we care neither about justice nor accountability, but merely want to stop the damage, we could subsidize the same corporations to repair damage they've already caused. Instead, for example, of the public paying Weyerhaeuser to deforest, as is currently the case, we could pay it to reforest. Not to make tree farms—virtual forests of genetically identical Douglas firs—but to use the inventiveness we talk so much about but rarely seem to use to life-serving ends in order to make life better for the forests and its other members with whom we share our home.

Of course this is a fantasy, as absurd as Fuller's notion of converting weaponry to livingry. Indeed, it's essentially the same fantasy. And not only is it an impossible fantasy for the reasons already discussed—a) weaponry (as well as massive public subsidies) being absolutely necessary to the unceasing flow of resources toward the center of empire, and b) Fuller's notion ignores violence to the natural world—but we face an even greater challenge to the possibility of ever living sanely, peacefully, or, saying much the same thing, sustainably. This impediment forms the tenth premise of this book, which I've described in previous books and which I'll explore more later on: *The culture as a whole and most of its members are insane. The culture is driven by a death urge, an urge to destroy life.*

❰ ❰ ❰

Here is how governments and people in this culture spend money. These make clear their priorities. In 1998, governments and people spent US $6 billion on basic education across the world; $8 billion on cosmetics in the United States; $9 billion on water and sanitation for everyone in the world; $11 billion on ice cream in Europe; $12 billion on reproductive health for all women in the world; $12 billion on perfumes in Europe and the United States; $13 billion on basic health and nutrition for everyone in the world; $17 billion on pet foods in Europe and the United States; $35 billion on business entertainment in Japan; $50 billion on cigarettes in Europe; $105 billion on alcoholic drinks in Europe; $400 billion on narcotic drugs in the world; $780 billion on military spending in the world. As the compiler of the list notes: "It would seem ironic that the world spends more on things to destroy each other (military) and to destroy ourselves (drugs, alcohol and cigarettes) than on anything else."[156]

❰ ❰ ❰

Most of my students at the prison are there at least partly because of drugs. Since the prison is a supermax, almost none of them are there for simple possession, or even dealing. Many are in for armed robbery committed to support their habits, or for murders committed under the influence or during drug deals gone bad.

Nearly all of them, as I mentioned before, hate prison with a passion I've rarely seen matched. They hate it partly because of characteristics that make prison really the quintessence of civilization: its routine dehumanization, its destruction of community, its isolation. My students are deprived of their families, with many knowing their children only through occasional letters and infrequent photos: they've shown me high school graduation pictures of children they've not seen since they were six and not held since they were infants. They've shown me pictures of wives and parents they'll never see again. Prisons also mirror and magnify the bureaucratic power structures and strict rules that characterize civilization. This is when you eat. This is what you eat. This is how many books you may have (which must have been sent directly from a bookstore or publisher). This is the sort of writing implement you may use. This is the sort you may not.

Those prisoners who do not hate prison generally fall into a very few categories. There are lifers and a few others—usually those who've already served decades—who've come to an enlightened sort of acceptance—the serenity to accept things they cannot change. There are people whose horrific childhoods make prison a comparative cakewalk. And there are J-cats, or crazy people (*J-cat* stands for category J, a prison classification meaning the insane).

Yet as I said before, when I ask my students whether they'll use again when they get out, even at the risk of coming back to prison, most say *yes*.

"It's very difficult," one said to me. "The first problem is the physical addiction. That can be hard to beat. And if you beat that, there's still the memory of how good it feels. Even though I've been clean now all these years in prison if you put drugs in front of me right now I'd want to take them, just so I could feel that good again. But these problems are nothing compared to the emotional addiction. So much of my identity has been wrapped up in drugs. Drugs became who I am. Without them I was nothing. But even kicking the emotional addiction still isn't the hardest part. It's all of my relationships. My wife and I used together—that was all bound up in our courtship, in our sex-life, in our daily activities. And she still uses. What am I supposed to do when I get out? Not only do I have to give up this thing that makes me feel so very good—or at least I think it makes me feel good—and not only do I have to step away from this thing that's been my identity for most of my life, but I'll have to change my whole web of friendships, and maybe even my family. I'm facing a third strike if I get caught again, which means I'd be in forever, but even facing that I just don't know if I can give up so much."

❰ ❰ ❰

One can be addicted to many things besides drugs, alcohol, tobacco. One can be addicted to television, sugar, coffee, low self-esteem, sex, authority, shopping, a specific (or specific type of) relationship. One can be addicted to a lifestyle. A whole culture, as we shall see (or perhaps as we already do), can be addicted to civilization.

My compact *Oxford English Dictionary* defines the verb *addict* (in excruciatingly tiny print that seems to get tinier with each passing year) as "to bind, devote, or attach oneself as a servant, disciple, or adherent." In Roman law, an *addiction* was "A formal giving over or delivery by sentence of court. Hence, A surrender, or dedication, of any one to a master."[157] It comes from the same root as *diction*: *dicere*, meaning *to pronounce*, as in a judge pronouncing a sentence upon someone. To be addicted is to be a slave. To be a slave is to be addicted. The heroin ceases to serve the addict, and the addict begins to serve the heroin. We can say the same for civilization: it does not serve us, but rather we serve it.

There's something desperately wrong with that.

☾ ☾ ☾

This might be a good time to remind readers of the necessary relationship between civilization and slavery, that in fact civilization originated in slavery, is based on slavery, requires slavery, would collapse without slavery. You needn't take my word for this, nor the word of anarchists, Luddites, or indigenous peoples. Nor do you merely need to take the word of pro-slavery philosophers or pro-technology CEOs. Nor do you merely need to take the word of Aristotle— propagandist extraordinaire—who wrote extensively in support of slavery and its necessity, indeed, its naturalness. Nor mainstream historians who recognize that, as Friedrich Engels (admittedly not a mainstream—i.e., pro-capitalist, pro-civilization—historian) wrote, "Without slavery, no Greek state, no Greek art and science; without slavery, no Roman Empire. But without Hellenism and the Roman Empire as the base, also no modern Europe. We should never forget that our whole economic, political and intellectual development has as its presupposition a state of things in which slavery was as necessary as it is universally recognized."[158] You don't even have to take the word of modern anti-slavery activists who point out that there are more slaves in the world today than came across on the Middle Passage. Just look around. Consider the immiseration inherent in the items surrounding you. Look for the slavery, both human and nonhuman, that went into their making. Just because you don't see the chains doesn't mean you don't benefit from their slavery, and from their deaths. How many salmon died to provide you electricity? How many rivers and mountains were enslaved to make this aluminum can? How many trees died to make this book? Further, how many people do you know who hate their jobs? On the other hand, how many people do you know who love their lives, and who live at least remotely integrated into the larger community that is their landbase?

ABUSE

We are going to fight them and impose our will on them and we will capture or . . . kill them until we have imposed law and order on this country. We dominate the scene and we will continue to impose our will on this country.

Paul Bremer, U.S. Administrator of occupied Iraq[159]

Something very unpleasant is being let loose in Iraq. Just this week, a company commander in the U.S. 1st Infantry Division in the north of the country admitted that, in order to elicit information about the guerrillas who are killing American troops, it was necessary to "instill fear" in the local villagers. An Iraqi interpreter working for the Americans had just taken an old lady from her home to frighten her daughters and granddaughters into believing that she was being arrested.

A battalion commander in the same area put the point even more baldly. "With a heavy dose of fear and violence, and a lot of money for projects, I think we can convince these people that we are here to help them," he said. He was speaking from a village that his men had surrounded with barbed wire, upon which was a sign, stating: "This fence is here for your protection. Do not approach or try to cross, or you will be shot."

Robert Fisk[160]

THE OTHER DAY, DEAR ABBY LISTED WARNING SIGNS OF POTENTIAL ABUSERS, saying, (in all caps, no less), "IF YOUR PARTNER SHOWS THESE SIGNS, IT'S TIME TO GET OUT." I followed her citation to the Projects for Victims of Family Violence, and was intrigued by what I saw. I was especially intrigued by the final sentence of the Projects' introduction: "Initially the batterer will try to explain his behavior as signs of love and concern, and a woman may be flattered at first. As time goes on, the behaviors become more severe and serve to dominate the woman."[161] This reminded me of something Robert Jay Lifton wrote in his extraordinary book *The Nazi Doctors*, about how before you can commit any mass atrocity, you must convince yourself that what you're doing is not in fact harmful but instead beneficial, so that, for example, Nazis weren't in their own minds committing genocide and mass murder, but instead purifying the "Aryan race." Of course we see the same on a daily basis, as we the civilized do not enslave the poor or indigenous but civilize them, and we do not destroy the natural world but instead develop natural resources. And I thought about this on a personal level: how very rare it is for someone to do something because he or she is a jerk. I know when I've treated people poorly, I've nearly always had my actions fully rationalized beforehand, and I've generally believed my rationalizations. That's one of the beautiful things about denial: by definition you don't know you're in it. Now, my own transgressions have been frankly pretty minor—a few hurt feelings here or there—but I've wondered about something of much greater consequence ever since I was a child: did my father believe the lies he told us about his own violence? Did he really think he was beating my brother because of where my brother parked the car? Or more seriously yet, did he really believe himself a day later when he denied the violence altogether? Similarly, do those in power believe their own lies? In their heart of hearts (presuming they still have them) do the scientists for the National Science Foundation really believe there's no connection between sonic blasts louder than nuclear explosions and the deaths of nearby whales? Do the National Academy of Sciences biostitutes really believe there's no connection between a lack of water in the Klamath and dead salmon? Does anyone really believe industrial civilization isn't killing the planet?

Now, to the list. I've greatly shortened (and in some cases modified) the

Projects' commentary, and although women sometimes do beat men (and certainly in this culture—where all of us are more or less crazy—women commit their fair share of emotional abuse, too), physical violence runs overwhelmingly enough from male to female to cause me to use the masculine pronoun for batterers. Nonetheless, if your partner is a woman and fits these characteristics, you, too, would be wise to follow Dear Abby's all caps advice.

The list begins with jealousy: Although the abuser says jealousy is a sign of love, it's instead a sign of insecurity and possessiveness. He'll question you about whom you talk to, accuse you of flirting, be jealous of time spent with family, friends, or children. He may call constantly or visit unexpectedly, prevent you from going to work because "you might meet someone," check the mileage on your car.

This leads to the second sign, controlling behavior: At first, the batterer will say he's concerned for your safety, your need to use time well, or your need to make good decisions. He'll be angry if you're "late" returning from the store or an appointment, will question you closely about where you went, whom you talked to. He may eventually not let you make personal decisions about your house or clothing; he may keep your money or even make you ask permission to leave the room or house.

The third characteristic is quick involvement. He comes on strong—"I've never felt loved like this by anyone"—and pressures you for an exclusive commitment almost immediately.

The pressure is because of the fourth characteristic: he needs someone desperately because he's very dependent, soon enough depending on you for all his needs, expecting you to be the perfect wife, mother, lover, friend. He then projects this dependence back onto you in an attempt to increase his control, saying, "If you love me, I'm all you need; you're all I need." You're supposed to take care of everything for him emotionally and in the home.

Because of his dependence he'll try to isolate you from all resources. If you have male friends, you're a "whore." If you have female friends you're a lesbian. If you're close to your family, you're "tied to the apron strings." He'll accuse people who support you of "causing trouble." He may want to live in the country without a phone, he may not let you use a car, and may try to keep you from working or going to school.

The sixth characteristic is that he blames others for his problems. If he's not successful in life, someone must be out to get him. If he makes a mistake, you must have upset him, kept him from concentrating. It's your fault his life isn't perfect.

And it's your fault he's not happy. It's your fault he's angry. "You make me angry when you don't do what I say." If he has to harm you, then, that, too, is your fault: you, after all, made him mad. And you certainly don't want to do that.

He gets upset easily. He's hypersensitive. The slightest setbacks are personal attacks.

He's often cruel, or at the very least insensitive to the pain and suffering of nonhuman animals, and also to children. He may beat them because they are incapable of doing what he wants: for example, he may whip a two-year-old for wetting a diaper.

He may conflate sex and violence. This may be under the guise of playfulness, wanting to act out fantasies that you're helpless, which serves the vital purpose of letting you know that rape excites him. Or he may simply drop the guise.

The next warning sign is that he may perceive and actualize rigid sex roles. You're supposed to stay at home and serve him. You must obey him, in great measure because women are inferior, less intelligent, unable to be whole without men.

He may verbally abuse you, saying cruel, hurtful, degrading things. He may run down your accomplishments, and may attempt to convince you that you cannot function without him. This abuse may come when you're surprised or vulnerable: he may, for example, wake you up in order to abuse you.

Sudden mood swings are another warning signal. He can be nice one minute, and explosively violent the next, which means of course he was never really nice to begin with.

You should watch out if he has a history of battering. He may acknowledge he hit women in the past, but will aver they made him do it. You may hear from ex-partners that he's abusive. It's crucial to note that battering isn't situational: if he beat someone else, he'll very likely beat you, no matter how perfect you try to be.

You should be very wary if he uses threats of violence to control you. "I'll slap your mouth off," or "I'll kill you," or "I'll break your neck." A batterer may attempt to convince you all men threaten partners, but this isn't true. He may also attempt to convince you you're responsible for his threats: he wouldn't threaten you if you didn't make him do it.

He may break or strike objects. There are two variants of this behavior: one is the destruction of beloved objects as punishment. The other is for him to violently strike or throw things to scare you.

The last characteristic on the Projects' list is the use of any force during an argument: holding you down, physically restraining you from leaving the room, pushing you, shoving you, forcing you to listen to him.

Now, I found this list very interesting in its own right, and given the rate at which women are abused (just in this country, a woman is beaten by her partner every ten seconds), it's also very important. But I found it even more interesting because it was immediately clear to me that these warning signs also apply to our culture as a whole. Let's go through them again.

Jealousy. The God of this culture has always been jealous. Time and again in the Bible we read, "I the LORD thy God am a jealous God, visiting the iniquity of the fathers upon the children unto the third and fourth generation of them that hate me,"[162] or "Ye shall not go after other gods, of the gods of the people which are round about you; (For the LORD thy God is a jealous God among you) lest the anger of the LORD thy God be kindled against thee, and destroy thee from off the face of the earth."[163] God today is just as jealous, whether he goes by the name of Science, Capitalism, or Civilization. Science is as monotheistic as Christianity, moreso really, since Science doesn't even have to *say* it's jealous: we've so internalized its hegemony that many of us believe the only way we can know anything about the world is through science: Science *is* Truth. Capitalism is so jealous it couldn't even allow the existence of the Soviet version of itself (they're both state-subsidized command economies,[164] the biggest differences being: a) the merging under the Soviet system of state and corporate bureaucracies into one huge bureaucracy that was even more inefficient and wasteful than the "capitalist" system of functionally separate bureaucracies working for the unified goal of production; and b) the Soviet Politburo was dominated by different factions of the Communist Party with more than 90 percent of the votes going to this party, while the American Congress is dominated by different factions of the Capitalist Party, with more than 90 percent of the votes going to *this* party). Civilization is just as jealous as science and capitalism, systematically disallowing anyone from perceiving the world in nonutilitarian terms, that is, perceiving the world not in terms of slavery, that is, not in terms of addiction, that is, perceiving the world relationally. Lots of so-called free thinkers like to comment on the tens of millions of people who have been killed because they refused to worship Christianity's God of Love—because God is after all a jealous God—but even they rarely mention the hundreds of millions of (indigenous and other) people who have been killed because they refused to worship Civilization's God of production, a God just as jealous as the Christian God, a God deeply devoted to the conversion of the living to the dead.

Control. I've thought for a couple of days now about what to put in this paragraph. I considered talking about the public school systems, which have as their primary function the breaking of children's wills—getting them to sit in one place for hours, days, weeks, months, years on end, wishing their lives away—in preparation for their lives as wage slaves. Then I thought about advertising, and more broadly television, and how through our entire lives we're manipulated by distant others who do not have our best interests at heart. I thought of the words of economist Paul Baran, "The real problem is . . . whether an economic and social order should be tolerated in which the individual, from the very cradle on, is so shaped, molded, and 'adjusted' as to become an easy prey of profit-greedy capitalist enterprise and a smoothly functioning object of capitalist exploitation and degradation."[165] But then I thought maybe I should write about face-recognition software, and of the implantation of ID chips first into pets, then into people. I thought of the words of a 1996 U.S. Air Force Scientific Advisory Board Report: "One can envision the development of electromagnetic energy sources, the output of which can be pulsed, shaped, and focused, that can couple with the human body in a fashion that will allow one to prevent voluntary muscular movements, control emotions (and thus actions), produce sleep, transmit suggestions, interfere with both short-term and long-term memory, produce an experience set, and delete an experience set. This will open the door for the development of some novel capabilities that can be used in armed conflict, in terrorist-hostage situations, and in training."[166] Of course one no longer needs to envision these sorts of weapons: many are already operational. I thought of the Joint Vision 20/20 Statement and the goal of "full-spectrum domination." I thought of the so-called Homeland Security Act of 2002, passed by the U.S. Senate by a vote of 90 to 9, that, in the words of even the conservative writer William Safire, means, "Every purchase you make with a credit card, every magazine subscription you buy and medical prescription you fill, every Web site you visit and e-mail you send or receive, every academic grade you receive, every bank deposit you make, every trip you book and every event you attend—all these transactions and communications will go into what the Defense Department describes as 'a virtual, centralized grand database.' To this computerized dossier on your private life from commercial sources, add every piece of information that government has about you—passport application, driver's license and bridge toll records, judicial and divorce records, complaints from nosy neighbors to the F.B.I., your lifetime paper trail plus the latest hidden camera surveillance—and you have the supersnoop's dream: a 'Total Information Awareness' about every U.S. citizen."[167] I thought of science, which has as its ultimate (and proximate) goal the conversion of the wild and wildly unpredictable natural world into something orderly, predictable, and controllable. There are simply too many examples of our culture's basis in the need for control for me to choose. You choose.

Quick involvement: I'm not sure how much quicker you can get than the choice offered to so many Indians as they were tied to stakes, piles of wood around their feet, of Christianity or Death. One Indian asked in response: If he converted to Christianity would he go to heaven? And if so, would there be other Christians there? When he found the answer to both questions was *yes*, he said he'd rather burn to death.

But there's something else about quickness. Civilization has only been on this continent a few hundred years. There are many parts of this continent, such as where I live, that became subject to civilization far more recently. Yet in this extremely short time this culture has committed us and the landscape to this technologized path, in so doing shredding the natural fabric of this continent, enslaving, terrorizing, and/or eradicating its nonhuman inhabitants, and giving its human residents the choice of civilization or death. Another way to say this is that prior to the arrival of civilization humans lived on this continent for at the very least ten thousand years, and probably much longer, and could drink with confidence from rivers and streams everywhere. After this culture's short time here, not only has it toxified streams and groundwater, but even mother's breast milk. That's an extraordinary and extraordinarily quick commitment to this technologized way of being (or rather non-being). Here's another way to say this: these days the decision to enslave or kill a river by putting in a dam is generally made in the several years it takes to write an Environmental Impact Statement and get funding. The process might drag on a decade or two at most. But such a decision, if it is to be made at all, should be made only after generations of observation: how can you possibly know what is best for any part of the land unless you interact with it long enough to learn its rhythms? For example, four days ago hooded mergansers landed on the pond outside my window. They stayed two days, and have now been gone two. They did this last year, only they arrived one day earlier, left one day earlier, and then came back a few days later and stayed a week. Will they come back next year? I don't know; I haven't been here long enough. And last year there were many rough-skinned newts living in the pond. I saw them almost every day. The mergansers ate some (rough-skinned newts are one of the most poisonous creatures around, but mergansers don't seem to mind). This year I haven't seen so many newts. Is that because of the mergansers, because of me, or because of something else entirely that I would only understand if I lived here long enough to start to know the place? I panicked two years ago because there weren't as many tadpoles as there had

been the year before. Was the population collapsing? Well, the next year the
frogs were quieter because there were fewer returning yearlings, and I was even
more worried. But these new males must have been especially virile, the females
especially fertile, because there were once again lots of fat babies. Many of these
tadpoles, however, were eaten by roving packs of backstriders, far more than
were eaten in the prior two years. Should I worry? The point is that I have no
idea, and I can have no idea till I've been here enough years, even generations,
to begin to know what is normal, expected, desirable. In the meantime, I'm a fool
if I do something grossly destructive.

Were we not abusive to the land, to each other, to ourselves, we would sit
back and see what the landscape gives willingly, what it wants us to have, what
it wants from us, what it needs from us. That's what you *do* in relationships, if
you're not abusive.

But we are abusive, so in the blink of a mountain's eye we have forced this continent (and the world) into an abusive relationship. The good news is that the
planet seems to be in the process of getting rid of the relationship.

Dependency. One of the advantages of not having to import resources is that
you need not depend on neither the resources' owners nor on the violence necessary to eradicate these owners and take what's theirs. One of the advantages
of not owning slaves is that you need not depend on them for either your "comforts or elegancies" or even the necessaries of life. We have at this point become
dependent on oil, on dammed rivers, on this exploitative way of being (or, once
again, non-being). Without it many of us would die, most all of us would lose
our identities.

Of course everyone is dependent. One of the great conceits of this way of
life is to pretend we're independent of our landbases, and indeed of our bodies: that clean streams (or clean breastmilk) and intact forests are luxuries.
We pretend we can destroy the world and live on it. We can poison our bodies and live in them. This is insane. The Tolowa were dependent on the
salmon, huckleberries, deer, clams, and so on who surrounded them. But these
others, too, were dependent on the Tolowa and on each other, as happens in
any long-term relationship.

I've spent a few days trying to figure out the differences between these forms
of dependency: the parasitic dependency between master and slave, between
addict and addiction on one hand, and the very real dependency on which all
life is based on the other. Sure, in some cases the difference is obvious: the
dependence is one-way. The natural world gets nothing out of our enslavement
of it, or at least nothing that helps it (dioxin doesn't count). While chattel slaves

generally receive food, clothing, and shelter, chances are good they could derive these without literally slaving away their lives. But in other cases the differences become more subtle. My students at the prison by all means gained something from drugs, else they would not have voluntarily taken them. Adults in abusive relationships obviously gain something from the relationships—or at least perceive they gain something from them—else they would walk away. But what? The backgrounds of many of my students are not exactly filled with love but rather the sort of extreme abuse that makes even my father seem a delight. Many were raised under conditions also of race and class oppression. For them perhaps these drugs neutralize, as they say, oppressive reality. But it goes even deeper: I know that many indigenous peoples the world over ritually (and for the most part very infrequently) use mind-altering practices or substances in order to gain insight. What is the relationship, if any, between my students' use of drugs and this mind-altering by indigenous peoples? I don't know. And so far as abusive relationships, I know that in my own family, my mother was convinced (by my father, and by society) that she had no other options, that to leave the person who was abusing her would be to suffer greatly. It would be to lose her children, and possibly her life. In exchange for suffering this physical and emotional abuse, however, she did get to live in a nice house. But there's something more.

All last week two words have kept coming to mind: *toxic mimicry*.

I used to believe that civilization is a culture of parodies. Rape is a parody of sex. Civilized wars are parodies of indigenous warfare, which is a relatively nonlethal and exhilarating form of play,[168] meaning civilized warfare is a parody of play. Abusive relationships are a parody of love. Cities are parodies of communities, and citizenship is a parody of being a member of a functioning community. Science—with its basis in prediction and extreme control—is a parody of the delight that comes from being able to predict *and meet the needs or desires* of one's friends and neighbors (this one came clear to me the other day on seeing my dogs' joy at guessing whether I was going to turn left or right on a walk, and feeling my own joy at guessing the same for them). This culture's recreational use of altered states is a parody of their traditional uses. Each of these parodies takes the form yet ignores the soul and intent of that which is being parodied.

But recently a friend convinced me that's not entirely accurate: the parody doesn't *ignore* the intent, but perverts and attempts to destroy it.[169] Rape is a toxic mimic of sex. War is a toxic mimic of play. The bond between slave owner and slave is a toxic mimic of marriage. Heck, *marriage* is a toxic mimic of marriage, of a real partnership in which all parties help all others to be more fully themselves.

I like the phrase *toxic mimic*, but it didn't quite help me uncover the relationship between these types of dependency. I asked my mom.

She gave me the answer in one word: "Identity."

"Really," I said. I had no idea what she was talking about.

"Abusers have no identity of their own."

I was going to ask what she meant, but I suddenly remembered a conversation I'd had years before with Catherine Keller, a feminist theologian and philosopher, and author of *From A Broken Web*. We'd been talking about how abuse communicates itself from generation to generation, and about what that abuse—on both personal and social levels—does to *who we are*. She talked about how not all cultures have been based on domination, then spoke of the rise of this culture, and the effects of this rise: "Within a group in which warrior males are coming to the fore and dominating the tribe or village, everyone in the group will begin to develop a sort of self that is different from that of earlier peoples, a self that reflects the defenses the society itself configures. . . . Another way to put this is that if people are trying to control you, it will be very difficult for you—in part because of your fear—to maintain an openness to them or to others. Quite often the pain you received you will then pass on to other people. Over and over we see the causing of pain—destructiveness and abuse— flowing out of a prior woundedness. We're left with an incredibly defensive fabric of selves that have emerged from this paradigm of dominance. And because the people who embody the defensive persona will dominate these societies, this kind of self-damaging and community-destroying and ecology-killing defensiveness tends to proliferate cancerously."

I'd asked her what she meant by defensiveness.

She'd responded, "Alan Watts said one of the prime hallucinations of Western culture—and I would add of the paradigm of dominance—is the belief that who you are is a skin-encapsulated ego. And just as the skin defends you from the dangers of the physical world, the ego defends you from the dangers of the psychic world. That leads to what I have termed the separative self. The etymology of the word *separate* is very revealing. It comes from the combination of the Latin for "self," *se*, meaning "on one's own," and *parare*, "to prepare." For this culture it is separation which prepares the way for selfhood."

This all made me think of my relationship with my mom. I live very close to her—three-eighths of a mile—and will live near her for the rest of her life. Part of this has to do with health problems on both my and her parts—I have Crohn's disease, she has vision problems—part of it has to do with the fact that she is family, and part of it has to do with the fact that I like her company. She presumably likes mine as well. Through my twenties and early thirties I took a lot of flak for this arrangement from some of my white acquaintances—never friends—who told me I was suffering from what they called separation anxiety, and that in order to grow up and become fully myself, I should move far away. I didn't really understand this, because I have a life of my own (as does she), and because the arrangement—at the time we lived probably five miles apart—works well for both of us on both practical and emotional levels, and because I knew that for all of human existence—save the last hundred years—it was expected that elders would live with or near one or more of their children. It's been a sudden shift. It struck me as significant that none of my indigenous or third world friends have ever found the arrangement anything but expected. In fact, when I'd tell my white acquaintances that part of the reason we can live so close is that I'm very clear about saying *no* to the things I don't want to do for her—for example, I dislike going to the grocery store so I don't usually take her—they'd nod and tell me what good boundaries I have. When I've told my indigenous or third world friends this same thing, they've looked at me, pained and disgusted, then asked, "With her vision problems, how does she get to the grocery store?"

Catherine continued, "There are many problems with the belief that separation prepares the way for self-hood, not the least of which is that it doesn't match reality. We know that on a physical level one is not 'on one's own,' that we have to breathe and eat and excrete, and that even on a molecular scale our boundaries are permeable. The same is true psychically. Life feeds off life, Whitehead says, and if we cut ourselves off from the way we psychically feed each other, the texture of our lives becomes very thin and flat. When we live in a state of defense, there is no moment-to-moment feeding from the richness of the endless relations in which we exist.

"For the system of dominance to perpetuate itself there must be clear rewards for those who manage to maintain a state of disconnection. People must be trained and initiated into that state, and they must be rewarded with a sense of dignity, indeed of manhood, if they are able to maintain a sense of self-control—as opposed to being present to their experience—and a sense of control over their surroundings, which would include as many people as possible.

"When you have a society organized so those at the top benefit from the labor of the majority, you have some strong incentives to develop the kind of self-hood that gets you there. The only kind of selfhood that gets you there is the kind of selfhood that allows you to numb your empathies. To maintain the system of dominance, it's crucial that the elite learns this empathic numbness, akin to

what Robert Jay Lifton calls 'psychic numbing,' so its members can control and
when necessary torture and kill without being undone. If its members are incapable of numbing, or if they have not been trained properly, the system of domination will collapse."

That's one of the reasons, she said, that civilization so often co-opts movements opposing domination. "Society as we know it may well need," she continued, "to live off of the energy of alternative movements. It needs to suck our
blood in order to feed itself, in part because a system of domination will always
be undernourished."

"How so?"

"Once we unplug from our vital connections—connections more like the
fiber of what we call nature where there aren't barriers between the relationships of things to each other—once we unplug from the way everything
branches into everything, and instead pursue the goals of civilization as we
know it, the energy source has to come from somewhere else. To some extent
it can come from sucking the labor of the poor, and to some extent it can come
from exploiting the bodies of animals and people treated like animals. The
exploiting of the bodies of women gives a lot of energy. But the parasitism of
the dominant culture is endless, because once you cut yourself off from the
free flow of mutually permeable life you have to get your life back somehow,
artificially."

I came back to the conversation with my mom, and heard her say, "That was
part of your father's problem. He had no solid identity of his own, which was
one reason he was so violent. Because he wasn't secure in his own identity, in
order to exist, he *needed* for those around him to constantly mirror him. When
you or I or your siblings didn't match his projections—when we showed any
spark of being who we actually were, thus forcing him to confront some other
person as someone different than himself—he became terrified, or at least he
would have become terrified if he would have allowed himself to feel that. But
to become terrified was too scary, and so he flew into a rage."

I just looked at her. I'd never heard this analysis before. It was very good. I
was thinking also that if my publisher were present he would probably be tearing his hair out at her penchant for making parenthetical comments, just as he
does with mine.

She continued, "His lack of a secure identity is also why he was so rigid. If
you're not comfortable with who you are, you have to force others to confront
you only on your own terms. Anything else is once again too scary. If you're
comfortable with who you are, however, it becomes no problem to let others be

their own selves around you: you have faith that whoever they are and whatever they do, you will be able to respond appropriately. You can be fluid and respond differently to different people, depending on what they need from you. He couldn't do that."

This same thing happens on a larger scale, of course. Deadened inside, we call the world itself dead, then surround ourselves with the bodies of those we've killed. We set up cityscapes where we see no free and wild beings. We see concrete, steel, asphalt. Even the trees in cities are in cages. Everything mirrors our own confinement. Everything mirrors our own internal deadness.

"One more thing," my mother said. "This lack of an identity is one of the reasons so many abusers kill their partners when their partners try to leave. They're not only losing their partners (and punching bags) but their identities as well."

That's also one of the reasons this culture must kill all non-civilized peoples, both human and nonhuman: in order to preclude the possibility of our escape.

Which brings us to the next category: abusers isolate their victims from other resources. I'm typing these words sitting in a manufactured chair staring at a manufactured computer screen, listening to the hum of a manufactured computer fan. To my left are manufactured shelves of manufactured books, written by human beings. Civilized, literate human beings who write in English (languages, many of them indigenous, are being destroyed as quickly as all other forms of diversity, and to as disastrous an effect: the language you speak influences what you can say, which influences what you can think, which influences what you can perceive, which influences what you can experience, which influences how you act, which influences who you are, which influences what you can say, and so on). To my right a window leads to the darkened outside and reflects back to me my uncombed dark hair surrounding the blur of my own face. I'm wearing mass-produced clothes, and mass-produced slippers. I do, however, have a cat on my lap. All sensory inputs save the cat originate in civilized humans, and even the cat is domesticated.

Stop. Think about it. Every sensation I have comes from one source: civilization. When you finish this paragraph, put down the book for a few moments, and check out your own surroundings. What can you see, hear, smell, feel, taste that does not originate in or is mediated by civilized human beings? Singing frogs on a *Sounds of Nature* CD don't count.

This is all very strange. Stranger still—and extraordinarily revealing of the degree to which we've not only accepted but reified this artificially imposed isolation, turned our insanity into a perceived good—is the way we've made a fetish and religion (and science, for that matter, as well as business) of attempting to define ourselves as separate from—different from, isolated from, in opposition to—the rest of nature. Abusers merely isolate victims from other resources. Far moreso even than this, civilization isolates all of us—ideologically and physically—from the source of all life.

We do not believe trees have anything to say to us (nor even that they can speak at all), nor stars, nor coyotes, nor even our dreams. We have been convinced— and this is the primary difference between western and indigenous philosophies—that the world is silent save civilized humans.

One of the most common and necessary steps taken by an abuser in order to control a victim is to monopolize the victim's perception. That is one reason abusers cut off victims from family and friends: so that in time victims will have no standard other than the abusers' by which to judge the abusers' worldviews and behavior. Abusive behavior—behavior that would otherwise seem extraordinarily bizarre (how crazy is it to rape one's own child? How crazy is it to toxify the air you breathe?)—can then become in the victim's mind (and even more sadly, heart) normalized. No outside influence must be allowed to break the spell. There can be only one way to perceive and to be in the world, and that is the abuser's way. If the abuser is able to mediate all information that reaches the victim, the victim will no longer be able to conceptualize that there is any other way to be. At this point the abuser will have achieved more or less total control.

This is, of course, the point we have reached as a culture. Civilization has achieved a completely unprecedented and nearly perfect monopolization of our perception, at least for those of us in the industrialized world. Fortunately, however, there do still exist people—mainly the poor, people from nonindustrialized nations, and the indigenous—who still have primary connections to the physical world. And fortunately, also, the physical world still exists, and all of us can at the very least reach out to touch trees still standing in steel and concrete cages. And we can see plants poking up through sidewalks, breaking cement barriers that keep them from feeling the sun. I would hope we can learn from these plants and break through these concrete and perceptual barriers.

The sixth characteristic is that abusers blame others for their problems. To make the jump to the cultural level it would be easy to simply list the ways our culture does this, and leave it at that. The capitalist media blames spotted owls and humans who love them for job losses in the timber industry, yet (surprise, surprise) ignores the greater number of jobs lost in the same industry to automation and raw log exports (as well as the cut-and-run nature of the industry). Politicians and other timber industry propagandists blame natural forests and environmentalists for fires, yet ignore the fact that logging is a significant cause of fires, and further, that fires burn hotter and more destructively in cutover forests and tree plantations than they do in natural forests. They ignore further the regenerative role fire plays in forests. We who care about the planet would be wise to not ignore this lesson about the destructive/regenerative powers of fire but learn it, and apply it when appropriate to the perceptual and physical barriers that monopolize our perception and that are killing the planet.

More blame: the bigot blames poor Mexicans when his employer's plant closes and moves to Mexico. The owner blames market conditions or damn unions for leaving him no choice but to move the plant. Go back in time and we have Israel's rulers, speaking through their God, blaming Canaanites because Israelites didn't want to follow "God's" (wink, wink) rules. Move forward and we have Crusaders blaming women for lack of success on the battlefield (sex, especially with an infidel, evidently displeases "God"). Then we have settlers blaming Indians for not giving up their land without a fight (as John Wayne later said, "I don't feel we did wrong in taking this great country away from them. There were great numbers of people who needed new land, and the Indians were selfishly trying to keep it for themselves"). Hitler and the Nazis blamed Communists and Jews for everything from world wars to defective dentures. Americans agreed at least so far as the Communists. Now it's terrorists who keep us from the Promised Land of Perpetual Peace and Prosperity™ (brought to you by ExxonMobil). There is always someone (else) to blame.

Something interesting happens when you combine an abuser's propensity to blame with the monopolization of the victim's perception: the victim comes to agree with the abuser, that all problems are actually the victim's fault. The wife tries tirelessly to make the perfect meal and if she's beaten it's because she's not a good enough cook, which means not a good enough wife, which means not a good enough person. Of course it's not because her husband is violent, abusive, insane. The child tries to perfectly clean the dishes, and violence comes to her because she is too sloppy. The teen tries to park the car in the right place— or rather not in the ever-shifting wrong place—so as to not be beaten. In an attempt to maintain control in a situation that is grievously out of control and that can never be in control so long as victims stay within the perceptual box created for them by their abuser, victims conspire with their abusers to focus on alterations of their own behavior in futile attempts to placate the abuser or at least delay or mitigate the inevitable violence, or at the very least shift this violence to another victim. Even worse than this self-focus being a mere tactic, it becomes a way of being (or rather non-being) in the world, such that victims come to *know* the fault is their own. Instead of stopping the abuse by any means necessary, they join with the abuser in doing violence to themselves.

They forget that assigning "blame" in this sense is a toxic mimic of the necessary task of assigning appropriate and accurate responsibility for the violence done to them, and doing something about it.

These same patterns are replicated on the larger social scale, at least among those who have been sufficiently enculturated. This is probably not the case among the primary victims of our culture, of course: those who remain free of civilization's perceptual box. I'm reasonably certain salmon, swordfish, and hammerhead sharks do not find themselves paralyzed by spasms of self-blame for their plight—*What could I do differently to placate these people? If only I were a better fish they would not hate me*—but instead know precisely who is killing them. The same can be said for the indigenous. You can't get much clearer than Sitting Bull, who said, when forced to speak at a celebration of the completion of a railroad through what had been his people's land: "I hate you. I hate you. I hate all the white people. You are thieves and liars. You have taken away our land and made us outcasts, so I hate you." It's important to note, by the way, that the white translator did not speak these words, but instead the "friendly, courteous speech he had prepared."[170]

And that's the problem.

Those of us whose vision has been defined by civilization, whose personalities have been formed and deformed in this particular crucible of violence, sometimes, like victims of childhood abuse, fail to adequately and accurately assign responsibility for the violence we suffer or witness, instead transforming raw impulses to assign responsibility—"You have taken our land and made us outcasts, so I hate you"—into friendly, courteous speech: some environmentalists even give training in "verbal nonviolence" so activists will be certain to not say "Fuck you" to police putting them, in copspeak, into "pain compliance holds," that is, torturing them. Abused children—and I know this from experience—generally are unable to face the fact that they have almost no power to stop the violence done to them and to those they love. As a consequence of this—and this dovetails nicely, or more accurately horrifically, with abusers blaming others for their own problems as well as abusers monopolizing victims' perceptions—victims often internalize too much responsibility, which in this case means any responsibility at all, for the violence they suffer or see. *I must have done something wrong, or my father would not hit me. I must be a slut or a temptress, and I must want him to do this to me—I know this because he tells me all of this—or he would not visit me at night.* This allows these children to pretend they have at least some power to halt or slow violence done to them, however illusory all evidence shows this power to be. That illusion can in fact be crucial to emotional survival. Of course when they're no longer children, the illusion becomes absurd and harmful.

Similarly, many of us trying to stop the destructiveness of this culture—and I know this not only from my own experience but from having worked with and talked to hundreds or even thousands of other activists—are routinely struck by the near-complete ineffectiveness of our work on any but the most symbolic levels. By almost any measure, our work especially as environmental activists is an appalling failure. Just today I spoke with a friend who for the past ten months has been sitting in an ancient redwood in Humboldt County, just south of here, in an attempt to keep the tree and the forest of which it is a part from being cut. Pacific Lumber is deforesting that watershed, as it is deforesting much of the state, and will eventually get to the tree in which she now lives. Previous cutting by this corporation has caused such severe flooding that local residents' homes have been destroyed. Some have put their homes on stilts. Once-pristine water supplies now resemble chocolate milk garnished with sticks, spiked with herbicides and diesel fuel. Years ago, in response to citizen outrage, the state's North Coast Region Water Quality Control Board—appointed by the governor, who is deeply beholden to big timber corporations—put together a scientific panel to study the problem, which is nearly always a good way to delay action while allowing primary destruction to continue. But the panel surprised the Board by unanimously declaring that cutting needs to be drastically reduced *now*, not only to protect local human residents, but for critically imperiled coho salmon and many other species. The Board's decision? You guessed it: ignore the citizens it purports to serve, ignore the scientific team it assembled, ignore everything but the "needs" of this grossly destructive corporation. This is democracy in action. This is the severing of reality from politics (or really, there's nothing to sever, since they've always been separated). This is the dismemberment of the planet. This is breathtakingly and obscenely routine.

The best and most courageous and most sincere of our efforts are never sufficient to the task of stopping those who would destroy.

Years ago, I wrote, "Every morning when I wake up I ask myself whether I should write or blow up a dam." I wrote this because no matter how hard activists work, no matter how hard I work, no matter how much scientists study, none of it really seems to help. Politicians and businesspeople lie, delay, and simply continue their destructive behavior, backed by the full power of the state. And the salmon die. I said back then, and I say now, that it's a cozy relationship for all of us but the salmon. Every morning I still make the decision to write, and every morning I think more and more I'm making the wrong damn decision. The salmon are in far worse shape now than when I first wrote that line.

I am ashamed of that.

We are watching their extinction.

I am ashamed of that as well.

To mask our powerlessness in the face of this destruction, many of us fall into the same pattern as those abused children, and for much the same reason. We internalize too much responsibility. This allows us activists to pretend we have at least some power to halt or slow violence done to us and to those we love, however illusory, once again, all evidence inevitably shows this power to be. And don't give me a lecture about how if we weren't doing this work the destruction would proceed even more quickly: of *course* that's the case, and *of course* we need to keep fighting these rearguard actions—I would never suggest otherwise—but do you realize how pathetic it is that all of our "victories" are temporary and defensive, and all of our losses permanent and offensive? I can't speak for you, but I want more than to simply stave off destruction of this or that wild place for a year or two: I want to take the offensive, to beat back those who would destroy, to reclaim what is wild and free and natural, to let it recover on its own. I want to stop in their tracks the destroyers, and I want to make them incapable of inflicting further damage. To want any less is to countenance the ultimate destruction of the planet.

But we all settle for less, and to make ourselves feel the tiniest bit less impotent we turn the focus inward. We are the problem. I use toilet paper, so I am responsible for deforestation. I drive a car, so I am responsible for global warming. Never mind that I did not create the systems that cause these. I did not create industrial forestry. I did not create an oil economy. Civilization was destroying life on this planet before I was born, and will do the same—unless I and others, including the natural world, stop it—after I die.

If I were to die tomorrow, deforestation would continue unabated. In fact, as I've shown in another book,[171] demand does not even drive the timber industry: overcapacity of very expensive pulp and paper mills (as well as, of course, this culture's death urge) determines in great measure how many trees are cut. Similarly, if I were to die, car culture would not slow in the slightest.

Yes, it's vital to make lifestyle choices to mitigate damage caused by being a member of industrial civilization, but to assign primary responsibility to oneself, and to focus primarily on making oneself better, is an immense copout, an abrogation of responsibility. With all the world at stake, it is self-indulgent, self-righteous, and self-important. It is also nearly ubiquitous. And it serves the interests of those in power by keeping our focus off them.

I do this all the time. *We're killing the planet,* I say. Well, no, I'm not, but thank you for thinking me so powerful. *Because I take hot showers, I'm responsible for drawing down aquifers.* Well, no. More than 90 percent of the water used by humans is used by agriculture and industry. The remaining 10 percent is split between municipalities (got to keep those golf courses green) and actual living breathing humans. *We're deforesting 214,000 acres per day, an area larger than New York City.* Well, no, I'm not. Sure, I consume some wood and paper, but I didn't make the system.

Here's the real story: *If I want to stop deforestation, I need to dismantle the system responsible.*

Just yesterday I caught myself taking on nonsensical responsibility. I was finishing a book with George Draffan about causes of worldwide deforestation. For one hundred and fifty pages we laid out explicitly and undeniably that this culture has been deforesting every place it touches at an ever-increasing pace for some six thousand years, and that current deforestation is driven by a massively corrupt system of interlocked governments and corporations backed, as always, by plenty of soldiers and cops with guns. (But you knew that already, didn't you?) Yet at the end, I found myself pleading with readers to drive the deforesters out of our own hearts and minds. I wrote, "We will not stop destroying forests until we have dealt with the urge to destroy and consume that hides in our hearts and minds and bodies." I cut the line. It's a fine first step—emphasis on *first*—because we surely cannot stop the destruction until we perceive it as destruction and not as "progress," or "developing natural resources," or even "inevitable," or "the way things are." But what about driving deforesters out of forests altogether? *That* is the real point. Anything less is far worse than just a waste of everyone's time: it paves the way for further destruction.

I recently saw an excellent articulation of the dangers of identifying with those who are killing the planet. It was in a "Derrick Jensen discussion group" on the internet. When I first heard of the group's existence, I was of course, flattered. People everywhere discussing me! Every guy's dream! My head swelled. Before this happened, I wasn't even convinced *I* would log on to discuss me. But I did. I followed the posts. My head swelled even more. I thought I'd give them a thrill, and posted something unpublished elsewhere. I considered the excitement they'd surely feel at this honor, and imagined how excited I'd have been when I was younger had the rock groups *UFO* or *Spirit* made some song accessible to only a few of us. I probably would have stayed up late that night listening to it over and over, and considering how special I was. Fortunately the response on the discussion group was more sedate. A few people wrote, "Nice essay." That's about it. Then they went back to discussing whatever they'd been discussing before. My head returned to normal size.

Now to the articulation I just read. A woman had commented that "We are going to go to war in Iraq." A man commented on her use of *we*, not realizing she was being ironic. His misunderstanding doesn't lessen the importance of his comments: "I find that many people (including myself when I'm not paying attention) slip into using the term 'we' when referring to actions of the U.S. government. I agree with Derrick's assertion that the government (I would say all governments) is a government of occupation, just as this culture is a culture of occupation. Though I'm coerced into participating in the system (by paying taxes, working, spending money in the economy) I do not consider myself one of the decision-makers. My choices are false choices, and my voice is not 'represented' by the government. A friend was wearing a great button the other day: 'U.S. out of North America.'"

He continued, "Those in power want us to associate ourselves with them, make us part of the 'we' so we become inseparable from them. This way they can not be challenged, questioned, or overthrown without attacking ourselves. This is the ultimate goal of nationalism, to fuse an entire nation into agreement with the leaders so no action, no matter how obscene, is questioned. Perhaps this is why when I bring up faults in the government, capitalism, the techno-industrial complex, or the culture as a whole, many people get extremely defensive, as if I'd just insulted their mother. The more we allow those in power to convince us we are to blame for their actions, the more we are unable to separate what we do from what we are forced to do or what rulers do in our name. The more all of this happens, the more power they gain and the more difficult any form of dissent becomes."[172]

❆ ❆ ❆

The phone rings. I answer. It's a friend. She asks, "How much longer do you think we're going to be in Afghanistan?"

She can't see this, but I look around, look outside at the redwood trees. I respond, "We're in Afghanistan? I thought we were in northern California."

Silence on the phone. A sigh, and finally she says, "How much longer do you think *our troops* are going to be in Afghanistan?"

I say, "I've got troops? Really? Will they do whatever I tell them? If I tell them to take out the dams on the Columbia River will they do that?"

More silence, until she says, "This is why I only call you every few weeks. I'll be in touch."

❰ ❰ ❰

We are no longer children. It is dangerous to us and to others to maintain the illusion that we are responsible for the destruction, an illusion that may have been appropriate when we were powerless. But we are not.

I remember the decision I made in my mid-twenties to pursue my life as a writer. I was scared to do this. I did not have sufficient self-confidence, I thought, to follow my dreams. I traced this lack of confidence to the abuse I'd suffered as a child. Part of my father's *modus operandi*—and I recognized this while very young—was that any time any one of us children (or our mother) revealed that something was important to us, one of three things would happen: he might use that thing as a form of payment for cooperation in his sexual abuse (I was interested in the Civil War as a child, and we took long trips to see battlefields, but at what cost?); he might use the promise of this thing to build up hopes so he could watch our faces as he dashed them; or he might simply destroy the thing itself in front of our eyes. I learned to not express my dreams.

I recognized in my mid-twenties that because of this abuse, I would have the best excuse in the entire world to not follow my dreams of becoming a writer. Who could blame me after what I'd been through? Mere emotional survival was triumph enough.

The choice quickly came to this: I could go the rest of my life with an airtight excuse for not doing what I wanted; or I could go the rest of my life doing what I wanted. It took me only a few months to decide which it would be.

❰ ❰ ❰

As a consequence of the belief that violence done to us is our own fault—or sometimes more simply because we do not want to be violated—we often become self-policing. I write this on an airplane flying home from giving talks. A friend took me to the airport. As we pulled into the parking lot we saw a uniformed man whose job it is, evidently, to search every car that enters.

I said, "I can't believe this."

"Do you want to not go in?"

I thought of the words I'd been told years before by a police officer when I'd commented that drivers licenses are in essence government "identity papers" we're "asked" to produce at least as often as people were in those old black-and-white movies of resistance against Nazis. He didn't appreciate my film reference, and told me, "If you don't like it, don't drive."

I also considered the checkpoints and travel limits heroes always faced in those movies, and the absolute necessity of such restrictions under repressive regimes. I thought of the comment I'd received more recently when I'd complained as an "airport security agent" put her fingers against the skin of my lower belly beneath the waistband of my pants. I'd asked her what she was doing.

She'd responded, "This is for your safety and the safety of others."

"You putting your hand inside my pants doesn't make anyone safer."

She'd said, "Flying is a privilege, not a right. If you don't like it, stay home."

I'd begun to disagree, and she'd motioned to a nearby cop. I'd had a plane to catch, and so I'd had a choice: I could make a scene, or I could get the hell out of Austin, Texas. I got the hell out of Austin, Texas.

Back at the airport parking lot, my friend said, "Let's just go ahead and park. Let them search the car. We have nothing to hide."

We looked at each other, shook our heads, and laughed.

This laughter kept us from cursing.

I'm not sure that's such a good thing.

《 《 《

I don't mean to suggest we should override every fear. I'm not sure we should override *any* fear. Fears should at least be listened to, whether or not we act on them. But I did not want to live a life based on fear. To live a life following my heart was important enough to me that I was willing to move into, through, and beyond this fear to my life on the other side.

There are certainly other fears I've not afforded the energy to move through. Because when I was a child there were beatings associated with water skiing and rapes associated with alcohol, to this day I carry powerful fears of both. But neither of those is particularly worth the effort to work my way through: I can happily live a life without water skiing or alcohol. I was not willing to live a life without my heart.

We can ask the same questions on the cultural level. Are we willing to live a life without clean air, clean water, wild animals: a livable planet? For what, precisely, will we face down our own fears?

We have the best excuse in the world to not act. The momentum of civilization is fierce. The acculturation deep. Those in power will imprison us if we effectively resist. Or they will torture us. Or they will kill us. There are so many of them, and they have weapons. They have the law. And many of them—probably in the final analysis nearly all of them—have no scruples, else they would never support the current system in the first place. Because of all this, there really is nothing we can do. We may as well admit that.

But the question becomes: would you rather have the best excuse in the world, or would you rather have a world?

((((

Here, once again, is the real story. Our self-assessed culpability for participating in the deathly system called civilization masks (and is a toxic mimic of) our infinitely greater sin. Sure, I use toilet paper. So what? That doesn't make me as culpable as the CEO of Weyerhaeuser, and to think it does grants a great gift to those in power by getting the focus off them and onto us.

For what, then, are we culpable? Well, for something far greater than one person's work as a technical writer and another's as a busboy. Something far greater than my work writing books to be made of the pulped flesh of trees. Something far greater than using toilet paper or driving cars or living in homes made of formaldehyde-laden plywood. For all of those things we can be forgiven, because we did not create the system, and because our choices have been systematically eliminated (those in power kill the great runs of salmon, and then *we* feel guilty when we buy food at the grocery store? How dumb is that?). But we cannot and will not be forgiven for not breaking down the system that creates these problems, for not driving deforesters out of forests, for not driving polluters away from land and water and air, for not driving moneylenders from the temple that is our only home. We are culpable because we allow those in power to continue to destroy the planet. Yes, I know we are more or less constantly enjoined to use only inclusive rhetoric, but when will we all realize that war has already been declared upon the natural world, and upon all of us, and that this war has been declared by those in power? We must stop them with any means necessary. For not doing that we are infinitely more culpable than most of us—myself definitely included—will ever be able to comprehend.

((((

To be clear: I am not culpable for deforestation because I use toilet paper. I am culpable for deforestation because I use toilet paper and I do not keep up my end of the predator-prey bargain. If I consume the flesh of another I am responsible for the continuation of its community. If I use toilet paper, or any other wood or paper products, it is my responsibility to use any means necessary to ensure the continued health of natural forest communities. It is my responsibility to use any means necessary to stop industrial forestry.

❨ ❨ ❨

The next characteristic of abusers is that they get upset easily. They're hypersensitive, and the slightest setback is seen as a personal attack. Much of the reason for this has to do with the fourth premise of this book, that violence in our culture flows only one way. This is true not only for violence, but for all control, all initiative. Those on top are allowed to have control and initiative. Those below must have them only insofar as control and initiative make them more effective proxies of those above.

Any breach of this etiquette must be dealt with swiftly, surely, and completely, so the hierarchy can remain seamless, safely unacknowledged, hidden from the possibility of change by either victim or perpetrator. That this is as true on the larger social scale as it is on the more personal or familial should be obvious, but I'll provide a couple of quick examples. Just last night I spoke with a group of students from San Marcos High School in Santa Barbara, California. The kids were delightful, intelligent, passionate, and defiant. One told me she had asked the school's administration for permission to put up posters containing these words from the Declaration of Independence: "That whenever any Form of Government becomes destructive of these ends [Life, Liberty, and the pursuit of Happiness], it is the Right of the People to alter or to abolish it." Far from rewarding her interest in history and politics (Who says kids these days don't know important historical documents?), administrators not only denied her request, but threatened her with "forced transfer" to another school should she post them anyway.

She asked my advice.

I suggested that since her request had already identified her to authorities, other students should put up the posters. Another student objected to this, saying that many students had already been threatened with expulsion.

"Why?"

She answered that they'd planned a one-period walkout to protest a school policy of administrators giving students' names and phone numbers to military recruiters. Teachers had infiltrated the organization planning the walkout, she'd said, under the guise of being advisors. When students rejected the teachers' advice to limit their protest to writing letters for the administration to ignore, teachers and administrators stood as one, telling students they'd be expelled if they walked out of any classes.

I told these kids I was proud of them, and that I was glad they had at such a young age experienced participatory democracy in action.

I wish I'd have told them another idea I had for the posters, but this didn't occur to me until much later: that they form alliances with students at other schools, so that other students put up posters of resistance at this school, and these students put them up elsewhere. Not only would this lessen the easy power of the administrators to harm those who speak out, but more importantly it would begin to make networks of organized resistance, cadres for the revolution we so desperately need.

No matter what they felt in their hearts, the teachers had probably been in a very bad position. My understanding of the school climate was that had they not gone along with this silencing of dissent, they could have lost their jobs. That's one of the ways the system works. If I complain about a woman in a uniform putting her hand in my pants, I miss my flight, and possibly get arrested. If these teachers do not stifle dissent, they possibly get fired.

This statement of course does not excuse their actions, but merely helps us understand them. Or maybe they had their actions fully rationalized, as presumably did the administrators.

The slightest real dissent—that not confined to places, times, and means designed or approved by those in power—must be perceived by those in power as an attack on the legitimacy of their rule.

Probably because it is.

It's a wondrous thing to get up off your knees, to stand again (or for the first time) on your hind legs, to say "Fuck you"—classes in "verbal nonviolence" notwithstanding—or to say "You have no right," or "No" to those in power, to choose where, when, and how you will express yourself, where, when, and how you will fight back, where, when, and how you will defend what and whom you love against those who exploit and destroy them.

You should try it some time. It's really fun.

❨ ❨ ❨

The next characteristic is that abusers are at least insensate to the pain of children and nonhumans. Bringing this to the larger cultural level requires, I think, only one word: vivisection. Okay, another: zoos. A couple more: factory farms. Okay, a few more: we're killing the planet. Correction: they're killing the planet, and they clearly do not hear the screams.

Do you?

❨ ❨ ❨

Abusers often conflate sex and violence. Rates of rape—so common as to be essentially normalized in the culture—make clear the conflation of sex and violence on the social level. Many films make it clear, too. So do many relationships. One can also say those magic words: breast augmentation surgery. Just yesterday I heard of a new fad in plastic surgery: reshaping the vulva to make it more visually pleasing, whatever that means (what about the notion that if you love a woman you will find her vulva beautiful, simply because it is hers?).

Really, though, this cultural conflation of sex and violence can be reduced to one word: *fuck*. It's an extraordinary comment on this culture that the same word that means *make love to* also means *do great violence to*.

❨ ❨ ❨

Abusers often actualize rigid sex roles. That this is true on the larger cultural level hardly needs remarking, and goes far beyond the stereotypically masculine values that dominate the culture. It also goes beyond the homophobia that's based on a fear of anything that confuses those rigid sex roles.

I've been thinking a lot lately about the seeming scientific obsession to artificially create or modify life, and also the obsession to search for life in outer space. It has always seemed profoundly absurd and immoral to me that billions of dollars are spent trying to discover life on other planets as trillions more are spent to eradicate life on this one. Were scientists to discover cute furry creatures on Mars with floppy ears and wriggly noses, Nobel prizes would soon be forthcoming (for the scientists, not the floppy-eared Martians). Yet when scientists on the real world see real creatures just like these, they reach for hair spray to put in the creatures' eyes for Draize tests (of course, the scientists would also leap to exploit the Martian bunnies faster than you can say Huntington Life Sciences).

Similarly, it makes no sense to me that we (read *they*) keep trying to recreate the "miracle of life" in laboratories as we (read *they*) daily the destroy the plenitude—we're learning it's not an infinitude—of miracles that surround us all.

But now I get it. It's those rigid sex roles combined with a devaluing of the feminine and a really bad case of womb envy, all topped with a heaping of sour grapes, boiling down to the fact that women have babies and men don't. If women are identified primarily or exclusively—rigidly—by their roles as creators of life, and if women are perceived as inferior (meaning whatever women do, men do better) then men, so as to not perceive themselves as less powerful than the women for whom they feel contempt, must figure out not only how to destroy the natural life they despise, but how to create some sort of life of their own.

A CULTURE OF OCCUPATION

Imagine if, for the last fifty years, we had sprayed the whole earth with a nerve gas. Would you be upset? Would I be upset? Yes. I think people would be screaming in the streets. Well, we've done that. We've released endocrine disruptors throughout the world that are having fundamental effects on the immune system, on the reproductive system. We have good data that shows that wildlife and humans are being affected. Should we be upset? Yes, I think that we should be fundamentally upset. I think we should be screaming in the streets.

Louis J. Guillette, Jr. [173]

I'm DRIVING THROUGH REDWOODS ON A FOUR-LANE HIGHWAY. A CAR materializes behind me, then speeds by so quickly I barely make out the sentence frosted on the rear window: *Drive it like you stole it.*

I laugh, then marvel at the boldness of this person seeming to beg police to give her tickets. But the longer I drive this ribbon of asphalt, the more significant the phrase seems. Let's change *it* in that sentence from a car to the land: *Live on this land like you stole it.* That's what members of this culture do. Probably because they did.

We should admit to ourselves, and this forms the eleventh premise of this book, that *from the beginning, this culture—civilization—has been a culture of occupation.*

What do occupiers do? They seize territory by force or threat of force. They take resources for use at the center of an empire. They degrade the landscape. They kill those who resist this theft. They enslave those whose labor is necessary for this theft, this degradation of the landscape. They eradicate those who are in the way—the humans and nonhumans whose land this is—and who must be removed so the occupiers can put the land to better use. They force the remaining humans to live under the laws and moral code of the occupiers. They inculcate future generations to forget their non-occupied past and to aspire to join the ranks of their occupiers, to actually join in the degradation of the landbase that was once theirs.

Because exploitation is so central to any culture of occupation—that's part of what defines it—this exploitation infects and characterizes every part of the culture.

This means any civilized government, by all means including the United States, is a government of occupation, set up to facilitate resource extraction (to bring resources from the country to the city, from colony to empire), a process these days called production, and to prevent interference in this process by those whose lives are diminished or destroyed by the devastation of their landbase, and also by those whose lives are diminished or destroyed laboring to serve production.

Any civilized economics, by all means including capitalism, is an economics of occupation, set up to rationalize resource extraction, and to pre-empt reasonable discourse about non-exploitative community relations.

Any civilized religion, including Christianity, Judaism, Islam, Buddhism, Confucianism, and so on, is a religion of occupation. A religion is supposed to teach us how to live, which if we're to live sustainably, means it must teach us how to live in place. But people will live differently in different places, which means religions must be different in different places, and must emerge from the land itself, and not abstract themselves from it. It's absurd to think that people will need the same guidance to live in the Middle East as in Tibet or the Pacific Northwest. And a transposable religion means that it could not have emerged from the particularities of that landscape. A religion is also supposed to teach us how to connect to the divine. Yet if a religion is transposed over space, it won't—can't—be so quick to speak to the divine in that particular place. The bottom line is that civilized religions lead people away from their intimate connection to the divinity in the land that is their own home and toward the abstract principles of this distant religion. How differently would we relate to trees if instead of singing "Jesus loves me, this I know, for the Bible tells me so," we, those of us who live in Tu'nes, were to sing, "I love these redwoods, and they love me. There's no finer feeling than to be loved by a tree"? On the other hand, what better words to get young slave children to sing in unison than "Little ones to Him belong, They are weak but He is strong." And finally, given the near ubiquitous belief among the overlords that their wealth and power is divinely ordained, how's this for a hymn to teach the poor? "I want to be a worker for the Lord, I want to love and trust His holy word, I want to sing and pray and be busy every day, in the vineyard of the Lord."

Any science of civilization will be a science of occupation, aiming toward ever more control of the occupied world, and toward the creation of ever more destructive technologies. Imagine the technologies that would be invented by a culture of inhabitation, that is, a sustainable culture, that is, a culture planning on being in the same place for ten thousand years. That culture would create technologies that enhance the landscape—what a concept!—and would decompose afterwards into components that help, not poison, the soil. The technologies would remind human inhabitants of their place in this landscape. The technologies would promote leisure, not production. The technologies would not be bombs and factory conveyor belts but perhaps stories, songs, and dances, and nets to catch set and sustainable numbers of salmon.

❨ ❨ ❨

The Squamish people, who live near what is now Vancouver, British Columbia, tell this story: A long time ago, even before the time of the flood, the Cheakamus River provided food for the Squamish people. Each year, at the end of summer, when the salmon came home to spawn, the people would cast their cedar root nets into the water and get enough fish for the winter to come.

One day, a man came to fish for the winter. He looked into the river and found that many fish were coming home this year. He said thanks to the spirit of the fish for giving themselves as food for his family, and cast his net into the river and waited. In time, he drew his nets in and they were full of fish, enough for his family for the whole year. He packed these away into cedar bark baskets, and prepared to go home.

But he looked into the river and saw all those fish, and decided to cast his net again. And he did so, and it again filled with fish, which he threw onto the shore. A third time, he cast his net into the water and waited.

This time, when he pulled his net in, it was torn beyond repair by sticks, stumps, and branches which filled the net. To his dismay, the fish on the shore and the fish in the cedar bark baskets were also sticks and branches. He had no fish, his nets were ruined.

It was then he looked up at the mountain and saw Wountie, the spirit protecting the Cheakamus, who told him that he had broken the faith with the river and with nature, by taking more than he needed for himself and his family. And this was the consequence.

And to this day, high on the mountain overlooking the Cheakamus and Paradise Valley, is the image of Wountie, protecting the Cheakamus.

The fisherman? Well, his family went hungry and starved, a lesson for all the people.

❆ ❆ ❆

Discourse under civilization is, as we see, a discourse of occupation, by which I mean there's lots of talk of bread and circuses to keep us occupied while we're systematically robbed of our landbase, our dignity, and our lives.

For example, I don't know about you, but sometimes I have what I've taken to calling Angelina Jolie moments. I'll be thinking about something else, and suddenly her image will pop up before me. I think it's because I'm so upset with how she was treated by Billy Bob Thornton, and how scandalous their whole relationship was. I'm sure you've heard that each carried around the neck a vial of the other's blood. And I'm sure you also heard that hubby Thornton sometimes said that when they were having sex he wanted to strangle her because he wanted them to be so close (see my previous discussion of the word *fuck*). But have you heard where she has his name tattooed? Ohmygosh, I'll bet when it was being done she was wishing his name was Ed.

Speaking of genitals, did you know that Nicole Kidman doesn't like to wear underwear? I read that in the newspaper, so it must be true. Nor did Marilyn Monroe. Nor, for that matter, did Tallulah Bankhead.

Stop.

Now, quick, what's the indigenous name for the place you live? Who are the indigenous people whose land it is? What are five species of plants and animals who live (or lived) within one hundred yards of your home, and who have been harmed by civilization? What are ten species of edible plants within one hundred yards of your home?

I find it odd and horribly disturbing that I can tell you—not from direct experience, mind you—what is on Angelina Jolie's genitals, and what is not on Nicole Kidman's, yet it took me two years of living in Tu'nes before I learned there was a massacre of several hundred Indians a few miles from my home at a place called Yontocket, another massacre nearby at a place called Achulet, and yet another at a place called Howonquet. I wonder how long it will take to learn of them all. Although I live in a riot of wildlife, I cannot name—or find—ten edible species outside my door (I think I should probably steer clear of those beautiful big red mushrooms with the small white spots). I've lived here almost four years, and it took me 'till last week to even learn of the existence of some new, or rather very old, neighbors: *Aplodontia rufa*, or mountain beavers (the oldest of the living rodents, a website says, also known for being hosts to the world's largest fleas!).

It is beyond passing strange—I would say obscene, as well as absolutely typical—that so much of our discourse concerns so many pieces of information that do not matter to our lives—I think I can state categorically that the knowledge that Angelina Jolie has a tattoo on her genitals and that Nicole Kidman doesn't wear panties will *never* make a tangible difference in my life—yet we know almost nothing of the land we inhabit, and of our living breathing neighbors who share this land.

That is a textbook example—*textbook*, as though a book written by someone far away carries more weight than my own direct idiosyncratic experience—of a discourse of occupation. Bring on the bread, and most especially bring on the circuses. And whatever you do, don't wake me until it's too late, until there's nothing I can do to resist, until I can in no way be held responsible for my failure to effectively act.

《 《 《

The conflict resolution methods of a culture of occupation will be different from those of a culture of inhabitation. The Okanagans of what is now British Columbia, to provide a counterexample, have a concept they call En'owkin, which means "I challenge you to give me your most opposite perspective to mine. In that way I will know how to change my thinking so I can accommodate your concerns and problems." The Okanagan writer and activist Jeannette Armstrong told me why her people developed this and similar technologies: "We don't have any fewer problems than you guys getting along. But we know that whomever we're having trouble with, their grandchild might marry our grandchild. So we have to accommodate one another. I have to ask myself how I can change to accommodate you. At the same time, because you, too, are Okanagan, you will be asking how you can change to accommodate me. We're going to be leaning toward one another." She talks of how all the people in her community share one skin. They share that skin with all of the people who came before, and all who will come after. This applies in a sense to their nonhuman neighbors as well.

In the dominant culture, familial and sexual relations are relations of occupation, not inhabitation. Rates of rape and child abuse reveal the degree to which the bodies of women and children are considered the property of their masters (husband: from Anglo-Saxon husbonda; hus, house, and honda, master). Vaginas become resources to be exploited (or at the very least husbanded), and those who live in the bodies containing these resources become pesky inhabitants to be terrorized into giving up the resource.

But something even more intimate than our family lives is infected by this complex of beliefs: our sense of what we consider a self. Who are you? Who, precisely, is the you that you consider you? Chances are good it's what Catherine Keller called the separative self, an isolated monad cut off from all others by psychological, spiritual, and existential barriers much stronger than skin. If your goal is to attempt to minimize acknowledging damage to yourself as you exploit others, this sort of self is just the ticket. If your goal is to inhabit relationships, this self is a really bad idea.

If you do believe you are a separative self, or act as though you believe you are a separative self, whom, exactly, are you cut off from? Do you consider your self to include your family? Your friends? The air you breathe? The *Aplodontia rufia* who live far closer to you than Angelina Jolie or Nicole Kidman? The solitary bees digging their nests in the dirt outside? The dirt itself, the living breathing dirt? The water that acts as intermediary between all of these? Are these all part of you? Are any of these part of you?

Or maybe you include only the parts of you that end at your fingertips. Or maybe you include even less than that. Maybe not even your emotions. Maybe not even your dreams. Maybe nothing but your thoughts. And maybe not even those.

I just got a note from a friend who put it well, "People never leave or even look outside the bubbles they create to meet their own immediate gratification. This is how we're taught to live: it's the city model on a micro level. These hollow beings (be they cities or people) suck in everything from around them and create a wall of aggression to keep outsiders outside. The more hollow and empty they realize they've become on the inside, the more fiercely they attack, disable, and devour their surroundings. It occurs to me that in a very real sense, we cannot hope to create a sustainable culture with any but sustainable souls."

She continued, "People see that the culture—and the same is true for many of our relationships—is broken in so many ways, and so unsustainable, but are terrified to probe too deep, because they think if it—civilization, their intimate relationship, whatever—crumbles, there might be nothing left. This is how we enter into these bubbles of perception—they form our earliest passage from a world of love to a world of fear and denial. It begins with wanting connection. And then we settle for something less, because we think the alternative is nothing at all. But our truth is still there—all of it is still there. We could wake up any time and reclaim the whole of our existence."[174]

❨ ❨ ❨

Precisely because those in power are so dependent (for their power, for their lives) on those they exploit, they must convince themselves and especially these others the opposite is true. Between open-mouthed kisses, fathers tell daughters no man could ever treat them so well. As carcinogens accumulate in our bodies (in our bones, organs, fat) movies, TV shows, magazines, and newspapers inculcate us to believe that without police (who count it among their jobs to protect the property and processes of polluters from the outrage and bombs of dying citizens) we would all be murdered in our sleep. As bombs (*their* bombs, never our bombs) fall on human beings around the globe (human beings who want to live and love and be loved and see their children grow to live and love and be loved) we are told by politicians that bombs (*their* bombs, never our bombs) are necessary to make the world safe for something they call democracy.

As forests are felled, rivers poisoned, soil toxified, as we see beautiful wild places we love destroyed, as we watch our grandparents, cousins, brothers, sisters, lovers, children, ourselves wasting away from cancer, the whole culture tells us time and again the same message: you cannot survive without this culture, without civilization.

All of these messages are feasible only because of outrageous narrowing and blurring of our ability to perceive and to think clearly. Safety must be made to seem dangerous, and danger must be made to seem safe. Benevolence comes to be called violence, and violence comes to be called benevolence. Fear feels like love, and love feels like fear.

I have experienced this. My father trained me well. I hated him when I was young, for the rapes I endured, the beatings I witnessed. But when he left, when I was maybe ten, I also felt deeply betrayed, and I hated him all the more for this further betrayal. At the time we did not talk about it, but I later learned from my sister that my father also raped her, and that she felt something similar when he left. She ran away. Later he came back. I hated him even worse for that. Later he left again. Still I hated him. I hated him for what he did to me, and I hated him for leaving behind a hollowed-out shell of me when he left.

All of this was precisely the sort of preparation I would need for a life of giving myself away, preparation for a process of schooling in which I was to give myself away to teachers, in preparation for a life of wage slavery, when I was to give myself away to the highest (monetary) bidder. I was similarly prepared to give myself away in personal relationships. The idea was that I should give myself away to those who held power over me until I had nothing left to give.

I am not unique.

That is what is expected of all of us.

That is what is expected of the world, that it give to those in power until it has nothing left to give.

But do we need to live like this? Do we need these masters? Do we need to give ourselves away to those who do not hold our best interests at heart, and do we need to allow them to hollow us out, and to hollow out the places we love?

It's very scary. Having been hollowed out, having been told time and again that we cannot exist without the social systems that lead to our degradation, it is very easy to come to believe we cannot live without them. No matter how much we hate our jobs, could we live without the capitalists who run the country? No matter how much we hate ExxonMobil, could we live without the oil it sucks from the earth and transforms into the very lifeblood of the industrial economy? No matter how much I hated my father, could I have lived without him? (Well, yes. I discovered quickly I could live without him, and have long done so. Since he no longer touches my life, I no longer even hate him.)

How deeply do we hold this belief that not only is civilization "a high stage of social and cultural development," but that we simply could not survive without it? How would we eat if we could not go to Safeway or Ray's Food Place (or KFC or Carl's Jr)? How would we clothe ourselves if we did not receive regular catalogs from J. Crew? I live now in the relatively stable climes of coastal northern California (average daily summer high, maybe sixty-five; average nightly winter low, maybe forty-five), but have lived most of my life where it gets cold: Colorado (last spring snow, June 16), northeastern Nevada (last spring freeze, July 4), North Idaho, and eastern Washington. Could I construct, using Stone Age tools, a shelter that would help keep me alive through a winter?

The answer to all of these is, not by myself.

But does that mean I—or you—could not survive without civilization?

That depends, first of all, on who you are. If you are a wild creature—although I doubt many Del Norte Salamanders will read this book, however much they may applaud (with their cute soft hands on stumpy little arms) my analysis—you could almost certainly live without civilization, and in fact almost certainly won't live if it's allowed to continue. I say "almost certainly" because while most nonhumans are harmed by civilization, nonhumans are by no means monolithic (part of our problem is so many of us consider "nature" to be something singular). Some—such as Norwegian rats, kudzu, and starlings—benefit mightily from civilization through the increase of their habitat and eradication of competitors and predators. Some microbes, too, benefit. Civilization has been such a boon to many microbes who feed off humans (especially overstressed humans in close quarters) that I've read persuasive arguments that microbes, not humans, are responsible for cities, which are in this perspective nothing more than microbe feedlots and factory farms. (These arguments always make me wonder if there are "human rights" activists among the microbes who complain about intolerable and "inmicrobane" living conditions humans are forced to endure in cities: "It's okay to eat them," say these viral activists, "but they should be allowed to live with dignity first!")

Nonetheless, for blue whales, spotted owls, hammerhead sharks, and Javan rhinos to survive, civilization has to go.

Soon.

But who cares about nonhumans, right? If they can't adapt to civilization, fuck 'em. We want to know about the only creatures who matter. Could humans survive without civilization?

Well, we have for more than 99 percent of our existence. But does that matter now? Could humans survive given current numbers? Perhaps more central to the concerns of most of the civilized, could we maintain our lifestyle (note that the question has not-so-subtly shifted from survival of living, breathing human beings to the capacity to maintain a capitalist, consumerist lifestyle where the rich buy second homes while the poor die of starvation and the world gets trashed)? Would taking down civilization cause massive deaths, massive suffering? Clearly more important to many, would we still be able to use the internet? I'll examine these questions later in greater detail, but for now let's break humans into quick subcategories, recognizing that humans are no more monolithic than cheetahs.

I think we—at least those of us who consider genocide a bad thing—can safely say traditional indigenous people living traditional ways would be better off if civilization disappeared tomorrow. They'd have been far better off if it had disappeared a long time ago. They could easily survive—and would survive better—without it.

The rural poor would also survive better without civilization. With no one to dispossess them, to use their land for cash crops, they could return to the subsistence farming that has supported them for a very long time. Recall the quote by the member of the tupacamaristas: "We need to be able to grow and distribute our own food. We already know how to do that. We merely need to be allowed to do so." The rural poor of the world know how to keep themselves alive. They merely need to be allowed to do so.

It seems pretty clear to me also that the rural rich—including, on a global scale, most rural people in the United States—would survive pretty well, too. They'd lose a lot of luxuries, like strawberries in January and shrimp year round. But because, as I've said several times, access to land means access to food, clothing, and shelter, these people would probably do well. Their relative wealth in material possessions—owning a gun, for example—would at least somewhat counterbalance their ignorance of how to feed themselves.

None of this alters the fact that there are too many humans for the land to permanently support. And we haven't yet begun to talk about cities.

The urban poor are in a much worse position than the rural poor. They obviously do not have access to land. In the long run, they would of course be far better off without civilization. The problem—and this is obviously a huge one—is that in the short run many of them would be dead: their food is funneled through the very system that immiserates them. Yet we need to remember that the continued existence of civilization and its extractive economies already guarantees the early deaths of many of them: these extractive economies are precisely how they became urban poor in the first place. I say this not to dismiss those deaths but to point out that we—or really, they—are in a double-bind of civilization's making: if we break down the distribution systems that feed them, many would probably die, yet those distribution systems are parts of a larger megasystem that cannot last, and that is quickly depleting the earth's capacity to support humans, a megasystem that already does these people great damage. This reveals the stupidity—and evil—of making people dependent on a system that exploits them, cutting off their direct connection to the real support for all life: the landbase.

But who cares about the poor, right? If they can't adapt to civilization, fuck 'em, and if they can't survive without it, fuck 'em twice. We want to know about the only humans who matter. What about the urban rich?

Well, I'm not too worried about them: they're the ones who got us in this mess. They can fend for themselves. And if they can't, fuck 'em.

〈 〈 〈

There are a number of reasons why my analysis of whether the urban poor could survive without civilization is bullshit. The first is that anytime anyone makes a prediction, that person should expect to be wrong. I can no more predict the outcome of such a complex set of actions as the end of civilization—whatever that means—than I could have predicted the Tampa Bay Devil Rays would lose more than a hundred games in 2002. Well, okay, I might have been able to predict the latter.

I do not know what will happen when civilization comes down, whether through ecological collapse or the efforts of those humans who resist it. Will the urban poor starve? With the removal of current power structures—which is certainly part of what I'm talking about—along with the cops who keep these power structures in place, will the poor take food from the rich? Will cops become even more violent than they already are? Will cities turn into battlegrounds? Or will the poor form collectives to take care of themselves and their neighbors, and take idle land from the rich to grow their own food? Will the poor be able to keep the food they grow? Will they be able to stay alive until their first crops come in? Will the rich hire (or convince) police to keep the poor from doing this? Will police do this simply on principle? Will police take the food for themselves? What will be the response on the part of the poor? Further, will violence against the natural world get worse? Will it shift its locus from the colonies closer to the heart of empire? I was recently in New England, and someone there commented that local trees had grown back over the last hundred years. He took that as a good sign: the people of the region had finally learned to not deforest their own backyards. I took it more as a sign of the increased reach of civilization: technological and social innovation have enabled these Yankees to deforest the globe—when they want wood fiber, they now come calling to someone else's backyard. The point is that when global trade collapses—global trade is another part of civilization that needs to go—if these people want fiber, they will once again cut the trees closest to them. But they won't be able to reach around the world. Will that inability be a good thing? I think so. But the *real* point is that I don't know what will happen.

Here's what I do know: the global industrial economy is the engine for massive environmental degradation and massive human (and nonhuman) impoverishment. The more this economy can be slowed, the less damage will be caused to the world, and the better the planet will be able to continue to support human (and nonhuman) life.

I also know that right now none of these urban poor die of starvation. They die of colonialism. As I mentioned before, while three hundred and fifty million people go hungry in India, former granaries in that country export tulips and dog food to Europe. While these same hundreds of millions starve, "their" government attempts to dump sixty million tons of grain into the ocean, because it cannot find export markets for that grain, and because it will not distribute food to those who cannot pay.

Seventy-eight percent of the countries reporting child malnutrition export food. During the much-publicized famine in Ethiopia during the 1980s, that country exported green beans to Europe. During the infamous potato famine, Ireland exported grain to England (and part of the reason the potato blight took hold in the first place was that the Irish were pushed to the poorest land).

Sure, there are too many people on the planet. Someday there will be fewer. But right now there is enough food to go around, enough, in fact, to make everyone fat: 4.3 pounds of food per person per day, around the world. This despite the exportation of non-food crops like coffee, tobacco, tulips, opium, and cocaine grown on land used for food production before the (often-forced) entry of the global economy, land that will be used again for local food production once the global economy collapses. This also despite the use of so much land for non-productive ends such as roads and parking lots. Pavement now covers over sixty thousand square miles just in the United States. That's 2 percent of the surface area, and 10 percent of the arable land.

Here's another reason my analysis of whether the urban poor would suffer more from civilization's crash than its continuation is bullshit, and this forms the twelfth premise of this book: *There are no rich people in the world, and there are no poor people. There are just people. The rich may have lots of pieces of green paper that many pretend are worth something—or their presumed riches may be even more abstract: numbers on hard drives at banks—and the poor may not. These "rich" claim they own land, and the "poor" are often denied the right to make that same claim. A primary purpose of the police is to enforce the delusions of those with lots of pieces of green paper. Those without the green papers generally buy into these delusions almost as quickly and completely as those with. These delusions carry with them extreme consequences in the real world.*

But really there are just people. None rich. None poor. Except in our minds. And so people starve.

When I predicted the urban poor might suffer under civilization's collapse, I may have been falling once again under the spell of the abuser who says we cannot survive without him. When civilization falls, many of those who die—or at least those who starve, which is what we're talking about right now—will be those who continue to believe what may be the central delusion of this culture, the delusion that there are rich and there are poor, that monetary wealth—and by extension food, and land (which means food)—is held by anything other than social contract and force. If the "poor" do not fall under this spell, and they can convince enough others it's not immoral to defend themselves from the hired guns of the (formerly) rich, there is a good chance they will survive.

❨ ❨ ❨

My statement that ownership is merely based on shared social delusion is not entirely accurate. First, we all know that the civilized notion of ownership is in truth based on force: the acquisition and maintenance of the property of the rich is the central motivating factor impelling nearly all state violence. But there's a deeper point to be made here, having to do with the mixing of one's body and the soil. When I say that I'm living on Tolowa land, I don't mean to imply that their ownership of this land is delusional, or even that it is based on social convention. Quite the contrary. They belong to the land, as the land belongs to them. It is still ownership, but not in the way that the civilized mean it. Typically when we the civilized speak of owning something, it means a person has the right to do what he wishes with it, to destroy it if he so pleases. It's my computer, so if I want to throw it off a cliff, nobody can stop me. But this other type of ownership has to do with responsibilities, and it has to do with the deal we spoke of earlier between predator and prey. If you live on a piece of land—if you own a piece of land—if you consume the flesh that is on that land, you are now responsible for the continuation of that land and its health. You are now responsible for the health of all the various communities who share that land with you. And because members of this community will consume your flesh, too, they will be just as responsible for the continuation and health of your community. At that point you will own the land, and it will own you.

$$(\quad (\quad ($$

Just as those who wish to dominate and exploit will use any excuse to maintain and expand their control, those who see themselves as victims will find any excuse to maintain their belief they could not survive without their exploiters. I've known many women who stay with men who beat them and their children because they do not know how they would otherwise pay the rent. This logic is insane: it is also all-too-common. I've known many people who sell their hours to jobs they hate for the same reason. This logic is just as insane, and, if anything, is even more common. I remained in those abusive relationships I mentioned earlier in part because I thought I could do no better. At the time it seemed to make sense: from the outside I now perceive my own insanity.

All of these are stupid reasons to stay in intolerable situations, and in all of the cases I know personally—the abused women, people who hate their jobs, my own ill-chosen relationships—the fears went entirely unrealized when the people had the courage to finally make a move.

How many of us recognize the atrocious nature of civilization, yet hang on because we fear we would not—could not—survive without it?

Recently a few members of the Derrick Jensen discussion group faced this subject head on. One wrote, "Although I may detest what civilization is doing, I am literally filled with it. I don't hunt, gather, or grow my own food, but buy it at diners and grocery stores. I clearly see how being civilized is like being in an abusive relationship, but I rely on this relationship for my food. And because civilization has taken over so much of the land and people, trying to live outside of it can be very hard and lonely."

Someone responded, "At one point, the civilized would, and often did, run away from civilization to join the indigenous, to become Pequots, or Lakota, or Goths, or Celts. Unfortunately, by now there is nowhere you can run to get away from civilization. When escape is not an option, what can you do if you're in an abusive relationship? *The Burning Bed* [a film about a woman who kills her abusive husband by burning him in his sleep] comes to mind."

Another person: "Those who think they can live outside of civilization are gravely mistaken. The only reason some who think they're living outside civ are left alone is they pose no real threat to the system and can easily be 'dealt with' if they do. They're 'allowed' to live their 'alternative' lifestyles. As Ted Kaczynski stated some time ago, 'You can have all the freedom you want as long as the authorities consider it unimportant.' In order for anyone to really have the chance to truly live 'outside' civilization anymore, civilization has to go."

This is all very true, and just another way of talking about civilization's monopolization of perception (and the world). But what will happen if we follow the example of *The Burning Bed*?

In answer, one person expressed the concern that "we've been caged in civilization so long our natural instincts and awareness have been dulled to the point we no longer trust we have the ability—or even the ability to learn how—to survive in a noncivilized environment."

Someone disagreed: "Perhaps I'm not seeing this clearly. I grew up in the country, and by the age of eight knew how to feed and shelter myself. If I could do it at eight, with no one teaching me, how much better could we all learn if we were being taught?"

The previous person responded: "I didn't mean to imply we've individually lost the ability to learn the skills needed to survive outside civilization. I should have emphasized our lack of deep-seated faith in our abilities, based on a lack of intimacy with the places we live (and, I would add, six thousand years of propaganda telling us nature is dangerous and civilization benign). For example, I've chosen to devote my energies primarily to learning, practicing, and perpetuating primitive survival and hunting skills, with the hope that when civilization collapses, such valuable knowledge will make it through to those trying to re-form more humane cultures. But here's my caveat: at one time I thought I'd become a competent survivalist. Then I had the opportunity to hunt with a half-Cherokee half-hillbilly who'd grown up in the backwoods of Appalachia, a man who could move with speed and stealth through thick underbrush, who could see and hear things in the forest I could not, who had a seemingly instinctual ability to know where to find animals based on the weather, season, time of day, and so on. This was a man possessing a deep-seated faith in his ability to survive in the wild. So while I think I have better survival skills and potential than most people, I have to admit I just don't have the deep-seated faith that can only come through spending the vast majority of your time in direct contact with the natural world, as my friend did in his youth.

"There's a broader problem, though, beyond individual survival, which is that of whole communities being able to learn to exist without the infrastructure of civilization. Again, I don't mean to imply we can't learn the needed skills. I just don't think in the near future we'll be able to master these skills with the grace evident in indigenous communities."

Someone else put in what became the final words: "I agree, but think that grace will come if (and when) we stick with it long enough. And I need to add something for people to think about as civilization comes down: don't leave out insects. They're abundant, self-cleaning (not much disease), tasty, and contain lots of protein and good fats. Bon Appetit!"

❨ ❨ ❨

After occupying Afghanistan, the United States invaded yet another country, Iraq.

The first night of the invasion I stood in the checkout line at Safeway, taking in the cover of a magazine—a picture of a fish with a human face—and pondering a question asked on another—whether after all this time Demi Moore will reconcile with Bruce Willis—when a man in his early twenties turned to me and said, deadpan, "So, we're at war."

I tried unsuccessfully to read his unshaven face beneath his baseball cap. I wanted to say, "*We're* not at war. *I'm* not at war. It's not my government. They're not my troops." But that would have required too much explanation. So I said, "Yes, the U.S. government is yet again bombing the shit out of poor brown people."

He nodded, and said, "Yes, it is. It certainly is." He turned away, and so did I.

❨ ❨ ❨

As civilization falls, we all—rich and poor alike—have far more to fear than starvation, even more than the dioxin that permeates our bodies. Those in power time and again show no hesitation at killing to gain and maintain access to resources or to otherwise increase their power. Indeed, as is being shown right now in Iraq, and has been shown repeatedly the world over, they show an absolute eagerness to do so (I was going to suggest those who think the U.S. invasion has nothing to do with oil should put the book down, but realized they've probably already tired of the big words).

But their eagerness to use violence to gain power is nothing compared to what awaits anyone within range when their power is threatened. Anybody who has ever been in a violent relationship knows that to leave is extremely dangerous, as abusers often kill their victims rather than let them escape (showing they'd rather kill than give up their control, and, as my mom said, give up their identities). They sometimes kill themselves as well, showing they'd rather die, too, than give up their control and identity.

This happens not only on a personal level. When Hitler finally realized his war was lost, he tried to take down all of Germany with him. Disobedience on the part of his lieutenants prevented Hitler from succeeding. Had the Nazis possessed a nuclear arsenal comparable to what is now wielded by the United States, Hitler would certainly have attempted to use it to destroy the world. If an abuser cannot control a thing, it shall not be allowed to exist. This is the quintessence of abuse.

Lately at talks I've begun commenting that if those who run the U.S. government were to find their power seriously threatened, whether through internal rebellion or ecological collapse, there's a good chance they wouldn't scruple at nuking L.A. or any other seat of resistance. Heck, they've nuked Nevada for decades without any threat to their power at all.

People nod when I say this. There are no gasps of shock or disbelief. People easily accept the very real possibility of "their" leaders using nuclear weapons on the people and landbase they purport to serve. People are often far ahead of me in their analysis and understanding that those in power will do anything to maintain that power, and will destroy everything under their control before they see it let free.

Starvation, frightening as it is, may not be our greatest fear.

〘   〘   〘

Given the radical obtuseness into which most of us—myself definitely included—are trained from infancy, I need to not be abstract but to be absolutely explicit. The United States government is a government of occupation. Capitalism is an economics of occupation. If a foreign power (or space aliens) were to do to us and our landbases what the dominant culture does—do their damnedest to turn the planet into a lifeless pile of carcinogenic wastes, and kill, incarcerate, or immiserate those who do not collaborate—we would each and every one of us—at least those of us with the slightest courage, dignity, or sense of self-preservation—fight them to the death, ours or far preferably theirs.

But we don't fight. For the most part we don't even resist.

How's it feel to be civilized? How's it feel to be a slave?

❰ ❰ ❰

Here's how it works. Those in power pass some law. It doesn't much matter how stupid or immoral the law is, it will now be enforced by people with guns: the police and the military. Or maybe some judge sets a precedent. Once again, it doesn't matter how stupid or immoral the precedent is, it will also be enforced by people with guns. This law or precedent may be that human beings are prop erty, that is, without rights (only responsibilities). It may be that corporations are persons, that is, with rights (and in this case, without responsibilities). It may be that corporate lies are protected free speech. It may be that corporate bribes are protected free speech. It may be that those who kill in the service of production are protected from accountability. It may be that those who destroy property "owned" by corporations face decades in prison as declared "terrorists."

Those in power often con the rest of us into being proud of being good, defined—by them and by us—as being subservient to their laws, their edicts. They con us into forgetting—and in time we become all too eager to con our-selves into forgetting—that those in power can and usually do legalize reprehen-sible activities that increase their power (for example, stealing land from the indigenous, invading countries with desired resources, debasing the landbase, all done legally, because those in power declare it to be so) and criminalize non-reprehensible activities that undercut their power (soon after the most recent invasion many people were arrested in New York City for pasting up pictures of Iraqi citizens, that is, humanizing the U.S.'s current targets; consider a law pro-posed in the Oregon legislature mandating twenty-five year minimum sen-tences for doing anything that would disrupt transportation or commerce, including standing in the street during an anti-war protest [I'm not kidding]).[175] Another way to say this is that those in power make the rules by which they maintain and extend their power. Of course. And then those in power hire goons—for when you take away the rhetoric of protecting and serving, the job of police and the military boils down to being muscle to enforce the edicts of those in power—to keep people in line.

When we forget that the edicts of those in power are merely the edicts of those in power, we lend these edicts a moral weight they do not deserve. Those in power (usually the rich) declare that those in power may under certain cir-cumstances kill those not in power (most often the poor), and the rest of us forget they're doing no more than using their power to get away with murder. Those in power declare that those in power may under certain circumstances devastate the landbases—oh, sorry, "develop the natural resources"—of distant communities, and the rest of us forget they're doing no more than using their power to get away with murdering communities and murdering the earth. Those in power declare that those in power may under certain circumstances destroy entire peoples, and the rest of us forget they're doing no more than using their power to get away with genocide.

Many of us do not effectively oppose the actions of the government that occupies our landbase because we're afraid of the consequences, afraid of being killed or imprisoned. That fear is, I think, one reason I have not yet taken out any dams. I am ashamed to admit that, but it is true.

If our fear drives us away from effective action, we should at least have the honor to not make a virtue of this cowardice. So often we pretend that to be a law-abiding citizen is to be a moral human being. Or we pretend the following is a position of moral superiority: to be under all circumstances opposed to all forms of violence (except, of course, that we do not seem to so much mind when it comes to using resources stolen by force from others and from the earth). Even to be opposed to using violence to stop violence done to ourselves and those we love is considered morally tenable, even desirable (and not, oddly enough, despicable). These rationalizations are essential to the maintenance of current power structures.

The thirteenth premise of this book: *Those in power rule by force, and the sooner we break ourselves of illusions to the contrary, the sooner we can at least begin to make reasonable decisions about whether, when, and how we are going to resist.*

❨ ❨ ❨

My friend who was tree-sitting down in Humboldt County—her name is Remedy—was recently pulled from the tree. With the assistance of the Humboldt County Sheriff's Department, Pacific Lumber sent several climbers after her. She locked down, which means she climbed far above the platform where she lived, the platform she shared with flying squirrels, crows, termites, ants, and tiny salamanders who live in rotted-out hollows high inside the trunk, wrapped her arms around the tree's woody flesh and put them inside metal sheaths, then locked her hands together. She did this so climbers would have to cut her away from the tree before they could pull her down.

The main climber is called Climber Eric. Pacific Lumber routinely hires him to take out tree-sitters. He climbs the trees, talks to the tree-sitters with the soft voice and smile of someone who knows he's backed by the full power of the state, and tells them things will go much better for them on the ground and in the courts if they come down now. If they don't come down on their own, he tells them he's going to bring them down, and still smiling says, "See those deputies on the ground? If you resist, or make even the slightest move against me, they'll shoot you." I do not know if he still smiles as he cuts tree-sitters from lockboxes. Nor do I know if he smiles as he puts them in pain compliance holds—that is, as he tortures them—until they go limp. Then he ties them and brings them to the ground. Because of Climber Eric's treatment, at least one tree-sitter is months later still unable to use his thumbs.

I need to say Climber Eric's use of violence is not limited to his professional life: he has twice been arrested for domestic violence.

As he cut Remedy out of her lockbox, he said to her, smiling (of course), "When you get down, to celebrate, you should get yourself a pearl necklace." *Pearl necklace* is a pejorative term for having a man's semen around your neck. I do not know if he was attempting to imply that *he* should be the one to ejaculate on her. I do know he was saying this to a woman whose hands were not free, and I know further that even if her hands had been available, and had she been able perhaps to slap him for the comment, she would still have had to hold back— to lie back and take his comment, as it were—for fear the deputies on the ground would have shot her.

Often when they pull down tree-sitters, the cops, goons, and loggers force the tree-sitters to watch as they cut the trees the sitters were trying to protect. In this case they didn't do that.

That night another tree-sitter—this one named Mystique—climbed back up. She was there a few days before Climber Eric came for her too. Mystique climbed higher and higher, far higher than Remedy had ever dared, to the top of this ancient redwood—which are the tallest of all trees—to where the trunk was smaller than her arm. Climber Eric followed. He reached for her. He told her that he and the other climbers, as well as the cops on the ground, had already agreed that if she fell they'd all say she committed suicide by jumping. He tied a single rope to her waist and lowered her upside down to the ground.

That night, another climber made his way up the tree. This climber has already withstood several assaults by Climber Eric. As I write this, the tree still stands.

❮ ❮ ❮

If Nazis or other fascists took over North America, what would we all do? What would we all do if they implemented Mussolini's definition of fascism: "Fascism should more appropriately be called Corporatism because it is a merger of State and corporate power"? And what would we do if they then instituted laws allowing them to put a significant portion—say one-third—of all Jewish males between the ages of eighteen and thirty-five into concentration camps? What if this occupied country called itself a democracy, but most everyone understood elections to be shams, with citizens allowed to choose between different wings of the same Fascist (or, following Mussolini, Corporate) party? What if anti-government activity was opposed by storm troopers and secret police? Would you fight back? If there already existed a resistance movement, would you join it? Substitute the word *African-American* for *Jewish* and ask yourself the same questions.

Now, would you resist if the fascists irradiated the countryside, poisoned food supplies, made rivers unfit for swimming (and so filthy you wouldn't even *dream* of drinking from them anymore)? What if they did this because . . . Hell, I can't finish that sentence because no matter how I try I can't come up with a motivation good enough even for fascists to irradiate and toxify the landscape and water supplies. If fascists systematically deforested the continent, would you join an underground army of resistance, head to the forests, and from there to boardrooms and to the halls of the Reichstag to pick off the occupying deforesters and most especially those who give them their marching orders?

Okay, so maybe your sense of kin, and your sense of skin, doesn't extend to the natural world. Maybe you don't yet love the land where you live enough that you will fight for it. But what if the fascists toxify not only the landscape but the bodies of those you love? What if their actions put dioxin—one of the most toxic substances known—and dozens of other carcinogens into the flesh of your lover, children, mother, brother, sister, father? Would you then fight back? What if the fascists toxify your own body? Would you still cling to the illusion that their edicts carry more weight than that brought to bear by their secret and not-so-secret police? Would you work for this regime? Would you teach others its virtues? Or would you fight back? If you will not fight back when they toxify your own body (and toxify your mind with propaganda leading you to believe their edicts carry moral weight), when, precisely, will you fight back? Give me—and more importantly yourself—a specific threshold at which you will finally take a stand. If you can't or won't give that threshold, why not?

None of these questions are rhetorical. The questions are real. They are, at this point, some of the most important questions there are.

《 《 《

How much closer must the culture cut before you will bring it down?

Prior to World War II, annual worldwide use of pesticides ran right around zero. By now it's 500 billion tons, increasing every year. Of course—ho hum—there are massive environmental problems associated with the fabrication and introduction into the environment of so many poisons. But I recently came across a study that might help shake the miasma from all but the already dead. Scientists compared children raised in an agricultural area in Mexico where chemical pesticides were used with those from nearby foothills where pesticides were not used (I'd like to say where pesticides were absent, but of course by now they're everywhere, some places are just not quite so saturated). Both the physical and mental growth of children exposed to pesticides were grossly retarded.

I've seen drawings by children of both groups. Those by children exposed to pesticides are pathetic, and I mean that in its deepest sense of raising pathos, except among the fully enculturated, who probably won't notice, or even noticing won't feel, or even feeling won't act.

Instead of the fully formed figures created by children four to six years old—stick figures with smiling faces or balloon men complete with belly buttons—the creations become instead unidentifiable scratches, as though a chicken stepped in ink then made its way across the paper.[176]

Let's be clear: Those in power are poisoning children, stealing their physical and cognitive health: making them weak, sick, and stupid.

How close must the culture cut before you will fight back?

WHY CIVILIZATION IS KILLING THE WORLD, PART I

The most potent weapon in the hands of the oppressor is the mind of the oppressed.

Stephen Biko, anti-apartheid activist tortured
to death by state police

WHY CIVILIZATION IS KILLING THE WORLD, TAKE ONE. Here are the words of Marine Corps Sergeant Sprague of Sulphur Springs, West Virginia, part of the U.S. force invading Iraq: "I've been all the way through this desert from Basra to here and I ain't seen one shopping mall or fast food restaurant. These people got nothing. Even in a little town like ours of twenty-five hundred you got a McDonald's on one end and a Hardee's at the other."[177]

❆ ❆ ❆

WHY CIVILIZATION IS KILLING THE WORLD, TAKE TWO. I received a note from a friend, Katherine Lo, a sophomore at Yale University. She set up a talk for me there. Soft-spoken almost to the point of shyness, she nonetheless possesses courage far beyond that held by most of us.

Her note: "I hung an upside-down American flag outside my window facing the main campus to express my dissent with the war the U.S. government is waging on the Iraqi people and the wars it has waged and is waging economically, politically, militaristically, and culturally on other countries and peoples.

"The next night several males carrying 2 x 4s entered my dorm suite without permission, then attempted to break into my bedroom, which was locked. After about ten minutes, they left the following note on my message board: 'I love kicking the Muslims ass bitches ass! They should all die with Mohammad. We as Americans should destroy them and launch so many missiles their mothers don't produce healthy offspring. Fuck Iraqi Saddam following fucks. I hate you, GO AMERICA.'"

She continued, "It is hard for me to fathom that people are capable of such malevolence. But this same hatred and racism is prevalent in the very policies of the U.S. government, the blind patriotism of many Americans, and the deeply sickening aspect of the dominant culture that has led some Americans to believe that an Iraqi life is somehow worth less than an American life. There is something seriously and fundamentally wrong here."

The incident in her room was not unique. She compiled a list of similar incidents that took place at Yale in just thirty-six hours. You could probably do the same for your own locale. The evening after the men entered her room, a group of undergraduates participated in a silent, non-violent vigil in the university's dining halls to mourn the deaths of Iraqi civilians. One participant, Raphael Soifer, was followed outside and spat on by a white male. That same evening, in response to an article Kat wrote, a number of anonymous, racist, and threatening posts were made on an online forum. Late that night, perhaps in response to posts on that forum signed by an African American, the following note was left on the door of the Afro-American Cultural Center: "I hope you protesters and your children are killed in the next terrorist attack. Signed Fuck You." Many undergraduates decided to fly flags upside down outside their windows as a sign of dissent, distress, and solidarity with Kat. At least one student's suite was illegally entered, her flag reversed. The next morning, students put up an art installation, permitted by the President's Office. The work included twenty-two American flags representing twenty-two U.S. invasions. One flag, in the center, hung upside-down. A group of husky white males confronted the activists, demanded to see the permit. When it was produced, the group ripped down the flags anyway (forming a parallel to the U.S. demanding that Iraq allow U.S. weapons inspectors into its country, and when Iraq acceded the U.S. invaded anyway). That morning, another upside-down flag outside a student's window was torn down and stolen.

❰ ❰ ❰

Smackyface.

What does that mean?[178]

We need to be explicit about interrogation techniques employed by the CIA and associated groups. I'm sure you've seen the CIA Torture Manuals—oh, sorry, Pain Compliance Manuals, oh, sorry, this time a real title (and I'm not making this one up) "Human Resource Exploitation Training Manual, 1983"—and I'm sure you can guess their contents. I'm sure you've seen the chapter from the 1963 CIA "KUBARK Counterintelligence Interrogation Manual" entitled *Coercive Counterintelligence Interrogation of Resistant Sources.* These manuals are explicit: "The following are the principal coercive techniques of interrogation: arrest, detention, deprivation of sensory stimuli through solitary confinement or similar methods, threats and fear, debility, pain, heightened suggestibility and hypnosis, narcosis, and induced regression." They go on to describe the advantages and disadvantages of each technique, and how each of them can be most effectively used to break their victims, that is, to cause three important responses, "debility, dependency, and dread," that is, to cause their victims to "regress," that is, to lose their autonomy. As one manual puts it: "these techniques . . . are in essence methods of inducing regression of the personality to whatever earlier and weaker level is required for the dissolution of resistance and the inculcation of dependence. . . . As the interrogatee slips back from maturity toward a more infantile state, his learned or structured personality traits fall away in a reversed chronological order, so that the characteristics most recently acquired—which are also the characteristics drawn upon by the interrogatee in his own defense—are the first to go. As Gill and Brenman have pointed out, regression is basically a loss of autonomy."[179]

In short and in vernacular, the point is to mindfuck victims (or as the manual also puts it: "Coercive procedures are designed not only to exploit the resistant source's internal conflicts and induce him to wrestle with himself but also to bring a superior outside force to bear upon the subject's resistance") until they give the perpetrators what they want. This is the essence of abuse. It is the essence of civilization. Every day we see these processes and purposes at work in the culture at large, whether it is teachers, bosses, cops, politicians, or abusive parents who try to exploit our internal conflicts to increase their control, safe in the knowledge that if we refuse to be so exploited they will use force to achieve the same ends.[180]

The manuals often describe the techniques with an absolute lack of attention to morality and humanity (and of course the same can be said for many manuals for teachers, bosses, cops, politicians, and [abusive] parents), as though they're talking not about the destruction of human psyches (and bodies), but about how best to get to the grocery store: "Drugs are no more the answer to the interrogator's prayer than the polygraph, hypnosis, or other aids." Or this: Techniques are designed "to confound the expectations and conditioned reactions of the interrogatee," and "not only to obliterate the familiar but to replace it with the weird." When victims have been hammered with "double-talk questions" and "illogical" statements long enough, all sensible points of reference begin to blur, and "as the process continues, day after day if necessary, the subject begins to try to make sense of the situation, which becomes mentally intolerable. Now he is likely to make significant admissions, or even to pour out his whole story, just to stop the flow of babble which assails him." Or this: "The manner and timing of arrest can contribute substantially to the interrogator's purposes. What we aim to do is to ensure that the manner of arrest achieves, if possible, surprise, and the maximum amount of mental discomfort in order to catch the suspect off balance and to deprive him of the initiative. One should therefore arrest him at a moment when he least expects it and when his mental and physical resistance is at its lowest. The ideal time at which to arrest a person is in the early hours of the morning because surprise is achieved then, and because a person's resistance physiologically as well as psychologically is at its lowest." Or this: "The effectiveness of a threat depends not only on what sort of person the interrogatee is and whether he believes that his questioner can and will carry the threat out but also on the interrogator's reasons for threatening. If the interrogator threatens because he is angry, the subject frequently senses the fear of failure underlying the anger and is strengthened in his own resolve to resist. Threats delivered coldly are more effective than those shouted in rage. It is especially important that a threat not be uttered in response to the interrogatee's own expressions of hostility. These, if ignored, can induce feelings of guilt, whereas retorts in kind relieve the subject's feelings. Another reason why threats induce compliance not evoked by the inflection of duress is that the threat grants the interrogatee time for compliance. It is not enough that a resistant source should be placed under the tension of fear; he must also discern an acceptable escape route." Or this: "1. The more completely the place of confinement eliminates sensory stimuli, the more rapidly and deeply will the interrogatee be affected. Results produced only after weeks or months of imprisonment in an ordinary cell can be duplicated in hours or days in a cell which has no light (or weak artificial light which never varies), which is sound-proofed, in which odors are eliminated, etc. An environment still more subject to control, such as water-tank or iron lung, is even more effective. 2. An early effect of such an environment is anxiety. How soon it appears and how strong it is depends upon the psychological characteristics of the individual. 3. The interrogator can benefit from the subject's anxiety. As the interrogator becomes linked in the subject's mind with the reward of lessened anxiety, human contact, and meaningful activity, and thus with providing relief for growing discomfort, the questioner assumes a benevolent role. 4. The deprivation of stimuli induces regression by depriving the subject's mind of contact with an outer world and thus forcing it in upon itself. At the same time, the calculated provision of stimuli during interrogation tends to make the regressed subject view the interrogator as a father figure. The result, normally, is a strengthening of the subject's tendencies toward compliance." Or this, "It has been plausibly suggested that, whereas pain inflicted on a person from outside himself may actually focus or intensify his will to resist, his resistance is likelier to be sapped by pain which he seems to inflict upon himself. In the simple torture situation the contest is one between the individual and his tormentor. . . . When the individual is told to stand at attention for long periods, an intervening factor is introduced. The immediate source of pain is not the interrogator but the victim himself. The motivational strength of the individual is likely to exhaust itself in this internal encounter. . . . As long as the subject remains standing, he is attributing to his captor the power to do something worse to him, but there is actually no showdown of the ability of the interrogator to do so."[181]

We need to bring this discussion to the real world. Twenty-four-year-old Ines Murillo was a prisoner in a secret army jail in Honduras, where she was interrogated by soldiers trained by these manuals, who gave reports on their interrogations to CIA officials who visited the prisons. For eighty days she was beaten, electrically shocked, burned, starved, exposed, threatened, stripped naked, and sexually molested. Her interrogators fed her raw dead birds and rats. To keep her from sleeping, they poured freezing water on her head every ten minutes. They made her stand for hours without sleep and without being allowed to urinate.[182]

She was not alone: just one of her captors has acknowledged that he himself tortured and murdered one hundred and twenty people.[183]

The CIA has aided torturers the world over. Indeed, the torturers and the CIA often work together (indeed, the torturers are often CIA "assets"). In the late 1940s, the CIA was central to creating Greece's secret police, the KYP, which soon began systematically torturing people. By the 1960s torturers were telling prisoners their equipment—such as a special "thick white double cable" whip that was "scientific, making their work easier," and the "iron wreath," a head screw progressively tightened around the head or ears—came as U.S. military aid.[184]

The CIA set up Iran's notorious SAVAK secret police, and instructed them in torture methods, with, for example, films on such topics as how to most effectively torture women.[185]

In 1950s Germany the CIA not only used normal methods to torture immigrants they suspected of being Soviet plants but they also used esoteric methods, such as applying turpentine to a man's testicles or sealing someone in a room and playing Indonesian music at deafening levels until he cracked.[186]

In Vietnam, the CIA set up its notorious Operation Phoenix, a systematic program of assassination, terror, and torture. It condoned confining prisoners in "tiger cages," five-by-nine-by-six-foot stone compartments, where three to five men would be shackled to the floor, beaten, mutilated. Their legs would wither, and they would become paralyzed, or at best reduced for the rest of their miserable lives to scuttling like crabs. Buckets of lime were emptied upon them.

Elsewhere in Vietnam, CIA assets applied electric shocks to victims' genitals, tapped six-inch dowels through victims' ears and into their brains, and threw victims out of helicopters in order to force their associates to talk.[187] More recently in Afghanistan, U.S.-backed troops loaded 3,000 prisoners into container trucks, sealed the doors, and left these to stand for days in the sun. A U.S. commander ordered an Afghan soldier to shoot bullets through the containers' walls to provide air holes. Soon enough, blood began to stream from the containers' bottoms. Those victims who survived so far were dumped in the desert and shot by Afghans who were watched over by thirty to forty U.S. soldiers. Often the Americans took more direct roles: as one Afghan soldier stated, "I was a witness when an American soldier broke one prisoner's neck. The Americans did whatever they wanted. We had no power to stop them." The bodies of their victims were left to be eaten by dogs.[188]

Latin America, Africa, Asia, Europe, Oceania, North America. There we find CIA-associated torture. Literally hundreds of thousands, if not millions, of human beings have been tortured or killed by people taught by these manuals.

Even *The Washington Post* has commented that CIA and U.S. Army Special Forces interrogators routinely beat prisoners. They hood them, deprive them of sleep, bombard them with light, bind them in painful positions with duct tape. As one agent put it: "If you don't violate someone's human rights some of the time, you probably aren't doing your job."[189]

This "doing their job" of course includes torturing children. In the same article where a CIA-agent spoke glibly of "playing smackyface" with victims, it was revealed that if "smackyface" doesn't work—and prisoners have died from what even a military coroner acknowledges is "blunt force injury"—the CIA has at its disposal other means to make victims "regress," or talk: the agent stated explicitly—and I have to say the capitalist journalist expressed no disapproval of the agent's stated option—that he had access to victims' young children. Surely their "regression"—the exploitation of these "human resources"—would make their father talk.

❨ ❨ ❨

We all know that agents of the United States government torture prisoners. We all know that this has been happening for a very long time. Part of the most recent response by those in power to this widespread understanding has been to redefine torture. A Department of Justice memo defined torture only as the intentional infliction of pain associated with "death, organ failure, or serious impairment of body functions." The President of the United States has insisted that the United States does not torture. In the process of not torturing, U.S. agents and their allies cuff prisoners' hands behind their backs, suspend them by these cuffs, and beat them with iron rods. They effectively liquefy their kneecaps. They force them to stand naked in freezing cells and douse them with water. They drown them time and again in a process used frequently enough to have a name: waterboarding, where "the prisoner is bound to an inclined board, feet raised and head slightly below the feet. Cellophane is wrapped over the prisoner's face and water is poured over him. Unavoidably, the gag reflex kicks in and a terrifying fear of drowning leads to almost instant pleas to bring the treatment to a halt."[190] They smother them to death inside of sleeping bags. They and their allies use electric drills to bore into their kneecaps, shoulders, skulls.

I wonder how the authors of that memo would define torture if they were not defining it in the abstract, if Premise Four did not reign supreme in this culture, if they knew that they and those they love could possibly receive the treatment they so blithely order.

❆ ❆ ❆

When are we going to acknowledge that those in power will scruple at nothing—have already scrupled at nothing—to increase their power? There is no limit to their obsession to control. This won't change because we ask nicely. It won't change because we live peaceably (just ask the indigenous).

What are you going to do about it?

❆ ❆ ❆

For years now I've been talking about blowing up dams to help salmon, but suddenly today I realized I've been all wrong.

This understanding came as I read a description of attempts by ancient Egyptians to dam the Nile, and the Nile's resistance to these attempts. It was all a pretty straightforward process. The Egyptians would erect a dam, and the river would shrug it off, probably with as little effort as a horse quivering the skin of its shoulder to get rid of a fly.

By now, however, the concrete straitjackets have become massive enough that rivers have a harder time sloughing them off, the equivalent, to extend the above simile, to encasing a horse in concrete, then leaving holes at the head and tail to allow food and water to pass. The rivers need our help. (I first wrote "They *may* need our help," but, even without my asking, a couple of rivers strongly requested I remove the qualifier.) They can't do it themselves, at least in the short or medium run.

I've always wanted to blow up dams in order to save salmon, sturgeon, and other creatures whose lives depend on wild and living rivers. But that's not right. We need to blow up dams for the rivers themselves, so they can again be the rivers they once were forever, the rivers they still want to be, the rivers they themselves are struggling and fighting to once again become.

☾ ☾ ☾

Liberating rivers, blowing up dams. The difference may seem semantic to you—like liberating versus invading Iraq, like "creating temporary meadows" versus clearcutting—but it doesn't to me, for a number of reasons.

The first, and probably most important, has to do with everything I've been talking about in this book. Rhetoric aside, both invading Iraq and clearcutting are motivated by the culture's obsession to control and exploit. The primary reason is to gain, maintain, and use resources—oil in the first case (as well as to provide a staging area for further invasions), trees in the second. Further, both invading and clearcutting damage landscapes, damage our habitat. They further enchain the natural world.

The primary motivation for liberating a river, on the other hand, isn't selfish, except insofar as it benefits oneself to live in an intact, functioning natural community (duh!), and insofar as doing good feels good.

This all leads to probably the most important question of this book so far: with whom or what do you primarily identify? A way to get at that question is to ask: whom or what do your actions primarily benefit? Whom or what do you primarily serve?

Who or what primarily benefits from the invasion of Iraq? Let me put this more directly: who/what benefits from U.S. access to Iraqi oil fields?

The U.S. industrial economy, of course. If you care more about and identify more closely with the U.S. industrial economy than you care about or identify with people killed by U.S. bombs or bullets (or by the "blunt force trauma" of smackyface)—people under whose land the oil resides—then you may support the U.S. invasion of Iraq.

I'm taking bets as to who's next on the list to be invaded. Smart money says Syria, but Lebanon and Iran aren't far behind. Here are the current odds, if you'd like to jump into the pool: Syria, 1:1; Lebanon, 3:1; Iran, 4:1; North Korea, 15:1 (North Korea actually having the ability to fight back reduces the odds tremendously); other 25:1; invade nobody 10,000:1 (and Colombia doesn't count, since the U.S. has already invaded [oh, sorry, is "advising"]; the same is true for the Philippines, and about a hundred and twenty other countries).

Similarly, if you identify more strongly with Weyerhaeuser or MAXXAM, or more broadly the industrial economy than you do with forests, you may support clearcutting.

Just today I saw an article in the local newspaper saying that local shrimp trawlers are complaining (accurately enough) about regulations California is (finally) putting in place to curtail the (extraordinary) damage done by trawling. Shrimp trawls are designed to maximize contact with the sea floor. They scrape away everything in their path, the undersea equivalent of clearcutting, picking up every living thing as they go. In some places 80 percent of the catch is "bycatch," that is, creatures the trawlers can't sell, and who are merely thrown overboard dead or dying.

Local trawlers say the regulations will force them out of business. Politicians say the regulations will hurt the local economy. This amounts to an explicit acknowledgment on both their parts that shrimping, and more broadly the local economy (and more broadly still the entire industrial economy) is predicated on harming and eventually destroying the landbase.

If you identify more closely with the local economy than the local landbase, it may make sense to you to support an economy that damages this landbase, your own habitat.

If, on the other hand, you identify more strongly with your landbase than with the economy, it may make sense to you to protect your landbase, your habitat. And since the industrial economy is poisoning us all, the same would be true for those who identify more closely with their own bodies and their own survival (and the survival of those they purport to love) than they do the industrial economy.

Who benefits from the removal of dams?

If you identify more closely with the Klamath River and its salmon, steelhead, lamprey, and other residents than you do with the agricorporations which primarily benefit from taking the river's water, it may make sense to you to help the river return to running free, to liberate it from its concrete cage, or rather, to help it liberate itself. The same would be true for the Columbia, Colorado, Mississippi, Missouri, Sacramento, Nile, and all other rivers who would be better off without dams.

With what/whom do you most closely identify? Where is your primary allegiance? Where does your sense of skin extend, and what does it encompass? Does it include ExxonMobil, Monsanto, Microsoft? Do you give them fealty? Do you give them time, money? Do you serve them? Does it include the U.S. government? Do you pledge it allegiance? Do you serve it? Does it include the land where you live? Do you act in its best interests?

⟨ ⟨ ⟨

I still haven't really gotten to the difference between liberating rivers and blowing up dams. It's one of focus and intent. I've written elsewhere that if I were once again a child faced only with the options of a child (i.e., no running away), but having the understanding I do now of the intractability of my father's violence, I would have killed him. But the point would not have been to kill him. The point would have been to liberate me and my family from the rapes and beatings, to stop the horrors.

Similarly, I don't have a thing for explosives. If I took out a dam, it wouldn't be so I could get off on the big kaboom. I'm not even sure it would be to help the salmon (although yesterday I saw seven baby coho in the stream behind my home, and fell in love with them all over again). It would be to help the river, which in turn would help the salmon. It would be to stop the horrors.

⟨ ⟨ ⟨

WHY CIVILIZATION IS KILLING THE WORLD, TAKE THREE. British scientists have at last discovered that fish do indeed feel pain.

Whether they admit it or not, everyone who has ever gone fishing knows this is the case. But for years an intense (and intensely stupid) debate has been carried on in all seriousness in scientific and fishing circles. In order to end the debate once and for all, scientists jabbed fish in the face with hot probes, and provided "mechanical" and "chemical stimuli" to the fish's faces as well. Sure enough, the fish "seemed" to feel pain.

Just to be certain, the scientists then injected bee venom or acetic acid into fish's lips. In the words of one researcher, "Anomalous behaviours were exhibited by trout subjected to bee venom and acetic acid." As a former beekeeper, I can attest to how much it hurts to have bee venom injected into one's lip, and how directly that leads to "anomalous behavior," in my case jumping up and down and cursing.

But evidently the (intensely stupid) debate isn't over. Dr Bruno Broughton, fisheries biologist for the United Kingdom's National Angling Alliance, fired a scientific salvo back, dismissing this research by saying one cannot "draw conclusions about the ability of fish to feel pain, a psychological experience for which they literally do not have the brains."[191]

This is of course a repetition of a line we've heard too many times, the equivalent of the National Science Foundation spokesman saying there's no causal connection between firing airguns at 240 db and whales beaching themselves, the equivalent of the National Academy of Sciences saying that salmon don't need water.

In order to maintain our way of living, we must tell lies to each other, and especially to ourselves.

From birth on we the civilized are systematically lied to, until in time we systematically lie to ourselves. We insulate ourselves from the pain of others (and from our own pain). We pretend it does not exist. Factory farmed chickens (and carrots) feel no pain. Dammed rivers feel no pain, no claustrophobia. Children made weak and stupid by pesticides feel no pain, no loss. Children with grotesque birth defects from depleted uranium feel no pain. But oh, I forgot, there has been no causal connection shown between the activities of those in power and any of these.

Nor has there been a causal connection shown between the systematic elimination of all wild creatures and the pain, terror, and despair these creatures must feel. But oh, I forgot, these creatures do not have the brains to feel any of these things: only humans feel these things. Only humans in power feel any of these things. Only humans highest on the hierarchy feel these things. Only humans highest on the hierarchy really exist.

And so it goes.

This is what science teaches us (*You will pull the vacuum-packed frog from its plastic shroud, or alternatively, you will scramble the brains of this live frog, make it as insensate as I am making you, as insensate as my elders made me*). It's what economics teaches us (*Money has value. Nonhuman life does not, except insofar as it can be somehow converted to cash. Among humans, because the rich have more money than the poor, and thus the capacity to make more money than the poor, the lives of the rich have more value than those of the poor.*)

This is what the military puts in place and the police enforce.

This is what is killing the world.

☾ ☾ ☾

I have seen tadpoles struggle when caught by backswimmers, and frogs flip frantically when held by the curved pincers of giant water bugs. I've reeled in fish fighting for their lives with hooks in their lips or throats or in the roofs of their mouths. I know these creatures feel pain. I do not need to burn or inject them with venom to know this.

Creatures eat each other. They cause pain to each other. That is part of life. That is part of death. That is part of eating. This causing of pain, this killing, happens whether or not we are vegetarians. It happens whether or not we choose to believe that others feels pain. I prefer to not cause pain, and must be reminded by my vegetarian friends when I accidentally step on a beetle or slug that I am a large mammal, and large mammals accidentally step on smaller creatures. But when I do cause pain, whether by accidentally squashing a sow bug, intentionally killing a fish or potato to eat, or pulling invasive scotch broom, I attempt to at least be honest about it.

❆ ❆ ❆

WHY CIVILIZATION IS KILLING THE WORLD, TAKE FOUR.
March 6.

That's why.

March 6, 1857, the United States Supreme Court rules in *Scott v. Sanford* that because blacks are "so far inferior" to whites, "they had no rights which the white man was bound to respect."

Fast forward.

March 6, 1974, Ayn Rand addresses West Point cadets, something she considered the greatest honor of her life. When someone has the impertinence to "express an unpopular view" and ask her about the United States' basis on the dispossession and genocide of Indians, she responds, "They didn't have any rights to the land, and there was no reason for anyone to grant them rights which they had not conceived and were not using. . . . What was it that they were fighting for, when they opposed white men on this continent? For their wish to continue a primitive existence, their 'right' to keep part of the earth untouched, unused and not even as property, but just keep everybody out so that you will live practically like an animal [and how else would she expect an animal—which is what we are—to live?], or a few caves above it. Any *white* person who brings the element of civilization has the right to take over this continent."[192]

Some things don't change.

(((

WHY CIVILIZATION IS KILLING THE WORLD, TAKE FIVE. In 1900, Senator Albert Beveridge of Indiana, who later won a Pulitzer Prize, and was much later included favorably in John F. Kennedy's immensely popular and influential *Profiles in Courage*, put forward his best arguments in favor of the United States invading—oh, sorry, liberating—the Philippines. I quote his argument at length because he articulates so perfectly and so guilelessly what is wrong with civilization, and because with a few minor changes his words could just as easily have been spoken two thousand years earlier or a hundred years later: "Mr. President, the times call for candor. The Philippines are ours forever, 'territory belonging to the United States,' as the Constitution calls them. And just beyond the Philippines are China's illimitable markets. We will not retreat from either. We will not repudiate our duty in the archipelago. We will not abandon our opportunity in the Orient. We will not renounce our part in the mission of our race, trustee, under God, of the civilization of the world. And we will move forward to our work, not howling out regrets like slaves whipped to their burdens, but with gratitude for a task worthy of our strength, and thanksgiving to Almighty God that He has marked us as His chosen people, henceforth to lead in the regeneration of the world.

"... For power to administer government anywhere and in any manner the situation demands ... is the power most necessary for the ruling provisions of our race—the tendency to explore, expand, and grow, to sail new seas and seek new lands, subdue the wilderness, revitalize decaying peoples, and plant civilized and civilizing governments all over the globe. ...

"Mr. President, this question is deeper than any question of party politics: deeper than any question of the isolated policy of our country even; deeper even than any question of constitutional power. It is elemental. It is racial. God has not been preparing the English-speaking and Teutonic peoples for a thousand years for nothing but vain and idle self-contemplation and self-admiration. No! He has made us the master organizers of the world to establish system where chaos reigns . He has given us the spirit of progress to overwhelm the forces of reaction throughout the earth. He has made us adepts in government that we may administer government among savage and senile peoples. Were it not for such a force as this the world would relapse into barbarism and night.[193] And of all our race He has marked the American people as His chosen nation to finally lead in the regeneration of the world. This is the divine mission of America, and it holds for us all the profit, all the glory, all the happiness possible to man. We are trustees of the world's progress, guardians of its righteous peace. The judgment of the Master is upon us: 'Ye have been faithful over a few things; I will make you ruler over many things.'

"What shall history say of us? Shall it say that we renounced that holy trust, left the savage to his base condition, the wilderness to the reign of waste, deserted duty, abandoned glory, forgot our sordid profit even,[194] because we feared our strength and read the charter of our powers with the doubter's eye and the quibbler's mind? Shall it say that, called by events to captain and command the proudest, ablest, purest race of history in history's noblest work, we declined that great commission? Our fathers would not have had it so. No! They founded no paralytic government, incapable of the simplest acts of administration. They planted no sluggard people, passive while the world's work calls them. They established no reactionary nation. They unfurled no retreating flag.

"That flag has never paused in its onward march. Who dares halt it now— now, when history's largest events are carrying it forward; now, when we are at last one people, strong enough for any task, great enough for any glory destiny can bestow? . . .

"Blind indeed is he who sees not the hand of God in events so vast, so harmonious , so benign. Reactionary indeed is the mind that perceives not that this vital people is the strongest of the saving forces of the world; that our place, therefore, is at the head of the constructing and redeeming nations of the earth; and that to stand aside while events march on is a surrender of our interests, a betrayal of our duty as blind as it is base. Craven indeed is the heart that fears to perform a work so golden and so noble ; that dares not win a glory so immortal.

"Do you tell me that it will cost us money? When did Americans ever measure duty by financial standards?[195] Do you tell me of the tremendous toil required to overcome the vast difficulties of our task? What mighty work for the world, for humanity, even for ourselves has ever been done with ease? . . .

"Do you remind me of the precious blood that must be shed, the lives that must be given, the broken hearts of loved ones for their slain? And this is indeed a heavier price than all combined. And yet as a nation every historic duty we have done, every achievement we have accomplished, has been by the sacrifice of our noblest sons.[196] Every holy memory that glorifies the flag is of those heroes who have died that its onward march might not be stayed. . . . That flag is woven of heroism and grief, of the bravery of men and women's tears, of righteousness and battle, of sacrifice and anguish, of triumph and of glory. It is these which make our flag a holy thing. Who would tear from that sacred banner the glorious legends of a single battle where it has waved on land or sea? . . . In the cause of civilization, in the service of the republic anywhere on earth, Americans consider wounds the noblest decorations man can win, and count the giving of their lives a glad and precious duty.

"Pray God that spirit never falls. Pray God the time may never come when Mammon and the love of ease shall so debase our blood that we will fear to shed it for the flag and its imperial destiny. Pray God the time may never come when American heroism is but a legend like the story of the Cid. American faith in our mission and our might a dream dissolved, and the glory of our mighty race departed.

"And that time will never come. We will renew our youth at the fountain of new and glorious deeds. We will exalt our reverence for the flag by carrying it to a noble future as well as by remembering its ineffable past. Its immortality will not pass, because everywhere and always we will acknowledge and discharge the solemn responsibilities to our sacred flag, in its deepest meaning, puts upon us. And so, Senators, with reverent hearts, where dwells the fear of God, the American people move forward to the future of their hope and the doing of His work.

"Mr. President and Senators, adopt the resolution offered, that peace may quickly come and that we may begin our saving, regenerating, and uplifting work. [Recall that the resolution he wishes to adopt is for the sacking, looting, and holding of the Philippines.] Adopt it, and this bloodshed will cease when these deluded children of our islands learn that this is the final word of the representatives of the American people in Congress assembled. Reject it, and the world, history, and the American people will know where to forever fix the awful responsibility for the consequences that will surely follow such failure to do our manifest duty. How dare we delay when our soldiers' blood is flowing?"[197]

Rhetoric aside, the ensuing American invasion left a large percentage of Filipinos dead. Massacres of every man, woman, and child encountered by American soldiers were commonplace, as was mass torture of combatants and noncombatants alike. The Philippines arguably continue to be, to this day, a colony of the United States.

❨ ❨ ❨

WHY CIVILIZATION IS KILLING THE WORLD, TAKE SIX.

Fast forward to the twenty-first century. Albert Beveridge is long dead, but the imperative, old as civilization, thrives. The flag has still not paused in its onward march, and no one has yet dared halt it.

Indeed, its pace is accelerating. Recall the stated goal of the U.S. military of "full-spectrum dominance." Or consider Michael Ledeen. The day I typed in the words of Albert Beveridge, I also came across the words of Ledeen, former consultant to President George W. Bush's national security adviser, and special adviser to the secretary of state. Considered a leading authority on intelligence and international affairs as well as one of the most influential advisors on U.S. policy in the Middle East, he has been profiled by *The New York Times* and *The Wall Street Journal.* One article lauded his "deep commitment to democracy [*sic*]," and stated that Ledeen "is a man who has helped shape American foreign policy at its highest levels." At least the latter is true: when Ledeen speaks, people like Vice President Dick Cheney and Secretary of Defense Donald Rumsfeld listen and act. People in the rest of the world die.

"Creative destruction is our middle name," Ledeen writes. "We do it automatically."[198] He speaks of "exporting the democratic [*sic*] revolution,"[199] which can be done through a process called "total war," best described by his colleague Adam Mersereau: "By 'total' war, I mean the kind of warfare that not only destroys the enemy's military forces, but also brings the enemy society to an extremely personal point of decision, so that they are willing to accept a reversal of the cultural trends that spawned the war in the first place. A total-war strategy does not have to include the intentional targeting of civilians, but the sparing of civilian lives cannot be its first priority. . . . The purpose of 'total' war is to permanently force your will onto another people. . . . Limited war pits combatants against combatants, while total war pits nation against nation, and even culture against culture."[200]

How does Ledeen suggest those in power prepare themselves psychologically to force their will onto another people? In an essay entitled "Machiavelli On Our War: Some Advice for Our Leaders," he states: "1. Man is more inclined to do evil than to do good." This of course says more about Ledeen's own proclivities and the proclivities of those in his circle than it does about human nature or the world at large. He continues, "Societies with a majority of good people are rare, and are constantly threatened by the evil-minded world outside. Peace is NOT the normal condition of mankind, and moments of peace are invariably the result of war. Since we want peace, we must win the war. Since our enemies are inclined to do evil, we must win decisively and then impose virtue on their survivors, so that they can't do any more evil to us. . . . 2. The only important thing is winning or losing. Don't worry about how the world will judge your strategy. Just worry about winning. Machiavelli tells us that if you win, everyone will judge your methods to have been appropriate. If you lose, they will despise you. 3. If you have to do unpleasant things, it is best to do them all at once, rather than to do a long series of little ones. Strike decisively, get it over with. Don't listen to your diplomats, who will try to convince you that you can achieve your goals with a little bit of nastiness and a whole lot of talking. . . . 4. It is better to be more feared than loved. You can lead by the force of high moral example. It has been done. But it's risky, because people are fickle, and they will abandon you at the first sign of failure. Fear is much more reliable, and lasts longer. Once you show that you are capable of dealing out terrible punishment to your enemies, your power will be far greater."[201]

All of this mirrors and brings up to date Caligula's favorite phrase, coined by the poet Lucius Accius, "*Oderint dum metuant*: Let them hate us so long as they fear us."[202] This line, now quoted regularly by those who run the United States government,[203] is perhaps the most important phrase in the history of civilization, and characterizes everything from childrearing practices to education to social regulation (the civilized term would be law enforcement) to relations with human neighbors to relations with the natural world. It characterizes civilization.

Ledeen more or less always urges politicians to go to war. And he more or less always urges them to do so quickly, ending many of his essays more or less the same way: "Peace in this world only follows victory in war. Enough talking, Mr. President. . . . Let's roll again. Faster, please."[204] Or, "One can only hope that we turn the region into a cauldron, and faster, please. If ever there were a region that richly deserved being cauldronized, it is the Middle East today."[205] Or, "Faster, please. What the hell are you waiting for?"[206] "Faster, please. Opportunity is knocking at our door."[207] "Iran is the heart of darkness. Enough already. Do it now."[208] "As in the war against Iraq, we have already waited far too long to get on with it. Faster, please!"[209] "No let's get on with the war. Faster, please."[210]

Those in charge of this culture are insane.

They are killing the world.

❆ ❆ ❆

WHY CIVILIZATION IS KILLING THE WORLD, TAKE SEVEN. During negotiations over the "Kyoto Protocol to the United Nations Framework Convention on Climate Change" (that's a lot of words to describe a document into which a lot of people put a lot of energy, and which is in the end nearly meaningless in terms of its effect in the real world; this was of course the point all along), Greenpeace activist Jeremy Leggett asked Ford Motor Company executive John Schiller how opponents of the Convention could believe there's no problem with "burning all the oil and gas available on the planet."

Schiller responded by first stating scientists get it all wrong when they say fossil fuels have been underground for millions of years: the Earth, he said, is just ten thousand years old.

How does he know this?

Because the Bible tells him so. Schiller says, "You know, the more I look, the more it is just as it says in the Bible." The Book of Daniel, he states, predicts that increased earthly devastation will mark the "End Time" and the return of Christ.[211]

All of this means that to many fundamentalists, the killing of the planet is not something to be avoided but encouraged, hastening as it does the ultimate victory of God over all things earthly, all things evil. Someone once asked Rick Santorum, this government's third most powerful Senator, why he consistently implements policies that harm the natural world. He replied that the natural world is inconsequential to God's plan, then referenced the impending rapture: "Nowhere in the Bible does it say that America will be here one hundred years from now."[212] (Now tell me you still believe the problems we face are tractable through reasonable discussion: tell me you believe these people will stop because we ask nicely, or because we make our cases through even the most impeccable logic.)

It's important to note that one hundred and seventy-eight members of the U.S. House of Representatives and forty-four members of the U.S. Senate are Christian fundamentalists, or are otherwise allied with the Christian right. The President of the United States and former attorney general are self-described fundamentalists.[213] The President of the United States has stated publicly his reason for bombing and invading Afghanistan and Iraq: "God told me to strike at al Qaida and I struck them, and then he instructed me to strike at Saddam, which I did."[214] And one of his advisors said, "George W. Bush really does seek information. He's very curious about the downing of a U.S. spy plane by China, and so he asked a lot of questions. He asked some detailed questions. Several times he asked, 'Do the members of the crew have Bibles?' 'Why don't they have Bibles?' 'Can we get them Bibles?' 'Would they like Bibles?'"[215]

It is quite possible, indeed likely, that the man with his "finger on the button" that could turn all of this planet into a radioactive wasteland (faster, rather than the slower way civilization is currently accomplishing this) could be actively and eagerly—rapturously—putting in place policies aimed at bringing an end to our time on earth, and the arrival of a mythical Prince of Peace.

This is why civilization is killing the world.

WHY CIVILIZATION IS KILLING THE WORLD, PART II

There have been periods of history in which episodes of terrible violence occurred but for which the word violence was never used. . . . Violence is shrouded in justifying myths that lend it moral legitimacy, and these myths for the most part kept people from recognizing the violence for what it was. The people who burned witches at the stake never for one moment thought of their act as violence; rather they thought of it as an act of divinely mandated righteousness. The same can be said of most of the violence we humans have ever committed.

Gil Bailie[216]

IT'S NOT JUST THOSE IN POWER WHO ARE INSANE. IT'S THE WHOLE culture. A national poll in 1996—and we see this sort of result all the time— showed that more than 40 percent of Americans believe the world in its present form will end at the battle of Armageddon in Israel between Jesus and the anti-Christ.[217] Presumably the evening's opening bout will be between the Virgin Mary and the Easter Bunny.

☾ ☾ ☾

We're fucked. We're so fucked.

Not in the good sense of the word.

☾ ☾ ☾

A reasonable definition of insanity is to have lost one's connections to physical reality, to consider one's delusions as being more real than the real world.

☾ ☾ ☾

WHY CIVILIZATION IS KILLING THE WORLD, TAKE EIGHT. Arrogance.

I have before me an advertisement for the University of California Berkeley Extension. It shows a picture of a man leaning back in his chair, arms folded behind his head, feet on his desk. He wears a white shirt and black tie. I can clearly see the soles of his business shoes. To the left of his shoes an artist has rendered four footprints. On the far left is a bird print. Then a small mammal's. Then a bear's. Then leading up to his shoes is a bare human footprint. The caption: "Evolution . . . doesn't have to take a million years."

The implication is clear: through "a million [sic] years," through birds, mammals, through all creatures, evolution has been leading toward businessmen, and more broadly toward this culture. We are the apex of all life on earth. We are the point. All of evolution has taken place so that we can wear uncomfortable clothes and sit at desks.

Flattering, isn't it?

It's not only Christians who believe the world was made for civilized humans.

☾ ☾ ☾

WHY CIVILIZATION IS KILLING THE WORLD, TAKE NINE. Each year Shell Oil corporation and the magazine *The Economist* hold an "international writing competition to encourage future thinking." The banner headline screams: "YOU WRITE A 2,000 WORD ESSAY. WE WRITE A $20,000 CHEQUE."

This year's topic: "Do we need nature?"

Remember the first rule of propaganda: if you can slide your assumptions by people, you've got them. Another way to say that—and every good lawyer knows this—is the person who controls the questions controls the answers. How would essays written in response be different if instead *The Economist*/Shell had asked one of the following: Does nature need us? Does nature need Shell Oil? Do humans need Shell Oil? Does nature need oil extraction? Do humans need oil extraction? Does nature need industrial civilization? Do humans need industrial civilization? Can nature survive industrial civilization? Can humans survive industrial civilization? What can we each do to best serve our landbases? Who is the *we* in *The Economist*'s/Shell's question?

Regarding this essay, here's probably the most important question of all: if our answers do not jibe with the financial/propaganda interests of Shell Oil and *The Economist*, do you think they'll still hand us a cheque for $20,000?

Just in case we've forgotten who precisely is cutting the cheque, the sponsors provide several questions to lead us on our (or rather their) way. Their first question is: "How much biodiversity is necessary?" This is an insane question, because it does not take physical reality (in this case biodiversity) as a given, but places it secondary to their mental constructs (in this case different people's opinions of "how much is necessary"). More sane questions, that is, questions more in touch with physical reality, would be "How much oil extraction, if any, is necessary? How many corporations, if any, are necessary? How can we help the landbase, on its own terms?"

The question is also insanely arrogant, because it presumes that we know better than the landbase how much biodiversity it needs. If you want to know how much biodiversity is necessary, don't ask me or any other human. Ask the land. And then wait a hundred generations, and your descendants will know the answer for that particular place where they have lived all this time.

And of course their question fails to ask, "How much biodiversity is necessary for *what*?"

Another of their questions: "Sustainable development sounds so natural and desirable that no one could possibly disagree with it. Yet technological advance makes today's definition of what is sustainable or unsustainable quickly obsolete. How can a concept purporting to look to the long term have any real meaning if technology keeps changing the parameters in the short and medium term?"

Once again, we must watch for insane premises leading to meaningless questions. What is their second sentence actually saying? What are its assumptions? A central assumption is that technological change is primary—the independent variable—and definitions of sustainability are secondary, dependent on technological change. Yet I fail to see how technological changes alter the definition of what is sustainable: an activity is sustainable if it does not damage the capacity of the landbase to support its members. Technology does not affect the "parameters" of sustainability or its definitions in the short, medium, or long term. Technologies can hinder—or, depending on one's definition of *technology*, help"[18]—one's ability to live in a place over a long time, but they do not affect what the term *means*. Of course living in place for a long time is not what this contest is about, nor is it what this question is about. It seems very clear to me that the real purpose of the "question" is to guide writers into calling into question the baseline nature of sustainability, which really is the bottom line of survival. Sustainability is and must be the independent variable, and the proper question to ask—if you're interested in surviving—is how any given technology helps or hinders your way of living's sustainability, that is, your survivability, that is, your viability, which means how it helps or hinders the health of the landbase to which you belong.

Another question, more of the same: "If man's [*sic*] success [*sic*] as a species, in terms of population growth and knowledge, is a natural phenomenon, how can man [*sic*] be said to threaten nature? Is the line between artificial and natural itself artificial?"

I'm sure by now you can parse for yourself the (insane) assumptions of these questions, and where they guide us. For example, they use the word *men* to encompass all humans, ignoring women (which is, says someone with a penis, how things of course should be). They use the word *men*—implying by the rest of the question civilized men—to encompass all cultures, ignoring the indigenous (which is, says someone born in a city, how things of course should be). They define *success* not as living in place over time but as conquering all other cultures and conquering the planet (this misdefinition of success is an old one. I believe the formative command was: "Be fruitful, and multiply, and replenish the earth, and subdue it: and have dominion over the fish of the sea, and over the fowl of the air, and over every living thing that moveth upon the earth[219]). They use runaway population growth as an example of success, something that seems grotesque in a conversation ostensibly about sustainability. Their use of the word *knowledge* in this context is as interesting as their word *success*. By *knowledge*, do they mean genetic engineering, or do they mean the thousands of languages being driven to extinction by the dominant culture, and along with them the knowledge of how to live in long-term relationships with the places where those languages were born? (I think it's safe to say the former, because another one of their questions is: "How do we balance the distrust of genetic modification with the needs of developing country farmers and people?" which implies not only that that genetic modification primarily helps the poor and not transnational chemical and oil corporations, but that resistance to genetic engineering is based on "distrust"—read unsophistication and stupidity—and not on the understanding that genetic engineering is bad for these farmers and for their landbases.) Having defined themselves as all of humanity—a fine use of the classic abuser's trick of monopolizing perception—their use of the phrase "how can man [*sic*] be said to threaten nature" becomes not only an attempt to naturalize the atrocious (*It's in our nature to terrorize, rape, exploit, and kill you, then steal your resources. We really had no choice*) but worse, an *explicit* statement that what is happening is not: It is an invitation to write an essay showing that the natural world is not in fact threatened (and don't give me any shit about that not being the case. If we saw a phrase like this on a high school or college exam, we'd know *exactly* what we'd need to write if we wanted to get an *A*. Now just multiply that incentive by $20,000). Sure, the logic goes, sharks may be getting hammered, as are marlins, flounders, salmon, whales, blacktailed prairie dogs, tiger salamanders,[220] spotted owls, marbled murrelets, Port Orford cedar, tigers, chimpanzees, mountain gorillas, orangutans, but "if man's success as a species, in terms of population growth and knowledge, is a natural phenomenon, how can man be said to threaten nature?"

It's the same old statement posed by the scientist from the National Science Foundation when he denied any link between air guns and beached whales. And to be honest, I want to respond the same way: If my success as a person, in terms of having the ability to purchase a gun and the knowledge on how to find you, is a natural phenomenon, and if death itself is a natural phenomenon, how can I be said to threaten you?

It's all insane. It's precisely the sort of nonsense the CIA extolled in their torture handbook—sorry, human resource exploitation manual. If you babble long enough, you can break people, get them to go along with almost any program.

But we still have one more part of this question: "Is the line between artificial and natural itself artificial?" We've all heard this argument before, usually put forward by those who wish to further exploitation: humans are natural, therefore everything they create is natural. Chainsaws, nuclear bombs, capitalism, sex slavery, asphalt, cars, polluted streams, a devastated world, devastated psyches, all these are natural.

I have two responses to this. The first I explored already in *The Culture of Make Believe*, where I said, "This is, of course, nonsense. We are embedded in the natural world. We evolved as social creatures in this natural world. We require clean water to drink, or we die. We require clean air to breathe, or we die. We require food, or we die. We require love, affection, social contact in order to become our full selves. It is part of our evolutionary legacy as social creatures. Anything that helps us to understand all of this is natural: any ritual, artifact, process, action is natural to the degree that it reinforces our understanding of our embeddedness in the natural world, and any ritual, artifact, process, action is unnatural to the degree that it does not."[221]

My second response to their question is: Who cares? I want to live in a world that has wild salmon and tiger salamanders and tigers and healthy forests and vibrant human communities where mothers don't have dioxin in their breastmilk. If you really want to argue that oil tankers, global warming, DDT, the designated hitter rule, and the rest of the massive deathcamp we call civilization is natural, well, you can just go off in a corner with your $20,000 cheque and your utilitarian-philosopher buddies and play your bullshit linguistic games while the rest of us try to do something about the very real problems caused by civilization. If you want to seriously propose these waste-of-time questions,[222] I've got nothing to say to you. I've got work to do. I've got a world to help save, from people exactly like you. I've got a civilization to help bring down before it does any more damage.

❨ ❨ ❨

WHY CIVILIZATION IS KILLING THE WORLD, TAKE TEN. It's 2003 and I read in the newspaper that "Industrial fishing practices have decimated every one of the world's biggest and most economically important species of fish.... Fully 90 percent of each of the world's large ocean species, including cod, halibut, tuna, swordfish, and marlin, have disappeared from the world's oceans in recent decades.... [F]ishing has become so efficient that it typically takes just 15 years to remove 80 percent or more of any species unlucky enough to become the focus of a fleet's attention."[223] Although these three sentences by themselves starkly reveal how and why civilization is killing the world, neatly tying together economics, technology, and planetary murder, there are other things about the article and others like it that reveal even more about what we are up against.

The first is the placement of the article, on page A13 (and taking up about one-fourth of the page, with the rest devoted to an ad for the new PCS Vision™ Picture Phone with BUILT-IN Camera). This is a point I've made before: if the murder of the oceans doesn't deserve to rank as front page news, I don't know what does.

The next is that, somewhat contradicting the first, I'm not sure this is really news at all. I told several activist friends about the article, and most responded, "I thought we already knew this."

They're right. Anybody who doesn't understand that industrial fishing is killing the oceans is either an industry stooge, a politician, or a bureaucrat. Or maybe a moron. But I repeat myself.

Time and again scientists put out studies showing how the natural world is being killed, and time and again the culture keeps killing the planet. I can guarantee that in three or four years another study will come out saying that the oceans are being killed. This study will make a big splash on page A13 of many papers. Ho hum. Wanna hand me the sports section?

For example, about thirty seconds of searching the internet revealed articles from 1996 and 1999 detailing how industrial fishing—in each case the technique of long-line fishing where lines thirty or more miles long holding thousands of hooks are strung behind boats—are killing the oceans (including seabirds such as albatross, who are getting absolutely hammered). 1996, 1999, 2003. Let's wait for 2006.

The world is not being destroyed because of a lack of information: it's being destroyed because we don't stop those doing the destroying.

The third is the entirely predictable yet still horrifying response by industry representatives. Linda Candler, speaking for the trade group International Coalition of Fisheries Associations, revealed that my conflation of industry stooges and morons was not in fact a slur by saying, "Research shows fisheries are more productive when fished." She noted that "fish populations respond by reproducing more" when a new predator, in this case the exact same long-line techniques decried in 1996 and 1999, doesn't overdo it.[224]

She's right, of course. Think of your own body. When you bleed, you obviously produce more blood to replace that which is lost. Using her logic, the more you bleed, the more you produce: QED, bleeding is actually good for you. Putting her logic in context, if someone were to drain 90 percent of Ms. Candler's blood, making sure, of course, to not overdo it, her body would presumably go into hyperproduction, and she would be even healthier than before.

Defending the indefensible makes anyone who tries it absurd.

The fourth is the entirely predictable yet still horrifying response by those other industry representatives, those who work for the government. Michael Sissenwine, director of scientific programs with the National Marine Fisheries Service and head of fisheries sciences at the National Oceanic and Atmospheric Administration, revealed that my conflation of bureaucrats and morons was not in fact a slur either when he responded to the death of the oceans by saying, "We shouldn't . . . conclude that a substantial reduction is a problem,"[225] and, further, that the "expected outcome of fishing is that stocks will decline. Even with very efficient sustainability [sic] plans in place you have to expect declines, sometimes of 50 percent or more. The issue is how much of a decline is reasonable and sustainable."[226]

Read this last sentence again. My dictionary defines decline as to slope downward. I learned in grade school math that if a line slopes downward, it eventually reaches zero. If a line slopes downward by 90 percent over fifty years (even assuming the line to be linear, while in this case the decline becomes ever-steeper as civilization approaches its endgame), this means in less than ten years the line will cross zero. My dictionary defines sustainable as "using a resource [sic] so that the resource [sic] is not depleted or permanently damaged."

I must be stupid. I cannot for the life of me understand what Michael Sissenwine, who is in charge of the two largest federal bureaucracies ostensibly tasked with protecting ocean fish, is saying. He seems to be saying that declines are sustainable, that declines of 90 percent are sustainable. And reasonable. And not a problem.

But he can't be saying that. Nobody can be that stupid. Or that brazen. Not even someone whose job it is to oversee the systematic murder of the oceans.

In a mere twelve words he has rendered the words decline, reasonable, and sustainable meaningless. Add his first sentence and he has destroyed the word problem. If the death—the murder—of the oceans isn't a problem, what is? Not only are these people vacuuming oceans, they are killing discourse. Defending the indefensible makes anyone who tries it absurd.

Ninety percent of the large fish in the oceans are gone. Those making decisions concerning the fate of the remaining fish do not consider this a problem. What are you going to do about it?

❦ ❦ ❦

WHY CIVILIZATION IS KILLING THE WORLD, TAKE ELEVEN. Targeted stupidity.

The interconnectedness of the global economic system is taken for granted. Most people understand that a downturn in one sector of the economy can lead to problems in another. The collapse of the Asian economies in 1997, for example, harmed the timber industry in the northwestern and southeastern United States, as corporations that had exported to Asia lost their markets. Yet many of the same people who natter endlessly about this form of interdependence somehow seem to believe that you can cut down a forest, replant with one species, and still have a forest. They will stare at you stupidly—or more likely scoff at you— if you talk about how harming voles harms Douglas firs. They see no problem with wiping out species after species, and cannot seem to grasp that species need habitat, and that habitat need species.

It is not that these people cannot understand interconnectedness. It is that their stupidity is targeted.

❦ ❦ ❦

WHY CIVILIZATION IS KILLING THE WORLD, TAKE TWELVE. Auschwitz. Treblinka. Bergen-Belsen. That's the reason. No, not because civilization turns the entire world into a labor camp, then a death camp, although that is the case. No, not because the endpoint of civilization is assembly-line mass murder, although that, too, is the case.[227] Instead it's because of the doctors at Auschwitz.

Here's why. Do you remember when I talked about how environmentalism is an abysmal failure, and I gave a reason or two for our ineffectiveness? I left off what I think is the most important reason, and it has to do with those doctors.

In his extraordinarily important book *The Nazi Doctors*[228] Robert Jay Lifton explored how it was that men who had taken the Hippocratic oath could participate in prisons where inmates were worked to death or killed in assembly lines. He found that many of the doctors honestly cared for their charges, and did everything within their power—which means pathetically little—to make life better for the inmates. If an inmate got sick they might give the inmate an aspirin to lick. They might put the inmate to bed for a day or two (but not for too long or the inmate might be "selected" for murder). If the patient had a contagious disease, they might kill the patient to keep the disease from spreading. All of this made sense within the confines of Auschwitz. The doctors, once again, did everything they could to help the inmates, except for the most important thing of all: They never questioned the existence of Auschwitz itself. They never questioned working the inmates to death. They never questioned starving them to death. They never questioned imprisoning them. They never questioned torturing them. They never questioned the existence of a culture that would lead to these atrocities. They never questioned the logic that leads inevitably to the electrified fences, the gas chambers, the bullets in the brain.

We as environmentalists do the same. We work as hard as we can to protect the places we love, using the tools of the system the best that we can. Yet we do not do the most important thing of all: We do not question the existence of this death culture. We do not question the existence of an economic and social system that is working the world to death, that is starving it to death, that is imprisoning it, that is torturing it. We never question a culture that leads to these atrocities. We never question the logic that leads inevitably to clearcuts, murdered oceans, loss of topsoil, dammed rivers, poisoned aquifers.

And we certainly don't act to bring it down.

❆ ❆ ❆

Here's an example. I recently gave a talk at a gathering of environmentalists called *Bioneers*. The speeches I listened to were quite good, with people speaking passionately and often very positively about the changes that need to be made, and the changes that are already being made. They spoke of the need for different models for farming, different models for community organization, different models for schooling. But no one spoke of power. No one discussed the self-evident fact that those in power destroy sustainable communities. No one spoke of the fact that even if farmers develop different models for how to live on their land more sustainably, those in power may decide that the farmers' land is needed for a Wal-Mart or should be drowned behind a dam, and those in power will simply take their land. And no one spoke of psychopathology. No one spoke of the dominant culture's need to destroy. No one spoke of the dominant culture's implacable destruction of indigenous cultures.

Not only our actions but our discourse remains inside the confines of this concentration camp we call civilization.

☾ ☾ ☾

WHY CIVILIZATION IS KILLING THE WORLD, TAKE THIRTEEN. I recently shared a stage with a dogmatic pacifist, who said there are no circumstances under which the shedding of human blood is appropriate. "Violence schmiolence," he said. "I wouldn't kill a single human being to save an entire run of salmon."

"I would," I shot back.

But I wasn't happy with my response. Here is what I wish I would have said, "Thank you for so succinctly stating the problem—why civilization is killing the world—which is the belief that any single human life (mine or anyone's) is worth more than the health of the landbase, or even that humanity can be separated (physically, morally, or any other way) from the landbase. The health of the landbase is everything. A run of salmon is worth far more than my life, or any other individual human life. The continuation of the existence of the great ocean fishes is worth more than any individual human life. The continuation of albatrosses is worth more than any individual human life. The continuation of leatherback sea turtles, redwoods, spotted owls, clouded leopards, Kootenai River sturgeon, all these are worth more than any individual human life. If we do not understand that, we can never hope to survive."

That is what I wish I would have said.

☾ ☾ ☾

WHY CIVILIZATION IS KILLING THE WORLD, TAKE FOURTEEN. The United States is currently planning to build at least three new bioweapons laboratories dedicated to the creation of new classes of toxins, including genetically engineered toxins.

This is, from the perspective of those in power, a good thing. From the perspective of the rest of us, this isn't quite so good.

How will they use these "bioweapons," and to what purposes?

Their own language provides a hint. They wrote about bioweapons, among other things, in the document *Rebuilding America's Defenses* [*sic*] put out by *The Project for the New American Century*, which, according to their website, is "a non-profit, educational organization whose goal is to promote American global leadership."[229] In other words, it's a right-wing think tank which has as its goals U.S. domination of the world. Who cares, right? It's just a few lunatics, right?

Well, yes, it is just a few lunatics. Unfortunately the lunatics include vice-president Dick Cheney, Secretary of Defense [sic] Donald Rumsfeld, the president's brother Jeb Bush, and Paul Wolfowitz, generally considered the mastermind behind the invasion of Iraq.

You really should get a copy of *Rebuilding America's Defenses* [sic].[230] Just don't read it late at night. But if you do get a copy, take a look at page sixty, where the authors state that "advanced forms of biological warfare that can 'target' specific genotypes may transform biological warfare from the realm of terror to a politically useful tool."[231]

Pretty clear, no?

These are the people with their fingers on the buttons. This is why civilization is killing the world.

❦ ❦ ❦

WHY CIVILIZATION IS KILLING THE WORLD, TAKE FIFTEEN. The Unabomber/Tylenol rule of threat perception.

I think about this rule every time I stand in line at the post office, which is fairly often. I live in a small town, where everyone seems to know everyone, and where the postal clerks enjoy chatting with all of us: one of the clerks has a son named Darrick with the same birthday as mine, another has a bad back, one spent his early years in the Detroit/Windsor area and likes Charlie Musselwhite, and . . . you get the idea. You also perhaps start to understand why the line so often extends past the double doors and well into the main lobby. Why are we all standing here? The Unabomber/Tylenol rule of threat perception.

After the Unabomber sent bombs through the mail that killed three people and injured twenty-three more, the United States Postal Service responded by instituting regulations banning any package weighing more than a pound from being dropped into a mailbox, instead forcing patrons to stand in line before (eventually) handing a package to a postal clerk. The good news is that I enjoy the conversations.

Now to the Tylenol half of it. In 1982 seven people died after taking Tylenol that had been laced with cyanide. Johnson and Johnson, the corporation that makes Tylenol, immediately recalled 31 million bottles of the pain reliever, at a cost of $125 million, and within a month and a half had designed new tamper-evident containers. The entire industry followed suit, until today nearly all consumables are packaged in similar containers.

What do these have to do with civilization killing the planet? Contrast the response to the Unabomber/Tylenol killings with the fact that air pollution from this country's coal-fired power plants causes 24,000 premature deaths each year,[232] or with the fact that global warming already kills tens of thousands of humans per year, or with the fact that dangerous products kill 28,000 Americans per year, exposure to dangerous chemicals and other unsafe conditions in the workplace kills another 100,000, and workplace carcinogens cause 28 to 33 percent of all cancer deaths in this country.[233] Contrast the Unabomber/Tylenol responses with the response by the government to the 240,000 Americans who will die over the next thirty years from asbestos-related cancers, the 100,000 miners who have died from black lung, the one million infants worldwide who died just in 1986 because they were bottle-fed instead of breastfed.[234]

Threats to a comparatively small number of people were responded to almost immediately. The threats were removed. Why? Because the threats were aberrations and not systematic. The solutions did not point toward problems that inhere in the system itself. Had the problems inhered in the system itself, not only would the problems not have been solved, but almost no one would even have noticed.

《 《 《

In related news, during the years since the September 11 bombings, the FBI has "reduced by nearly 60% the number of agents assigned to white-collar crime, public corruption and related work,"[235] transferring these agents to terrorism investigations, despite the fact (or perhaps because of the fact) that corporate crimes cost orders of magnitude more—both in lives and in dollars—than either street crime or "terrorism."

《 《 《

Instead of the Unabomber/Tylenol rule, I could have called it the Fantasy Football rule, or maybe the Rotisserie League rule. The Earth Liberation Front and the Animal Liberation Front are considered by the FBI to be together the nation's number one domestic terrorist threat, even though they've never hurt anyone. The feds' rationale is that the ELF and ALF have caused significant financial loss to corporations. And it is true that some members of the ELF—elves—seem proud of the fact that the ELF has cost corporations and the government tens of millions of dollars through "economic sabotage." I hate to break it to both the elves and the G-men, but that's comparatively trivial compared to the real terrorists. I am of course describing those who play fantasy football and baseball. According to a scoop in today's *San Francisco Chronicle*, "America's addiction to fantasy sports could cost the nation's businesses $36.7 million daily"[236] as people who "should" be working are instead checking the internet to see how their favorite players fared (I'll bet you wish you'd picked up Johan Santana after his first few starts). If the FBI really cared about stopping serious economic sabotage, they would crack down immediately on websites that encourage such behavior. They would shut down rototimes.com, rotoworld.com, hardballtimes.com, and even ESPN.com. It's a travesty that such sites are allowed to operate openly, without harassment! They're encouraging terrorist behavior!

Maybe this means that if members of the ELF *really* want to cause economic damage to those in power, instead of burning SUVs they should just play fantasy baseball.

Or maybe not.

❆ ❆ ❆

Instead of the Unabomber/Tylenol rule, I could have called it the Terrorism rule. Although members of governments around the world and members of the capitalist press like to talk a lot about terrorism, the numbers aren't that high. Using their definitions of terrorism,[237] there have been about 1,300 people killed per year by terrorists since the September 11, 2001 attacks, and precisely zero in the United States. Contrast that with the numbers above. But the politicians talk incessantly about terrorism (or at least terrorism by enemies of states), and they do not talk about these other deaths. This is partly because of premise four of this book, and partly because of the Unabomber/Tylenol rule.

Think of that whenever you hear those in power mention the word *terrorism*.

❆ ❆ ❆

Abusers are volatile. They may be pleasant one moment, and violent the next. I go back and forth on whether I believe their volatility is real.

Argument in favor: Abusers are fragile. They're frightened. Because they have no identities of their own (which also means that they could never identify with their bodies nor with the landbases that give them life) they have no capacity to react fluidly to whatever circumstances arise. They must then control their surroundings. So long as those surroundings remain perfectly under control abusers can maintain at least an exterior calm. But threaten that control (or their perceived entitlement to control and exploit) and the fury that forever seethes beneath their surface bursts full-blown into the world.

Argument against: I strongly suspect, based on my own experience of abusers, that their volatility is at least quite often fabricated for manipulative purposes, making the volatility of abusers akin to the planned "outbursts" of CIA interrogators when victims refuse to fall into the trap of abusing themselves, refusing, for example, to stand for days at a time. In other words, the volatility may not be real at all, but part of a calculated strategy to keep victims off guard, to get them to police themselves.

But there's another argument for the fundamental falsity of an abuser's volatility, which refers instead to the first half of the statement: it is possible that an abuser's pleasantness is never real pleasantness, instead being a mere temporary (and probably tactical) lessening of the relentless tightening of attempted control. Instead of an abuser being like a jug of gasoline—noxious enough, but often not immediately fatal until and unless some spark sets it off, meaning ultimate responsibility for your own immolation rests on you for being silly enough to ever let flint strike steel—perhaps it's more accurate to say that to enter or to be forced to enter into a relationship with an abuser is more like being bound tightly by ropes tied by someone trained in the Japanese art of *hojojutsu*, about which one expert wrote: "Knots were developed that could hold almost anybody in any position. The knots were so designed that if a person tried to wiggle free the rope around the neck would tighten, restricting the airflow and choking the victim."[238]

This, for me, is the experience of being in a relationship with an abuser: if you do not struggle but only lie motionless, the abuser merely confines you, but every slightest movement in any direction on your part—and I want to emphasize *every* movement in *any* direction—tightens the abuser's hold over you.

Given all this, how real is the "pleasantness" of an abuser? Only very stupid or very desperate abusers—and this is as true on the larger social scale as it is on the familial—are *always* oppressive. Unrelenting oppression is not nearly so effective at control as is intermittent oppression mixed with rewards. If the oppressor were *only* oppressive, victims would realize they have nothing left to lose. Those who believe they have something left to lose are ever-so-much-more manipulable. Those who realize they have nothing left to lose have nothing left to fear, and they can be extremely dangerous to their victimizers.

I go back and forth on this question—is an abuser's volatility real?—on the cultural level, too, and for the same reasons. Certainly those in power have always hated the indigenous and have always reacted with rage toward those who threaten their perceived entitlement (as I put it in *The Culture of Make Believe*, "[I]f the rhetoric of superiority works to maintain the entitlement, hatred and direct physical force remain underground. But when that rhetoric begins to fail, force and hatred wait in the wings, ready to explode"[239]).

In addition to this hatred and rage that undergirds so many of the actions of those in power and the culture in general, I strongly suspect that much of the moral outrage and righteous indignation expressed by those in power before they invade yet another (probably defenseless) country containing resources they want or need, or before they punish those who try to stop their depredations, is so much playacting. I know, you're shocked—shocked!—at the implication that those in power may be sometimes less than honest about their true motivations and feelings. But it's pretty clearly true.

The question remains: are they then volatile, or do they just pretend to be volatile. Or both?

Not that any of this necessarily makes a difference in the real world. Whether those in power blow you up because they hate you for wanting to defend your landbase or because they want your resources doesn't much matter. You're just as dead.

But there *still* remains the second part of this question: is this culture's *niceness* real?

Here's why I'm belaboring this point: people who haven't thought about these issues at all—especially those who are aware of neither history nor current events, which means a hell of a lot of people—sometimes ask, if industrial civilization (or occasionally more specifically the U.S.) is so awful, why does everyone want to be "like us"? Well, the truth is, they generally don't, at least not until their landbase, and thus culture, has been destroyed. As J. Hector St. John de Crévecoeur commented in his *Letters from an American Farmer,* "There must be in the Indians' social bond something singularly captivating, and far superior to be boasted of among us; for thousands of Europeans are Indians, and we have no examples of even one of those Aborigines having from choice become Europeans! There must be something very bewitching in their manners, something very indelible and marked by the very hands of Nature. For, take a young Indian lad, give him the best education you possibly can, load him with your bounty, with presents, nay with riches, yet he would secretly long for his native woods, which you would imagine he must have long since forgot; and on the first opportunity he can possibly find, you will see him voluntarily leave behind all you have given him and return with inexpressible joy to lie on the mats of his fathers."[240] Here's how Benjamin Franklin put it: "No European who has tasted

Savage life can afterwards bear to live in our societies."[241] He also wrote, "When an Indian Child has been brought up among us, taught our language and habituated to our Customs, yet if he goes to see his relations and make one Indian Ramble with them, there is no persuading him ever to return, and that this is not natural [to them] merely as Indians, but as men, is plain from this, that when white persons of either sex have been taken prisoners young by the Indians, and lived a while among them, tho' ransomed by their Friends, and treated with all imaginable tenderness to prevail with them to stay among the English, yet in a Short time they become disgusted with our manner of life, and the care and pains that are necessary to support it, and take the first good Opportunity of escaping again into the Woods, from whence there is no reclaiming them."[242] These descriptions are common. Cadwallader Colden wrote in 1747 of whites captured by Indians, "No Arguments, no Intreaties, nor Tears of their Friends and Relations, could persuade many of them to leave their new Indian Friends and Acquaintance[s]; several of them that were by the Caressings of their Relations persuaded to come Home, in a little time grew tired of our Manner of living, and run away again to the Indians, and ended their Days with them. On the other Hand, Indian Children have been carefully educated among the English, cloathed and taught, yet, I think, there is not one Instance that any of these, after they had Liberty to go among their own People, and were come to Age, would remain with the English, but returned to their own Nations, and became as fond of the Indian Manner of Life as those that knew nothing of a civilized Manner of living."[243] At prisoner exchanges, Indians would run joyously back to their families, while white captives had to be bound hand and foot to not run back to their captors.[244]

The civilized who chose to stay among the Indians did so because, according to historian James Axtell, summarizing the stories of whites who wrote about their lives among Indians, "they found Indian life to possess a strong sense of community, abundant love, and uncommon integrity—values that the European colonists also honored, if less successfully. But Indian life was attractive for other values—for social equality, mobility, adventure, and, as two adult converts acknowledged, 'the most perfect freedom, the ease of living, [and] the absence of those cares and corroding solicitudes which so often prevail with us.'"[245]

Because Indian life was more enjoyable, pleasant, and non-abusive than life among the civilized, the conquistador Hernando de Soto had to place armed guards around his camps, not so much to keep Indians from attacking, but to keep European men and women from defecting to the Indians.[246] Likewise,

Pilgrim leaders made running away to join the Indians an offense punishable by death.[247] Other colonial rulers did the same. When, to provide one example among many, in 1612 some young Europeans in Virginia "did runne away unto the Indyans,"[248] the governor ordered them hunted down, tortured, and killed: "Some he apointed to be hanged Some burned Some to be broken upon wheles, others to be staked and some to be shott to deathe."[249] We can ask ourselves whether the governor was actually outraged and acting out his volatility, or whether he simply preferred that his subjects fear him, even if that meant they hate him. The reasoning was straightforward: "all theis extreme and crewell tortures he used and inflicted upon them to terrify the rests for Attempting the Lyke."[250]

When even this failed to stem the flood of desertions—and who can blame the deserting colonists?—the civilized saw no option but to slaughter the Indians and thus eliminate the possibility of escape. (The aforementioned governor, for example, in another case of runaway white folks, sent his commander and some troops "to take Revendge upon the Paspeheans and Chiconamians [Chickahominies]," Indians unfortunate enough to live closest to the whites. This "Revendge" consisted of going to where the Indians lived, killing about fifteen of them, capturing their "quene" and her children, and making sure to "cutt downe their Corne growing about the Towne." On the boat ride home, the soldiers of civilization "begin to murmur because the quene and her Children weare spared." Not wanting to upset his soldiers, the commander threw the children overboard before "shoteinge owtt their Braynes in the water." The Governor, displeased at the sparing of the "quene," ordered her burned at the stake. But the commander, "haveinge seen [sic] so mutche Bloodshedd that day," convinced his boss to let him merely stab her to death instead.[251]

The elimination of the possibility of escape has, of course, been from the beginning one of the central motivators for nearly all actions perpetrated by civilization.

So, given the choice between Christianity or death, capitalism or death, slavery or death, civilization or death, is it any wonder that at least some do not choose to die? I recently watched some old movie about Alcatraz, and Art Carney, playing the Birdman of Alcatraz, says something that goes to the heart of this: "The only thing worse than life in prison is no life at all."[252] We may as well face up—and fess up—to the prevailing logic: if we're stuck with a system that is based on rigid hierarchies, where those at the top systematically exploit those below—and this is as true on the personal and familial levels (wanna talk about rates of rape and child abuse?) as it is on the grand social level—a system that is killing the planet, that is toxifying our bodies, that is making us stupid and insane, that is eliminating all alternatives, we may as well have a nice car. If I can't live in a world with wild salmon and egalitarian social relations, and in a body free from civilization-induced diseases (choose your poison: mine is Crohn's disease), I may as well belly up to the bank and surround myself with as many luxuries as possible. If I'm going to be encased in an 880-by-90-foot steel-walled luxury prison called the *Titanic*, and that prison will soon become my icy tomb, it's better, I suppose, in the meantime to be riding first class than to be scrubbing the toilets of "my betters."

My point, however, is that these goodies that make up the bulk of the system's "pleasantness" are entirely conditional on your subservience to those above you on the hierarchy. What happens to you if you act on a disbelief in the property rights of the rich? What happens if you act on a belief that police (and more broadly the state, and more broadly still those at the top of the hierarchy) do not have a monopoly on violence, and that violence perpetrated by those in power may (and sometimes will) be met by violence perpetrated by those considered to have no power at all? What happens if you act on a disbelief that those in power have the right to toxify the planet? What happens when you become convinced that violence from the powerless cannot be disallowed given the magnitude and relentlessness of the violence of the powerful?

You are, in a word, dead.

BRINGING DOWN CIVILIZATION, PART I

It IS possible to get out of a trap. However, in order to break out of a prison, one first must confess to *being in a prison. The trap is man's emotional structure, his character structure.* There is little use in devising systems of thought about the nature of the trap if the only thing to do in order to get out of the trap is to know the trap and to find the exit. Everything else is utterly useless: Singing hymns about the suffering in the trap, as the enslaved Negro does; or making poems about the beauty of freedom *outside* of the trap, dreamed of *within* the trap; or promising a life outside the trap after death, as Catholicism promises its congregations; or confessing a *semper ignorabimus* as do the resigned philosophers; or building a philosophic system around the despair of life within the trap, as did Schopenhauer; or dreaming up a superman who would be so much different from the man in the trap, as Nietzsche did, until, trapped in a lunatic asylum, he wrote, finally, the full truth about himself—too late. . . .

The first thing to do is to find the exit out of the trap.

The nature of the trap has no interest whatsoever beyond this one crucial point: WHERE IS THE EXIT OUT OF THE TRAP?

One can decorate a trap to make life more comfortable in it. This is done by the Michelangelos and the Shakespeares and the Goethes. One can invent makeshift contraptions to secure longer life in the trap. This is done by the great scientists and physicians, the Meyers and the Pasteurs and the Flemings. One can devise great art in healing broken bones when one falls into the trap.

The crucial point still is and remains: to find the exit out of the trap. WHERE IS THE EXIT INTO THE ENDLESS OPEN SPACE?

The exit remains hidden. It is the greatest riddle of all. The most ridiculous as well as tragic thing is this:

THE EXIT IS CLEARLY VISIBLE TO ALL TRAPPED IN THE HOLE. YET NOBODY SEEMS TO SEE IT. EVERYBODY KNOWS WHERE THE EXIT IS. YET NOBODY SEEMS TO MAKE A MOVE TOWARD IT. MORE: WHOEVER MOVES TOWARD THE EXIT, OR WHOEVER POINTS TOWARD IT IS DECLARED CRAZY OR A CRIMINAL OR A SINNER TO BURN IN HELL.

It turns out that the trouble is not with the trap or even with finding the exit. The trouble is WITHIN THE TRAPPED ONES.

All this is, seen from outside the trap, incomprehensible to a simple mind. It is even somehow insane. *Why don't they see and move toward the clearly visible exit?* As soon as they get close to the exit they start screaming and run away from it. As soon as anyone among them tries to get out, they kill him. Only a very few slip out of the trap in the dark night when everybody is asleep.

Wilhelm Reich[253]

OFTEN WHEN I MENTION AT TALKS THAT I'M WRITING A BOOK ABOUT bringing down civilization, people interrupt me with cheers. They shout, "Hurry up and finish," or "Sign me up" (the exception to this, for reasons that escape me, is New England, where people are more likely to stroke their chins, furrow their brows, and murmur, "What a strange and interesting idea"). Indeed, at one talk in Kansas someone introduced me by saying, "We brought Derrick here because he's got the balls to say we need to take down civilization." Presumably were I a woman he would have said *ovaries.* Hundreds of people show up, and we talk into the wee hours about the whys and hows of bringing it down.

Yet not everyone is happy. Recently, for example, an attorney volunteered to be on my legal team when I get arrested under the Patriot Act.

"That's nice," my mom said when I told her, "But the Feds have bigger things to worry about."

"Like what?" I responded, somewhat hurt.

"Like making up excuses to lock up poor brown people."

"Good point."

I got compared to Hitler once simply because I suggested that someday the population will be smaller than it is now. I told the woman—who also said, "You seemed like such a nice man until you opened your mouth"—that I failed to see how bringing together a very simple ecological understanding with an intense opposition to genocide and the centralization of power could put me in the same camp as one of civilization's sterling examples.

Then a few days ago I hit the trifecta. Someone—a dogmatic pacifist, not that you asked—compared me in one breath to Stalin, Mao, and Pol Pot. She was a bit fuzzy on the first two—especially considering each killed tens of millions of people to industrialize their economies—but her reasoning on Pol Pot was that he wanted to deindustrialize, and so do I, ipso facto, I must be for genocide, mass murder, and the killing of anyone who wears eyeglasses. I didn't say much in response, in great measure because she had the bit between her teeth, and nothing I could have said would have made the slightest difference. Had she stopped to take a breath, however, here is what I would have said to her, "All morality is particular. Everything is particular. Taking down civilization is not a monolithic act, as if I could snap my fingers and suddenly the lazy-boy recliners and ergonomic computer chairs would disappear, leaving so many millions of people hanging surprised in the air for one long instant before they fall to the soil that still lives beneath their recycled carpet, floorboards, and the concrete of their suddenly disappeared foundations."

Bringing down civilization first and foremost consists of liberating ourselves by driving the colonizers out of our own hearts and minds: seeing civilization for what it is, seeing those in power for who and what they are, and seeing power for what it is. Bringing down civilization then consists of actions arising from that liberation, not allowing those in power to predetermine the ways we oppose them, instead living with and by—and using—the tools and rules of those in power only when we choose, and not using them only when we choose not to. It means fighting them on our terms when we choose, and on their terms when we choose, when it is convenient *and effective* to do so. Think of that the next time you vote, get a permit for a demonstration, enter a courtroom, file a timber sale appeal, and so on. That's not to say we shouldn't use these tactics, but we should always remember who makes the rules, and we should strive to determine what "rules of engagement" will shift the advantage to our side.

Bringing down civilization is not about being morally pure—morality defined, of course, according to those in power—but instead it is about defending our own lives and the health and lives of our landbases.

Bringing down civilization is millions of different actions performed by millions of different people in millions of different places in millions of different circumstances. It is everything from bearing witness to beauty to bearing witness to suffering to bearing witness to joy. It is everything from comforting battered women to confronting politicians and CEOs. It is everything from filing lawsuits to blowing up dams. It is everything from growing one's own food to liberating animals in factory farms to destroying genetically engineered crops and physically stopping those who perpetrate genetic engineering. It is everything from setting aside land so it can recover to physically driving deforesters out of forests and off-road-vehicle drivers (and manufacturers and especially those who run the corporations) off the planet. It is destroying the capacity of those in power to exploit those around them. In some circumstances this involves education. In some circumstances this involves undercutting their physical power, for example by destroying physical infrastructures through which they maintain their power. In some circumstances it involves assassination: At a talk someone asked me what, given the opportunity, I would have said to Hitler, and I immediately responded, "Bang, you're dead." She then asked what, given the opportunity, I would say to George W. Bush . . .

All morality is particular, which means that what may be moral in one circumstance may be immoral in another. And the morality of any action must be put into the context of a system—civilization—that is killing or immiserating literally billions of human beings, killing our collective future, killing our particular landbases, killing the planet. In other words, our perception of the morality of every particular act must be informed by the certainty that to fail to *effectively* act to stop the grotesque and ultimately absolute violence of civilization is by far the most immoral path any of us can choose. We are, after all, talking about the killing of the planet.

Just last night I shared a stage with Ward Churchill, a Creek/Cherokee/Métis Indian, and author of more than twenty books (I asked how many, and he laughed and then said it's a bad sign when he no longer remembers the precise number). Ward is known for his militancy, as you can probably guess from some of his titles—*Struggle for the Land: Indigenous Resistance to Genocide, Ecocide, and Expropriation in Contemporary North America*, and *Pacifism as Pathology: Reflections on the Role of Armed Struggle in North America* come to mind—and he's known as well for his clarity of thought and expression on issues of resistance. So it came as no surprise when he said onstage, "What I want is for civilization to stop killing my people's children. If that can be accomplished peacefully, I will be glad. If signing a petition will get those in power to stop killing Indian children, I will put my name at the top of the list. If marching in a protest will do it, I'll walk as far as you want. If holding a candle will do it, I'll hold two. If singing protest songs will do it, I'll sing whatever songs you want me to sing. If living simply will do it, I will live extremely simply. If voting will do it, I'll vote. But all of those things are allowed by those in power, and none of those things will ever stop those in power from killing Indian children. They never have, and they never will. Given that my people's children are being killed, you have no grounds to complain about whatever means I use to protect the lives of my people's children. And I will do whatever it takes."

The crowd gave him a standing ovation.

Let's just hope they convert his words into actions.

❨ ❨ ❨

I think it would be virtually impossible for even the most dogmatic pacifist to make a moral argument against immediately taking down every cell phone

tower in the world. Cell phones are, of course, annoying as hell. That might be
a good enough reason to take down the towers, but there are even better reasons. There is of course the very real possibility that tower transmissions cause
cancer and other health problems to humans and nonhumans alike. Even
ignoring this, however, there's the fact that towers—cell phone, radio, and television—act as mass killing machines for migratory songbirds: 5 to 50 million
per year.[254] These birds die so the jerk at the table next to you can yammer at
full volume (of course) about his latest financial conquest (thank god this time
the conquest isn't sexual or you might soon be arraigned for murder). Now, I'm
sure some hypothetical pacifist could assemble some hypothetical situation
where cell phones save lives. For example, a woman is alone on a dark country
road. Her car breaks down. She dials 911, then turns on the radio to pass the time
while she waits for a cop to show up. She hears a report of a homicidal madman who escaped from a local prison (he was in prison because budgets for
mental hospitals were gutted during the Reagan era, and no, silly, she doesn't
hear that on the radio: radio stations are owned by large corporations, and
would *never* provide useful political analysis). He—the madman, not Reagan—likes to kill women on dark lonely roads (Reagan preferred killing poor
brown people, and those at a distance, and no, that analysis doesn't come from
the radio either). He has only one hand, the other being a hook he uses for
awful purposes at which the radio only hints. She shivers. Finally the cop
arrives, approaches her driver's side window. She checks his hands before she
rolls it down just a fraction. She pops her hood from inside, he fiddles a
moment, the car miraculously starts. He drops the hood, gets back in his car.
She drives away, feeling a slight tug on her car as she does. When she gets home
she looks at the passenger side, and finds, of course, a bloody hook stuck to the
handle of her car door. Saved by a cell phone!

I recognize that we can construct much less fabulous cases: almost a third of
911 calls (almost 50 percent in big cities) come from cell phones.[255]

My point, however, is that we can just as easily construct hypothetical situations that will keep us from doing *anything*. The same woman, for example,
driving alone down a dark country road, picks up her cell phone to call her
dear elderly mother. Her mother shuffles to answer, falls down the stairs, and
breaks her neck, but is able to grab the phone and gasp, "Dial 911." Her daughter picks up her second cell phone (you do have multiple cell phones, don't
you?), begins to dial, and because she's not paying attention to her driving,
plows into three orphan waifs huddling for warmth, security, and comfort by
the side of the road, leaving them all paralyzed from the neck down. (Because

they have no health insurance, and because politicians steadfastly refuse to put in place universal health coverage, they all soon die). Her car hurtles across a ditch, wiping out the last population of a highly endangered salamander, then smashes into a tree. She hears her mother's dying gasps, and as she loses consciousness she sees a hook shining in the moonlight outside her passenger window. The madman, by the way, did not have a thing for children or salamanders, meaning that they had previously been safe.

Are cell phones beneficial to human and nonhuman life? What are the effects of cell phones on the landbase?

We'd have an even harder time rationalizing our inaction in allowing television and radio towers to exist (and I hope you're not going to suggest it would be immoral to take out television towers, that migratory songbirds should die so we can watch *The Best Damn Sports Show, Period* on Fox Sports Net).

To the direct killing of birds we can add as a cost of cell phones the effect of speeded-up business communications, which decreases the quality of individual lives in a culture addicted to speed ("People who work for me should have phones in their bathrooms," said the CEO of one American corporation[256]), and which decreases the ability of the natural world to sustain itself (the activities of the economic system are killing the planet: the higher the GNP, the more quickly the living are converted to the dead).

☾ ☾ ☾

The question becomes, how do you take out a cell phone tower?

I need to say up front that I'm a total novice at this sort of thing. I am, to slip into the language of the mean streets, a goody two-shoes. My whole life I've rarely done anything illegal, not out of an equation on my part of morality and obedience (or subservience) to laws—at least I hope not—but instead partly because many illegal activities such as using illegal drugs repulse or scare me while others such as insider trading simply do not hold my interest. Even with those that do hold my interest—e.g., taking out dams, hacking, destroying (or otherwise liberating) corporate property—I'm not only almost completely ignorant of how to do it but fairly nervous about getting caught. Don't get me wrong: I've raised a little hell in my time. Sometimes I go crazy and turn right on red without coming to a complete stop, and I routinely drive four or sometimes even nine miles over the speed limit. A few anarchist friends were trying to set up a talk where I'd share the stage with a couple of former Black Panthers. One of them did time for robbing a bank, the other for hijacking a plane. I thought a moment, then confessed, "I once shoplifted dog food from Wal-Mart." High fives were exchanged around the table.

I have to add that were I more attracted to illegal activities I would probably curtail them because of what I write. I presume, my mom's reality checks notwithstanding, that I've drawn at least a little attention from the powers-that-be, and the last thing I want to do is give them an excuse to pop me for something non-political (and frankly I'm not too keen on getting popped for something political either). If they want to come after me because of what I write, I'll take them on, and if someday I have the courage to quit writing and take out dams (note the plural, *dams*: I don't agree with the Plowshares tactic of turning yourself in if you destroy property belonging to the occupiers), they can try to catch me. But in the meantime, I'm not going to give them any cheap opportunities.

All of which is to say I'm a coward. I'm going to write about how I would take down a cell phone tower here in town, but I'm not going to do it. If I were going to do it, I wouldn't be so stupid as to write about it, or even talk about it with anyone I didn't know and trust literally with my life. And all of *that* is to say that you FBI agents reading this book (and the ones tracking my strokes on my keyboard) can go ahead and lose your erections. This book isn't a confession. And even if your CIA buddies decide to play smackyface with me there isn't much I can confess (unless you count the survey stakes I've removed, but I've already written about that, and besides, removing survey stakes is a fundamental human duty).

Recon is always the first step in any military action, so I drive my mom's car to the cell phone tower behind Safeway. I take her car not out of some fiendishly clever plot to make it so that if anything happens she'll get sent up the river instead of me, but because my car has been sitting on blocks in her driveway for more than a year now (I never knew, by the way, that moss can grow along the weather stripping around the rear window).

There are two towers I know of in Crescent City. There's the one behind Safeway, and another off in the woods a quarter mile north. The one closest to the grocery store is in the open, which would obviously make taking it down more problematic. The tower is enclosed in a chain-link fence topped by barbed wire. The two sides of this fence farthest from Safeway face thick woods, which would provide cover. I'm certain the fence could be cut easily and quickly.

The problem is that I wouldn't know what to do next. There are a couple of sheds inside, and I'd imagine that some gasoline and matches could render the whole thing inoperable. That may be great for (temporarily) stopping the guy at the restaurant from bothering his neighbors, and would slow the destructive march of the economic system, if only ever so slightly, but it wouldn't do a damn thing for the birds. Unfortunately, the tower itself is probably three feet in diameter, hollow with a two-inch shell of some sort of metal.

I sit in my car and look at it. I'm nervous, as though even *thinking* about how I would do this is enough to draw cops to me. (The same is true now as I write this.) Of course if I *were* going to bring this down I would never have driven here for reconnaissance. At least not during the middle of the afternoon. I would have parked far away and walked. And there's no way I would have done it in this town, either. Crescent City is too small and I'm too well known. For crying out loud, at the (excellent) Thai restaurant two blocks south of this tower they know me well enough to always bring me a huge glass of water without me asking, and they like me well enough to pack my salad rolls full to bursting (of course after they read this book my future salad rolls may be limp and wrinkly). I'm almost surprised no one has stopped by while I'm sitting in this car, just to say hi and pass the time of day.

I don't know what to do. I'm a writer. I wouldn't know how to take down this tower any better than I would know how to write a computer virus, or how to perform brain or heart surgery. Worse, I'm spatially and mechanically inept—probably a couple of standard deviations below the norm—with a heavy dose of absent mindedness thrown in for good measure (and it seems that absent mindedness would be a tremendous curse to *any*one contemplating anything deemed illegal by those in power).

An example of the spatial ineptitude: whenever I pack for a road trip, my mom always takes a look at my suitcase, sighs, and repacks everything in about half the space.

An unfortunate experience in eighth-grade woodshop class highlights the mechanical problems. For our final project, we got to build whatever we wanted. I chose a birdhouse. I was excited. From close observation I knew the birds in our area (though I no longer live in a region with meadowlarks, recorded versions of their songs still make me smile), and from reading books I knew their habits and preferences. In some cases I knew their Latin names. I cut each piece of wood as meticulously as I could, nailed them together as tightly as they would go (admittedly there were a fair number of gaps where my cuts hadn't *quite* been straight), then put putty in the nail holes. I stained it all (an irregular) dark brown. On the final day of class we each brought our projects to the front, one at a time. The other pieces looked pretty good and I got increasingly nervous as my turn approached. For good reason. When I held up my birdhouse, the entire class burst into laughter. One of them—I still remember your name, David

Flagg, and you're still not on my short list of people to invite over to dinner—pointed at the lumps of still-white putty and shouted, "It looks like the birds have already been on it." Even the teacher laughed so hard he had to remove his shop glasses and wipe his eyes.

The infamous shower curtain episode makes clear my absent-mindedness. My shower curtain was hanging too far into the tub. It floated when I showered, and I often stepped on or even tripped over it. After about a year of this I decided to fix it and cut off the bottom of the shower curtain. Only later did I remember that the bar (which I had purchased and installed) was spring-loaded, and it was a simple matter to just raise it a few inches.

The point is that when it comes time for us to start taking out dams, I'm not sure I'm the one you want holding the explosives.

That said, here's what I'm thinking as I look at the cell phone tower. Basic principles. There are, I'd think, maybe six major ways to take down anything that's standing. You can dismantle it. You can cut it down. You can pull it down. You can blow it up. You can undermine it until it collapses. You can remove its supports and let it fall down on its own. This is all as true for civilization as it is for cell phone towers.

In the (smaller) case before us, I think we can out of hand dismiss dismantling and digging. So far as the former, the tower is constructed of two or three huge pieces, and is obviously not a candidate for dismantling. And the big parking lot (as well as presumably deep footings) would certainly eliminate digging.

Pulling it down can be dismissed just as easily, unless you've got some big earthmoving equipment and a hefty cable to attach fairly high up on the tower. I don't think my mom's car has the horsepower to move it (and I know mine sure as hell doesn't). I keep picturing that scene from *The Gods Must Be Crazy* where they attach one end of a cable to a tree and the other to a jeep, and end up winching their vehicle into the air. *Oh, hello, officer. What am I doing up here? That's a very good question. My cell phone reception has been really crappy lately, and I thought I'd get better reception if I got closer to the antenna. And say, would you mind helping me down?*

Cutting would probably work, so long as we're clear that we're not talking about hacksaws. In that case I may as well ask my friends the *aplodontia* to come gnaw it down. This tower is *big*. A grinder wouldn't work either in this case. There are lots of cell phone and other towers out in the mountains, and so long as you had lookouts, grinders might work out there, but that much noise here in town seems contraindicated. *Oh, hello, officer. What am I doing here? That's a very good question.* . . . But an acetylene torch might do the trick, although once again here in town there's a good chance it would draw some attention. And so far as me doing it, I *have* used acetylene torches, but you don't *even* want to hear about my experiences in metal shop class (and yes, David, I still remember you from there, too).

Explosives would have the advantage of rendering moot whether anyone notices, because timers are easy enough to make that even I could use them. By the time the tower comes down I could easily be in another state (not quite so dramatic as it sounds since I live about twenty minutes from the border). Additionally, in this case explosives would be safe. Although I've been saying that this tower is "behind Safeway," it's *way* behind Safeway, in an old abandoned parking lot. The problem, once again, is that I know nothing about explosives. I was certainly a nerd in high school, college, and beyond, but evidently the wrong kind of nerd for the task at hand. While the science geeks were busy seeing what bizarre ways they could combine chemicals to blow things up and dropping M-80s down toilets in (usually unsuccessful) attempts to get school cancelled (though, being geeks, I was never quite sure why they wanted to cancel school), my friends and I were reading books and playing Dungeons & Dragons (and a hell of a lot of good that does me now: if only a +3 Dwarven War Hammer could bring down civilization, I'd be in great shape).

Ah, the pity of a misspent youth.

This all makes me wish I would have joined the Navy Seals and learned how to blow things up (I probably would have learned how to kill people too: strange, isn't it, how when the system's soldiers are taught to kill, that's banal — the final night at boot camp drill instructors sometimes christen their students' new lives by saying, "You are now trained killers"[257]—but when someone who opposes the system even *mentions* the *k* word, it's met with shock, horror, the fetishization of potential future victims, and the full power of the state manifesting as those who've been trained to kill in support of the centralization of power). Or better, it makes me wish I had a friend who was a Navy Seal and who shared my politics.

This brings us to removing the tower's supports and letting it fall on its own. That may be the easiest, and something even I could handle. The other tower, in the woods to the north, has about twenty guy wires. Everything I've read suggests these wires are even more deadly to birds than the towers themselves. Some places you can pick up dead birds by the handful beneath the wires. Their necks are broken, skulls cracked, wings torn, beaks mangled. But I also know what happens when high-tension wires are severed: those opposed to their own decapitation ought to be far away.

But there's good news in all of this. There are giant bolts surrounding the base of the tower behind Safeway. I'd imagine they're very tight, but for one of the few times in my life my physics degree might come in handy. Of course you don't really need a physics degree to understand that if you want to unscrew a tight bolt all you need is a long lever arm on your wrench. Just as Archimedes said, "Give me a long enough lever and a place to stand and I can move the world," I'll go on record as saying that if you give me a long enough lever arm I can unscrew any bolt in the world—oh, okay, maybe just a lot of bolts that are pretty damn tight. So a huge pipefitters wrench with a long metal pipe over the end to extend your lever arm might be enough to get you the torque you'd need to loosen the base (failing that, you could always cut the bolts instead of the tower itself: remember, always attack the weakest point!). Then walk away and wait for the next windstorm to do the trick.

Emboldened by the realization that this just might be doable, I make my way through the dense forest to the northern tower. I quickly find a path, which opens into a large meadow. The only problem is that this is the wrong meadow: no tower. So it's back into the woods, this time on a game trail. Note that I didn't say *big* game. Sometimes I crawl on my belly. I cross a mucky streambed and see prints of (very small) deer. Often I stop to pull Himalayan blackberry thorns from my shirt. A few times from my arms, hands, fingers, face. I realize that somehow a thorn has lodged in my heavy denim pants at the—how do I say this delicately?—very top of the inseam. With every step it scrapes against my, well, let's just say *extremely* high on my thigh. Finally the path opens out again, and I'm there.

The first thing I do is thank the gods for making turnbuckles (actually that's the second thing I do after taking the thorn out in my pants). Loosening the wires, and even undoing them, would be simplicity itself. There's a lot of them, but security would be no problem here: forest surrounds this tower on all sides. Even the tower itself could be easily attacked: it's made of a spindly grid of metal tubing. I could cut through the thing in an hour or two with a hacksaw. Someone with a torch could do it in minutes.

All this talk of taking down towers makes me wish I was a farmer, not only because the farmers I've known have generally been crackerjack mechanics—I was a farmer (commercial beekeeper) in my twenties, and learned to my dismay that most farmers spend far more time with machines than animals—but also because back in the 1970s a group of farmers called the Bolt Weevils were pioneers in the art and science of taking down towers. They specialized in towers with high-tension electrical wires.

It all started when the United Power Association and the Cooperative [*sic*] Power Association decided to put a 400 mile transmission line across Minnesota farmland between coal-fired generating stations in North Dakota and the industry and homes of the Twin Cities.[258] As always, the poor would be screwed so the rich could benefit. First, as with water, most of this electricity would not be used to benefit human beings, but industry. Second, the utility corporations chose to put the power lines across lands belonging to politically powerless family farmers rather than across huge corporate farms with political clout.

One of the farmers, Virgil Fuchs, became aware of the plan, and went door-to-door informing his neighbors. He was just in time: representatives from the utility corporations were right behind him trying to get farmers to sign easements. After Virgil's warning, not one farmer signed.

What follows is a story we've heard too many times, of local resistance overwhelmed by distant power, of politicians and bureaucrats who go out of their way to feign community interest while going just as far out of their way to stab these communities in the back. In essence, it's the story of civilization: of human beings and communities harmed so cities and all they represent may grow.

Local townships passed resolutions disallowing the power lines, and county boards refused permits for construction. The response by the corporations was to ignore local concerns and turn to the state for help. The farmers also turned to the state for help, speaking to their purported representatives. The response by the state government's Environmental Quality Council was predictable: public hearings were held, people voiced their opinions, and after discovering that opinions ran overwhelmingly against the power lines, the state doctored the transcripts of the meetings (dropping out unfavorable testimony), then went ahead and granted the permits. One county sued, but the case was dismissed.

Government representatives promised they would at least let farmers know when construction would begin, but they lied. Suddenly one day surveyors showed up in Virgil Fuchs' fields.

Here is why in many ways I respect at least some family farmers more than most environmentalists: Fuchs fought back. He drove his tractor over the surveyors' equipment, and rammed their pickup truck.

It must be said, however, that Fuchs was in some ways risking less by doing this than if he had committed the same actions as an environmentalist. He was sentenced to community service, and eventually even the record of his arrest and conviction was expunged. You and I both know that any environmentalist who did this to equipment belonging to any extractive corporation would probably get charged with attempted murder and receive at least fifty years in prison: remember that environmental activist Jeffrey Luers is serving more than twenty-two years for torching three SUVs in the middle of the night when no one was around, and three environmentalists face up to eighty years for allegedly torching an unoccupied logging truck. Similarly, when gun-wielding farmers in the Klamath Valley stood off sheriffs and sabotaged public dams to force water to be diverted away from salmon and toward their (publicly subsidized) potato farms, sheriffs joined the fun and no one was arrested, let alone indicted, let alone prosecuted, let alone sent to prison, let alone shot. And they got the water. If you or I re-sabotage those dams to keep water for salmon (water for fish: what a quaint notion!), and we pull guns on sheriffs as we're doing so, we, too, wouldn't go to prison: we would go to the cemetery.

Farmers began gathering at Fuchs's farm and at others across several counties. They fought the surveyors wherever and however they could. They'd suddenly, for example, gain permission from the county to dig a ditch across a road (to prevent vehicles from driving across them) for this reason or that. One farmer stood next to the surveyors and ran his chainsaw so the workers couldn't communicate.

Local sheriffs did the right thing, or at least didn't do the wrong thing. One said, "As sheriff of this county, I became involved when the landowners and other concerned citizens objected to trespasses of their property [by the power companies]. In the meantime the power companies expect my department to use unlimited force, if necessary, to accomplish their survey and ultimately the routing of the power line. In my opinion this is a situation that began with the Environmental Quality Council, at the request of the power companies, and that's where the problem should be remanded for resolution. I will not point a gun at either the farmer or a surveyor. To point a gun is to be prepared to shoot, and this situation certainly does not justify either. It does justify a review of the conditions that bring about such citizen resistance."

Where is this sheriff when environmentalists need him? Would that sheriffs would always defend local humans against distant corporations, or at the very least not enforce the ends of these corporations through violence.

The governor also refused to intervene. That's where things stood when a new governor took office that winter. Things looked good for the farmers: the new governor considered himself a populist. As one farmer said, "He thought of himself as representative of the people, with a capital P, not of the bureaucracy or the bigwigs or the business people, and so he had, I think, a great hope and belief that he could get people together and solve the problems."

But when politicians present themselves as representatives of regular people it's time to start packing (either your luggage so you can flee, or a pistol, so you can, well, you know . . . You choose which).

The governor took to slipping off in secret to visit farmers at their homes. He told them he sympathized, and said, "You really got stuck in this case."

Philip Martin, head of United Power Association, sympathized too. He'd grown up on a farm, and he even knew and loved Virgil's mother—"She reminded me somewhat of my own mother," he said—but as from the beginning of civilization the demands of this deathly economic system trumped all human cares, feelings, and needs. Demand for electricity was growing by 10 percent per year, construction of the lines had already begun, and the clock was ticking on interest on a $900 million federal loan. The logic was, "I may love my mother, but if the economic system—and more broadly civilization—demands it (or hell, even hints at it) I'll screw her over and leave her for dead."

Martin was clear on the source and solution of the problem: "We built all the way across North Dakota and we had one person protesting it. That was solved when the law enforcement—he did some damage—and the law enforcement there initiated the action to put him in prison, or jail. And pretty soon he said, 'I'll be a good boy, I won't do anything more,' and they let him out, and we built a transmission line. We didn't have any problem in North Dakota."

But, he continued, in Minnesota, "The law enforcement refused to enforce their own laws. We would go out and try to survey, and they would simply pull up all our stakes, they would destroy everything we had out there.[259] And there was never anything done. President Norberg, who was president of the cooperative, and I were out there at many meetings. I drove a car with an escort in front of it and back of it with guns going off, sticking out the windows."

The farmers said the transmission lines would come in over their dead bodies. They filed more lawsuits, which went to the Minnesota Supreme Court. The Supreme Court decided against them. This journey through the courts radicalized many of the farmers, who up to that point had believed in the system. One farmer stated: "I had the feeling that it was all decided. The courts weren't acting as courts at all, they were just a front. And it was just a terrible, terrible shock to me. I thought, gee, this can't be."[260]

That November, construction started in western Minnesota. When farmers protested, the corporations filed $500,000 lawsuits against them.

The farmers found allies, from former Vietnam War protesters to Quakers to musicians. The corporations, of course, already had allies in the court system, and now the governor, and through him police with guns. For all of his rhetoric, when push came to shove, the governor, as representative of the state's economic system, shoved the power lines down the farmer's throats. He said, "You know, this is a nation of laws. And there are a lot of things that I don't like, you know, and I'm sure there's many things that you don't like, but there's a process that we can work, it's a process that's open. It's a process that people in November go and they make that mark on that ballot." Let me translate: "It does not matter whether this or any other particular law or action is good for humans or the landbase. It does not matter whether you like what happens to your landbase, to your children, or to you. It does not matter whether I like it. It does not matter if the laws were designed by and for the rich, and the same is true for the courts and law enforcement. It does not matter if we lie to you and put you through processes of sham public participation. Your participation in processes that affect your life, the lives of your children, and your landbase begins and ends with a checkmark on a ballot in a meaningless election. The only thing that matters is the growth of the economic system. If you don't like it, we will send in people with guns to put down resistance."

Farmers broke up construction sites and corporate representatives said construction would not continue without police protection. The governor sent in state troopers, with up to ten cars and twenty cops protecting individual dump trucks.

The state legislature considered a moratorium on construction until further health studies could be performed. It was already known that electrical lines can lower conception rates and milk production in dairy cows. And the state's own guidelines warned farmers against refueling their vehicles under the transmission lines, and warned school bus drivers against picking up or discharging children under them.

Across the state, people overwhelmingly favored the farmers over the utility corporations. But, as a corporate attorney argued, "The critical question for you as legislators is, is this a government of law, or of men?"

Think for a moment about that question, and think about its implications.

The legislators thought about it long enough to kill the moratorium.

By now the cops (who may have sympathized, but who were too enthralled to the machinery of civilization to follow their human hearts) were behind the power lines one hundred percent. They told farmers they couldn't assemble, couldn't drive county roads, couldn't stop on township roads, couldn't speak. When a farmer asked why cops were stopping farmers on county roads, the officer responded, "We will do whatever we can to get that power line through." The farmer made the point that the officer did not say, "We are there to protect you," nor even "We are there to protect the workers."

In August, someone loosened the bolts on one of the 150-foot steel transmission towers. Soon after, it fell, and soon after that so did three more. People cut guard poles in half, they cut bolts three-quarters of the way through, then replaced them, waiting for someone to step on and break them.

The governor called out the FBI. A helicopter soon guarded the power line, presaging the sort of surveillance that is now familiar to the poor in many parts of the country. There were more than seventy arrests in one county alone. But home-cooked justice prevailed this time, as even the two people convicted of felonies were sentenced only to community service. In some cases, everyone refused to testify against the farmers.

A reporter asked one farmer whether he agreed with those who were bringing down towers. The farmer responded, "I wish a few more would come down, and I think they will, as time goes on. They shouldn't have done this to us in the first place. We did everything we could lawfully. We went to Minneapolis, got lawyers, went through the courts. But either the judges are paid off, or they just don't realize what's going on here. I think there's a lot of different laws and ways you can look at it. There's moral laws, too. I don't know, I don't figure it's wrong what we're doing out here. Sure, people think you gotta stay with the law, but what is the law? Who makes it? We should have more of a say with what goes on in this state too, you know. They can't just run over us like a bunch of dogs."

Although the farmers ultimately lost—the power lines have been operating for two decades now—over the next two years they knocked down ten more towers, and shot out thousands of insulators.

Dissatisfied even with victory, the power corporations wanted to make sure no one would ever again challenge their hegemony. In the words of Philip Martin, "We got the federal government to pass the law" that it's a federal crime to take down a tower transmitting electricity across state lines.

❈ ❈ ❈

I'm sitting again by the cell phone towers, and this time I'm thinking, *I could do this*. There are, as with so many activities we may find intimidating, several categories of barriers to action. There's the intellectual: I must convince myself it's necessary. There's the emotional: I must feel it's necessary. There's the moral: I must know it's right. There's the consequential: I must be willing and prepared to deal with the effects of my actions. Related to this, there's the fearful: I must be willing to cross barriers of fear, both tangible, real, present-day fears and conditioned fears that feel just as real and present but are not (e.g., if I wanted to go waterskiing, which I don't, I would have to face not only whatever fears I might have of speeding behind a boat, but my visceral repulsion to waterskiing based on beatings associated with it when I was a child: there is no longer any danger of my father hitting anyone if I were to go waterskiing, but it still *feels* like there is. How many of our other fears have been inculcated into us by our families or the culture at large?). There's the technical: I must figure out how best to proceed. There are undoubtedly others I can't think of.

For someone to act—and this is a generic process, applying as much to asking someone out as to weeding a garden as to writing a book as to removing cell phone towers as to dismantling the entire infrastructure that supports this deathly system of slavery—each of these barriers to action must be overcome or sometimes simply bypassed in moments of great embodiedness, identification, and feeling (e.g., if someone were attempting to strangle me [with bare hands, as opposed to the toxification of my total environment] my movement through these various barriers to action would of necessity be visceral and immediate: no pondering, just reaching for the pen to stab into his eye.)

Sure, I don't know how to take down a cell phone tower. But that's not why I don't act. A purpose of this book is to help me and perhaps others examine and, if appropriate, move past these other barriers to leave us only with the technical questions of *how to*, because so often *how to* is actually the easiest question, the smallest barrier.

I could take out a cell phone tower. So could you. We're not stupid (I'm presuming no members of the current Administration have made it this far in the book). And while our first few attempts may not be pretty—you'll notice I don't show you the first stories I ever wrote (at the time, my mother said they were good, yet now we both laugh when she says, "They were terrible, but I could never tell you that") and even now I don't show you my first drafts—but we would learn, just as we learn to do any technical task. I'm certain that if I made as many birdhouses as I write pages, not even David Flagg could laugh at them.

Practice makes perfect. This is as true of taking down cell phone towers as of writing. And fortunately, there are a lot of cell phone towers (I bet you never thought you'd see me append *fortunately* to a statement like that!). According to some estimates there are 138,000 cell phone towers in the U.S. (more than 48,000 of which are over two hundred feet tall[261]), plus radio and television towers. And the number of American cell phone users went up another 23 million between 2000 and 2001, leading to the erection of 20,000 new towers.[262]

That's a lot of practice. If we just put our hearts and minds and hands to it, it probably won't take very long before we get pretty good at it, so that taking down towers becomes something natural, like breathing, like taking long deep breaths of cool fresh air. Soon enough, we'll wonder what took us so long to get started.

❰ ❰ ❰

A teenager approached me after a talk. His eyes were on fire with intelligence and eagerness. He said, "I want to help you bring down civilization. I want to burn down factories."

Sometimes when people say things like this to me I distance myself from them. This is partly in case they're feds trying to entrap me—it's a classic trick: the feds suggest the action, entice you into doing it, provide the materials, and when you acquiesce you find yourself saying good-bye to your life for the next sixty years. It's partly because I don't know these people, and they could very well be crazy: the last thing I'd want to do would be to associate myself with some pyro who gets off on the flames, and who masturbates in the corner as the building crumbles (well, that's actually the second to last thing I'd want to do: the last thing would be to associate myself with a fed agent provocateur who gets off on putting people in little concrete cages). And it's also partly to protect myself from people with bad boundaries: to come up and semi-publicly tell a complete stranger you want to burn down a factory would seem at the very least to be a fundamental breach of security.

But I immediately fell in love with this kid's fierce sincerity. I thought a moment. There was no one around. I said, "Now, I would never want to discourage you or anyone from burning down a factory. But at the same time I want to emphasize that you have to be smart. One stupid mistake can cost you a lot."

He nodded.

"How old are you?"

"Sixteen."

"Can I ask you a personal question?"

He nodded again.

"Have you ever had sex?"

He shook his head.

"If you do this, and you get caught, you won't be having sex for at least twenty years. I'm not saying don't do it. I'm just saying this isn't a game, and there are real consequences for acting against the wishes of those in power, for effectively opposing production. That doesn't mean we should be afraid of those in power.

It means we should be very, very smart. Think it through, and then think it through a hundred more times. And then follow your heart."

He nodded again.

☾ ☾ ☾

I don't always respond that way. Sometimes, as I said, I get as far away from them as I can. But once I was approached by someone who said, "I know how destructive dams are, and I know what's at stake. My people are people of the salmon. Our entire way of life is centered around them. If you can get me the explosives I'll take out a dam."

I'd never met this man before, but I knew him by reputation. He wasn't a fed. Nor was he crazy. Nor did he have bad boundaries. Nor was he young and inexperienced. He knew what he was talking about, and he knew what he would be risking.

He said, "I have young children, so I can't do it for a few years. But when they're old enough, I'll do it."

Unsaid, but hanging in the air between us, was the fact that once his children were old enough to understand, he would be prepared to die or go to prison to help the river run free.

"I don't know how to do it," I said. "And I don't know how to get explosives."

He nodded and smiled wryly, then said, "That's okay. You've got a few years."

A HISTORY OF VIOLENCE

Few of us can easily surrender our belief that society must somehow make sense. The thought that the state has lost its mind and is punishing so many innocent people is intolerable. And so the evidence has to be internally denied.

Arthur Miller[263]

WHEN I WROTE ABOUT THE CIA'S *HUMAN RESOURCE EXPLOITATION Training Manual*, I forgot to mention that the Agency also put out instruction manuals on how to commit murder. The manuals make pretty fascinating reading in a ghoulish sort of way, if you can force yourself to forget that the book belongs not in the fiction section of the CIA bookshelf (along with their press releases and their analyses of the threats posed by other countries) but in the how-to section.

I think the words from *A Study of Assassination: A CIA Manual* describe the culture and the government far more starkly and elegantly than I ever could, so I'll quote at length: "TECHNIQUES: The essential point of assassination is the death of the subject. A human being may be killed in many ways but sureness is often overlooked by those who may be emotionally unstrung by the seriousness of this act they intend to commit. The specific technique employed will depend upon a large number of variables, but should be constant in one point: Death must be absolutely certain. . . . Techniques may be considered as follows:

"1. Manual.

"It is possible to kill a man with the bare hands, but very few are skillful enough to do it well. Even a highly trained Judo expert will hesitate to risk killing by hand unless he has absolutely no alternative. However, the simplest local tools are often much the most efficient means of assassination. A hammer, axe, wrench, screw driver, fire poker, kitchen knife, lamp stand, or anything hard, heavy and handy will suffice. A length of rope or wire or a belt will do if the assassin is strong and agile. All such improvised weapons have the important advantage of availability and apparent innocence. The obviously lethal machine gun failed to kill Trotsky where an item of sporting goods succeeded. . . .

"2. Accidents.

"For secret assassination . . . the contrived accident is the most effective technique. When successfully executed, it causes little excitement and is only casually investigated. The most efficient accident, in simple assassination, is a fall of 75 feet or more onto a hard surface. Elevator shafts, stair wells, unscreened windows and bridges will serve. Bridge falls into water are not reliable. In simple cases a private meeting with the subject may be arranged at a properly cased location. The act may be executed by sudden, vigorous [excised] of the ankles, tipping the subject over the edge. If the assassin immediately sets up an outcry, playing the 'horrified witness,' no alibi or surreptitious withdrawal is necessary. In chase cases it will usually be necessary to stun or drug the subject before dropping him. Care is required to insure that no wound or condition not attributable to the fall is discernible after death.

"Falls into the sea or swiftly flowing rivers may suffice if the subject cannot swim. It will be more reliable if the assassin can arrange to attempt rescue, as he can thus be sure of the subject's death and at the same time establish a workable alibi. . . .

"Falls before trains or subway cars are usually effective, but require exact timing and can seldom be free from unexpected observation.

"Automobile accidents are a less satisfactory means of assassination. If the subject is deliberately run down, very exact timing is necessary and investigation is likely to be thorough. If the subject's car is tampered with, reliability is very low. The subject may be stunned or drugged and then placed in the car, but this is only reliable when the car can be run off a high cliff or into deep water without observation.

"Arson can cause accidental death if the subject is drugged and left in a burning building. Reliability is not satisfactory unless the building is isolated and highly combustible. . . .

"3. Drugs.

"In all types of assassination except terroristic, drugs can be very effective. If the assassin is trained as a doctor or nurse and the subject is under medical care, this is an easy and rare method. An overdose of morphine administered as a sedative will cause death without disturbance and is difficult to detect. The size of the dose will depend upon whether the subject has been using narcotics regularly. If not, two grains will suffice. . . .

"4. Edge Weapons.

"Any locally obtained edge device may be successfully employed. A certain minimum of anatomical knowledge is needed for reliability. Puncture wounds of the body cavity may not be reliable unless the heart is reached. The heart is protected by the rib cage and is not always easy to locate. Abdominal wounds were once nearly always mortal, but modern medical treatment has made this no longer true. Absolute reliability is obtained by severing the spinal cord in the cervical region. This can be done with the point of a knife or a light blow of an axe or hatchet.

"Another reliable method is the severing of both jugular and carotid blood vessels on both sides of the windpipe. . . .

"5. Blunt Weapons.

"As with edge weapons, blunt weapons require some anatomical knowledge for effective use. Their main advantage is their universal availability. A hammer may be picked up almost anywhere in the world. Baseball and [illegible] bats are very widely distributed. Even a rock or a heavy stick will do, and nothing resembling a weapon need be procured, carried or subsequently disposed of. Blows should be directed to the temple, the area just below and behind the ear, and the lower, rear portion of the skull. Of course, if the blow is very heavy, any portion of the upper skull will do. The lower frontal portion of the head, from the eyes to the throat, can withstand enormous blows without fatal consequences.

"6. Firearms.

"Firearms are often used in assassination, often very ineffectively. The assassin usually has insufficient technical knowledge of the limitations of weapons, and expects more range, accuracy and killing power than can be provided with reliability. Since certainty of death is the major requirement, firearms should be used which can provide destructive power at least 100% in excess of that thought to be necessary, and ranges should be half that considered practical for the weapon. . . .

"The .300 F.A.B. Magnum is probably the best cartridge readily available. . . . These are preferable to ordinary military calibers, since ammunition available for them is usually of the expanding bullet type, whereas most ammunition for military rifles is full jacketed and hence not sufficiently lethal. . . .

"An expanding, hunting bullet of such calibers as described above will produce extravagant laceration and shock at short or mid-range. If a man is struck just once in the body cavity, his death is almost entirely certain. Public figures or guarded officials may be killed with great reliability and some safety if a firing point can be established prior to an official occasion. The propaganda value of this system may be very high. . . .

"The sub-machine gun is especially adapted to indoor work when more than one subject is to be assassinated. An effective technique has been devised for the use of a pair of sub-machine gunners, by which a room containing as many as a dozen subjects can be 'purifico' in about twenty seconds with little or no risk to the gunners. It is illustrated below. . . .

"A large bore shotgun is a most effective killing instrument as long as the range is kept under ten yards. It should normally be used only on single targets as it cannot sustain fire successfully. The barrel may be 'sawed' off for convenience, but this is not a significant factor in its killing performance. . . .

"The sound of the explosion of the proponent in a firearm can be effectively silenced by appropriate attachments. . . . The user should not forget that the sound of the operation of a repeating action is considerable, and that the sound of bullet strike, particularly in bone is quite loud. . . .

"A small or moderate explosive charge is highly unreliable as a cause of death, and time delay or booby-trap devices are extremely prone to kill the wrong man. In addition to the moral aspects of indiscriminate killing, the death of casual bystanders can often produce public reactions unfavorable to the cause for which the assassination is carried out.

"Bombs or grenades should never be thrown at a subject. While this will always cause a commotion and may even result in the subject's death, it is sloppy, unreliable, and bad propaganda. . . .

"Homemade or improvised explosives should be avoided. While possibly powerful, they tend to be dangerous and unreliable. Anti-personnel explosive missiles are excellent, provided the assassin has sufficient technical knowledge to fuse them properly."[264]

And so on.

（ （ （

Another warning sign of abusers, from that list adapted from Dear Abby, is a history of violence: "He may acknowledge he hit women in the past, but will aver they made him do it. You may hear from ex-partners that he's abusive. It's crucial to note that battering isn't situational: if he beat someone else, he'll very likely beat you, no matter how perfect you try to be."

In other words, as we saw earlier, abusers generally don't change ("there is no cure," is how *The Guardian* put it), and unless you want to be abused you should probably take past as prologue.

Likewise, we can read the culture's past as prologue. "Civilization originates," as I've quoted Stanley Diamond before, "in conquest abroad and repression at home."[265] So we can ask ourselves, Will civilization and the civilized commit genocide? To answer, let's first ask, Where are the indigenous of the Middle East, the Levant, the Mediterranean, Europe, Africa? Where are the intact and unthreatened indigenous elsewhere? Given the relentless fervency of the prologue (and main body), can we expect the denouement to be different?

Next, Will civilization and the civilized commit ecocide? To answer, just ask, Where are the forests of the Middle East, the Levant, the Mediterranean, Europe, Africa? Where are the other intact biomes in these or other places? How stupid or delusional must we be to expect some sort of magical reduction in the destructiveness?

Next, What does this culture's past tell us to expect about the treatment of women? Members of this culture—read male members of this culture—have routinely raped, killed, mutilated, enslaved, and otherwise abused women from its beginning. This abuse does not seem to be abating, and there is no good reason to think it will.

A classic line used by abusers and their codependents is that while things may have been bad in the past, now we must move on, start fresh, forget these atrocities that are no longer applicable in these brave new circumstances. This amnesia serves both parties well by allowing them to continue their disturbing and destructive dance of victimization. The abuser gets to continue to act out his (or her) hatred and self-hatred by hurting the victim (and thus himself through destroying the relationship, as well as that with which he has come to identify), and the victim gets to continue to act out her (or his) hatred and self-hatred by allowing herself (or himself) to be hurt. A loss of amnesia would sorely threaten their cozy relationship and reveal the enforced stupidity required on both parts to believe the convenient lies promising future change, promising some future utopia when the violence will no longer have to be.

We hear and too often believe the same lies on the cultural level. We nod our heads solemnly when timber industry spokespeople tell us they've reformed their methods of cutting, and *this time* they'll do it right. Meanwhile rates of deforestation continue to accelerate. Biodiversity collapses. The world burns. We breathe a sigh of relief that at least all the states in the United States have rescinded the bounty rewards they gave to the civilized for bringing in the scalps of dead Indians, and are thankful that at least John Ford is dead and can no longer put out his propaganda, yet we look away as languages and cultures disappear down a memory hole.

I suppose this is when I'm supposed to cite Santayana, that those who forget the past are condemned to repeat it. And that quote is certainly true so far as it goes. But it won't remain true very much longer. The pace of everything is increasing: the destruction is becoming more outrageous and omnipresent, extending now from the militarization (and trashing) of space to the changing of the weather to the toxification of the deepest oceans to the manipulation and pollution of our genetic materials; the frantic distractions as attempts to avoid seeing the destruction—have you watched any movies lately, or how about the Home Shopping Network?—are becoming ever more trivial, ever more obscene (as obscenities become trivialized and trivia becomes our staple). Civilization has entered its endgame, reached the end point of its exponential journey on a finite planet. It is consuming the world. It is consuming all of us. It will not last.

It may be possible to save some specific places or peoples or plants or animals or fungi or rocks or other natural life from being devoured and destroyed by this deathly culture (if the 138,000 cell phone towers, for example, kill 27.6 million migratory songbirds per year [roughly mid-range of the estimates] each collapsed cell phone tower saves an average of two hundred migratory songbirds per year). There's a world to be liberated. What are you going to do about it?

☾ ☾ ☾

WHY CIVILIZATION IS KILLING THE WORLD, TAKE SIXTEEN. Polar bears: "About half a mile upriver, I came to a very strong shoot of water, from thence I saw several white-bears fishing in the stream above. I waited for them, and in a short time, a bitch with a small cub swam close to the other shore, and landed a little below. The bitch immediately went into the woods, but the cub sat down upon a rock, when I sent a ball through it, at the distance of over a hundred and twenty yards at the least, and knocked it over; but getting up again it crawled into the woods, where I heard it crying mournfully and concluded that it could not long survive.

"The report of my gun brought some others down, and another she bear, with a cub of eighteen months old, came swimming close under me. I shot the bitch through the head and killed her dead. The cub perceiving this and getting sight of me made at me with great ferocity; but just as the creature was about to revenge the death of his dam, I saluted him with a load of large shot in his right eye, which not only knocked that out, but also made him close the other. He no sooner was able to keep his left eye open, than he made at me again, quite mad with rage and pain; but when he came to the foot of the bank, I gave him another salute with the other barrel, and blinded him most completely; his whole head was then entirely covered with blood. He blundered into the woods; knocking his head against every rock and tree that he met with.

"I now perceived that two others had just landed about sixty yards above me, and were fiercely looking round them. The bears advanced a few yards to the edge of the woods, and the old one was looking sternly at me. The danger of firing at her I knew was great, as she was seconded by a cub of eighteen months; but I could not resist the temptation."

The author, a Captain George Cartwright, really the first person to solidly establish civilization on the shores of Newfoundland, then moved toward another part of the river. "I had not sat there long, ere my attention was diverted to an enormous, old, dog bear, which came out of some alder bushes on my right and was walking slowly towards me, with his eyes fixed on the ground, and his nose not far from it. I rested my elbows, and in that position suffered him to come within five yards of me before I drew the trigger; when I placed my ball in the centre of his scull, and killed him dead: but as the shore was a flat reclining rock, he rolled around until he fell into the river.

"On casting my eyes around, I perceived another beast of equal size, raised half out of the water. . . . I crept through the bushes until I came opposite to him, and interrupted his repast, by sending a ball through his head; it entered a little above his left eye, went out at the root of his right ear, and knocked him over, he then appeared to be in the agonies of death for some time; but at last recovered sufficiently to land on my side of the river, and to stagger into the woods.

"Never in my life did I regret the want of ammunition so much as on this day; as I was by the failure interrupted in the finest sport that man ever had. I am certain, that I could with great ease have killed four or five brace more.[266]

Eskimo curlews: "Hunters would drive out from Omaha and shoot the birds without mercy until they had literally slaughtered a wagonload of them, the wagon being actually *filled*, and with the sideboards on at that. Sometimes when the flights were unusually heavy and the hunters well-supplied with ammunition, their wagons were too quickly and easily filled, so whole loads of the birds would be dumped on the prairie, their bodies forming piles as large as a couple of tons of coal, where they would be allowed to rot while the hunters proceeded to refill their wagons with fresh victims."[267]

Wilson snipe: "The birds being only in the country for a short time I had no mercy on them and killed all I could, for a snipe once missed might never be seen again."[268]

Golden plover: "The gunners had assembled in parties of from 20 to 50 at places where they knew from experience that the plovers would pass. . . . Every gun went off in succession, and with such effect that I several times saw a flock of a hundred or more reduced to a miserable remnant of five or six. . . . The sport was continued all day and at sunset when I left one of these lines of gunners they were as intent on killing more as they were when I arrived [before dawn]. A man near where I was seated had killed 63 dozens. I calculated the number [of hunters] in the field at 200, and supposing each to have shot only 20 dozens, 48,000 golden plovers would have fallen there that day."[269]

Ivory-billed woodpeckers: As the state of Louisiana tried desperately in the early 1940s to buy the habitat of the last of these birds in the United States, the board chair of Chicago Mill and Lumber responded, "We are just money grubbers. We are not concerned, as are you folks, with ethical considerations." The company argued that cutting this habitat would provide jobs (where have we heard that argument before?) but they lied (where have we seen corporate executives lie before?): their labor force consisted of German POWs, who themselves were "incredulous at the waste—only the best wood taken, the rest left in wreckage." The trees were used to make chests to hold tea.[270]

Northern spotted owls: Just to show how much things have changed in the last sixty years, I need to say that, coincidentally, the very day I wrote the previous paragraph, the Canadian Broadcasting Corporation carried a news story entitled "B.C. Court OKs Logging in Endangered Owl Habitat." There are, it seems, only twenty-five pairs of northern spotted owls still living in British Columbia, indeed in Canada. The birds are going extinct in the United States as well. The article stated, "The B.C. Court of Appeal has upheld a lower court ruling permitting old-growth logging in the last remaining habitat for the bird, saying economic interests can be weighed against the interest of the species."[271]

Remember that one working definition of insanity is to have lost one's connections to physical reality, to consider one's delusions as being more real than the real world. The judges (and other industry representatives) in this case are insane, attempting to "weigh" the needs of an intellectual and philosophical system against living beings.

Of course environmentalists are just as insane. As part of their pathetic and necessarily ineffective "defense" of these and other creatures, environmentalists have been reduced to saying, "If the logging industry gets [sic] a reputation for having killed a [sic] species, they're not going to benefit because worldwide markets aren't going to buy wood from B.C. if they know that B.C. logging companies are killing owls to get it."[272]

Another reasonable working definition of insanity is that it is insane to keep acting in the same way and to expect different results. Apart from the appalling stupidity of this environmentalist's statement, it has no basis in historical fact. Destroying the habitat of ivory-billed woodpeckers obviously did not harm the U.S. timber industry. Destroying the habitat of creatures never harms corporations, or at least it doesn't harm them because of public perception (if people within this culture loved the natural world, they would stop its destruction): corporations can certainly destroy the landbase and thus undercut their own eventual profitability, but of course by then the damage is done. Within this culture, the fantastic and ever-changing "needs" of the economic system will always "outweigh" the needs of physical reality (in exactly the same way that the fantastic and ever-changing "needs" of abusers always "outweigh" the needs of everyone else). If we do not understand this, we have no chance of surviving.

Halibut: "The fishermen of Newfoundland are much exasperated whenever an unfortunate halibut happens to seize their baits: they are frequently known in such cases to wreak their vengeance on the poor fish by thrusting a piece of wood through its gills and in that condition turning it adrift. The efforts which are made by the tortured fish to get its head beneath the water afford a high source of amusement."[273]

I have before me a photograph of—what do I call this?—a mound of fish inside a rolled up commercial fishing net. The pressure from the tons of fish inside the net forces the faces of those on the outside of the mass through the net. Their eyes bulge from the pressure, their mouths gape. In the background a man looks off to the side, presumably working the machinery that tightens the net around the wild fish. If this catch is typical of commercial catches, most of these fish will be thrown back overboard, dead.[274]

Prairie dogs: If you have internet access, you might do a Google search for "red mist," or go to www.seekersoftheredmist.com/. You will discover that when someone shoots a prairie dog or other "varmint" with a high powered rifle, the creature explodes into fine red mist. This provides "varmint hunters" with what they call "instant visual gratification." Oftentimes the "hunters" sit in chairs, scoped rifles attached to specially made tables, and then attempt to create red mist. They will also try for what they call "flipper shots"—also called "The Olga Korbut"—where the creature is sent flying end over end; or "the Chamois" in which the creature's entire skin is removed with one shot; or "Hoover Time," a head shot on a prairie dog peeking out of its den.[275]

Sometimes the "hunters" do not rest their guns on tables. Here is an account—not unusual in the least—I saw just today: "I had to run up to the Caprock yesterday for an errand. I took the opportunity to stow the .223 and a can of ammo in the truck before I took off. I got the chance to take my 3 year old with me, so when we got in the truck, he found my earphones and played with them. I talked to him about wearing them and leaving them on, how important it was, and that he had to do what I told him. It takes about 1.5 hours to get from here to there, so we had several chances to go over the rules. When we got to the first PD town, I ordered him to put on his ears, and I rolled down the window and grabbed the gun. Those PDs must have been shot recently because that was the last I saw of them. Further down the road, we pulled over again. I checked his ears several times before I finally pulled the trigger. I was very impressed. He watched, followed orders, kept his ears on and handed me rounds, one at a time. Very cool. I didn't feel like I had a lot of time, so I only got 7 pups and 1 barbed-wire fence (oops! Dad or I will fix the neighbor's fence . . . again). After a couple of hits, Gavin said, 'Cool! Fly!' Oh, yeah! It was a good day, even though I only got to shoot for about 10 minutes. I've got a 'hunt' planned for next weekend, so I'm excited."[276]

❨ ❨ ❨

WHY CIVILIZATION IS KILLING THE WORLD, TAKE SEVENTEEN. My only experience of military boot camp comes from movies, and thus is fictitious. I'd probably know more about them if I had never seen these movies (all writers, remember, including writers of movies, are propagandists). Here is what a former Marine sergeant says about boot camp: "Deceit and manipulation accompany the necessity to motivate troops to murder on command. You can't take civilians from the street, give them a machine gun, and expect them to kill without question in a democratic society; therefore people must be indoctrinated to do so. This fact alone should sound off alarms in our collective American brain. If the cause of war is justified, then why do we have to be put through boot camp? If you answer that we have to be trained in killing skills, well, then why is most of boot camp not focused on combat training? Why are our privates shown videos of U.S. military massacres while playing Metallica in the background, thus causing us to scream with the joy of the killer instinct [sic] as brown bodies are obliterated? Why do privates answer every command with an enthusiastic 'kill!' instead of, 'yes, sir!' like it is in the movies? Why do we sing cadences like these?: 'Throw some candy in the schoolyard, watch the children gather round. Load a belt in your M-60, mow them little bastards down!!' and "We're gonna rape, kill, pillage and burn, gonna rape, kill, pillage, and burn!!' These chants are meant to *motivate* the troops; they enjoy it, salivate from it, and get off on it. If one repeats these hundreds of times, one eventually begins to accept them as paradigmatically valid."[277]

HATRED

Alienation as we find it in modern society is almost total: it pervades the relationship of man to his work, to the things he consumes, to the state, to his fellow man, and to himself.[278] Man has created a world of man-made things as it never existed before. He has constructed a complex social machine to administer the technical machine he has built. Yet this whole creation of his stands over and above him. He does not feel himself as a creator and center, but as the servant of a Golem, which his hands have built. The more powerful and gigantic the forces are which he unleashes, the more powerless he feels himself as a human being. He is owned by his creations, and has lost ownership of himself.

Erich Fromm[279]

IF YOU RECALL, THE TENTH PREMISE OF THIS BOOK IS "THE CULTURE as a whole and most of its members are insane. The culture is driven by a death urge, an urge to destroy life." The fourteenth premise, somewhat related to the tenth, is, "*From birth on—and probably from conception, but I'm not sure how I'd make the case—we are individually and collectively enculturated to hate life, hate the natural world, hate the wild, hate wild animals, hate women, hate children, hate our bodies, hate and fear our emotions, hate ourselves. If we did not hate the world, we could not allow it to be destroyed before our eyes. If we did not hate ourselves, we could not allow our homes—and our bodies—to be poisoned.*"

This hatred can be more or less overt, in such manifestations as the Seekers of the Red Mist, the KKK, or the military (called "peacekeepers" by those in power, and "trained killers" by those who teach them their cadences). Sometimes the hatred is harder to see. As I tried to show exhaustively in *The Culture of Make Believe*, any hatred felt long enough no longer feels like hatred, it feels like what passes in this culture for religion, economics, tradition, the erotic (each of these being toxic mimics of what they would be in a human culture). It feels like science. It feels like technology. It feels like civilization. It feels like the way things are.

When you somehow extricate yourself from these iron cages of hate, what do you see?

I'm standing in line at a Safeway checkout counter, holding torment in my hands—torment I will soon enough take into my body—holding in my hands the processed flesh of plants and animals who were systematically enslaved and tortured, who were not merely killed—we all have to kill to eat: as a tree said to me, "You're an animal, you consume, get over it"—but who were denied their very nature, disallowed from ever simply existing, from being who they are, free and wild.

I look at the magazines, so many processed women, artificial models showing others, by contrast, their own inadequacies—including the attractive flesh-and-blood woman standing right in front of me, who is nowhere near as attractive (can *never* be as attractive) as these distant women neither she nor I shall ever meet—teaching them first and foremost to hate themselves, to hate their own never-good-enough bodies.

The checkout guy hates his job. Or at least he would if he allowed himself to feel in his body the slipping away of his own precious lifetime. Perhaps, though, it's more accurate to say "his own no-longer-precious lifetime," since if it were really precious he would not—could not—sell it so cheaply, nor even sell it for money at all. But he has been trained to never think of that, and especially to never feel it. If he thought of that—if he felt himself spending the majority of his life doing things he did not want to do—how would he then act? Who would he then be? What would he then do? How would he survive in this awful, unsurvivable system we call civilization? How, too, would we all respond if we fully awoke to the effects of the drip, drip, drip of hour after hour, day after day, year after year sold to jobs we do not love (jobs that are probably destroying our landbase to boot), and how would we respond, too, if we paid attention to the effects of other incessant drippings such as airbrushed photo after airbrushed photo on something so intimate as what—not whom, *never* whom—we find attractive?

Two days ago I was at a meeting of local grassroots environmentalists. One longtime activist approached me to say, "I read your books, and even if your facts are true and your analysis is correct—and it really seems they are—I cannot allow myself to go there, because I would not survive in this system. I need denial, even if I know that's what it is, and I need to hope that the system will change on its own, even if I know it won't."

A high school student bags the groceries. She's been through the mill. Twelve years of it, not counting her home life, twelve years of sitting in rows wishing she were somewhere else, wishing she were free, wishing it was later in the day, later in the year, later in her life when at long last her time—her life—would be her own. Moment after moment she wishes this. She wishes it day after day, year after year, until—and this was the point all along—she ceases anymore to wish at all (except to wish her body looked like those in the magazines, and to wish she had more money to buy things she hopes will for at least that one sparkling moment of purchase take away the ache she never lets herself feel), until she has become subservient, docile, domestic. Until her will—what's that?— has been broken. Until rebellion against the system comes to consist of yet more purchasing—don't you love those ads conflating alcohol consumption (purchased, of course, from major corporations) and rebelliousness?—or of nothing at all, until rebellion, like will, simply ceases to exist. Until the last vestiges of the wildness and freedom that are her birthright—as they are the birthright of every animal, plant, rock, river, piece of ground, breath of wind—have been worn or torn away.

Free will at this point becomes almost meaningless, because by now victims participate of their own free will—having long-since lost touch with what free will might be. Indeed, they can be said to no longer have any meaningful will at all. Their will has been broken. Of course. That's the point. Now, they are workers. They are productive members of this great and benevolent structure of civilization that brings good to all it touches. They are happy, even if this happiness requires routine chemical assistance. There is no longer any need for force, because the people—or more precisely those who were once people—have been fully metabolized into the system, have become self-regulating, self-policing.

Welcome to the end of the world.

She wears around her neck a cross, symbol of Christianity, symbol of dying to the flesh so she can be reborn to the spirit, symbol of perceiving the world—the body, her own body—as an evil place, a vale of tears where the enemy death constantly stalks, a place that is not and can never be as real as the heaven where bodies—these wild and uncontrollable *things* we've come to see as so flawed—no longer exist, a place that can never be home. (Would Christians object to the systematic exploitation, toxification, and despoliation of heaven as I object to the same on earth?)

I have friends who are Buddhists. They, too, are trained away from their bodies, away from the real, away from the primary, away from the material, away from their experience, away from what they call samsara (literally *passing through* in Sanskrit: what my dictionary calls "the indefinitely repeated cycles of birth, misery, and death caused by karma,"[280] and what one Zen Buddhist calls "the hellish world of time and space and the shifting shapes which energy assumes, the fluctuating world which is apprehended by the senses and presided over by the judgmental ego,"[281] all of which sounds like an awful drag, and really, to be honest, does not sound in the slightest like life as I experience it), away from what they call illusion, and toward what they tellingly and pathetically call "liberation" from this earth. As Richard Hooker puts it on his "World Civilizations" web pages, "If the changing world is but an illusion and we are condemned [sic] to remain in it through birth after birth, what purpose is there in atmansiddhi? The goal became not an eternity in a blissful afterlife, but moksha, or 'liberation' from *samsara*. This quest for liberation is the hallmark of the *Upanishads* and forms the fundamental doctrine of both Buddhism and Jainism."[282]

In short, Buddhism and Christianity both do what all religions of civilization must do, which is to naturalize the oppressiveness of the culture—get people (victims) to believe that their enslavement is not simply cultural but a necessary part of the existence to which they've been "condemned" (what does it say about them and the lives they lead that they perceive life not as a beautiful gift from the world, something for them to cherish and be grateful for, but as something to which they've been condemned?)—and then to point these people away from their awful (civilized) existence and toward "liberation" in some illusory better place (or even more abstractly, no place at all!). How very convenient for those in power. How very convenient for those who enslave human and nonhuman alike. These are religions for the powerless. These are religions to *keep* people powerless.

There are many Buddhist stories I love (as there are many Christian stories I love). In one of them, set during Japan's feudal period, an army sacked a neighboring shogun's village. Most of the villagers had already fled, but when the general of the attacking troops entered a Zen monastery, he found the master meditating. The general raised his sword. The master did not respond. The general sputtered, "Don't you realize I'm the man who could cut off your head without blinking an eye?"

The Zen master responded, "Don't you realize I'm the man who could have my head cut off without blinking an eye?"[283]

Since hearing this story I've admired the Zen master's equanimity in the face of certain death, and when the time comes I pray I manifest the same serenity. But the more I've thought about this story the more I've realized that the Buddha not only is always killed on the road, as Tom Robbins wrote ("Ideas are made by masters, dogma by disciples, and the Buddha is always killed on the road"[284]) but, and I'm sort of inverting his language here to emphasize a similar point a different way, the Buddha *must* be killed on the road, by each and every one of us, each and every day.

It all has to do with something I've been hammering on throughout this book: that all morality is dependent on a particular context, as is effective action. What may be appropriate and moral in one circumstance may be inappropriate or immoral in another. This means that while it's often useful to look to others for models on how we might behave under certain circumstances, it's foolish to the point of being potentially fatal to consider these models as applicable in all (or sometimes even in any) other circumstances. It is crucial to this story of the Zen master, for example, that the master faced down a shogun's general who was steeped in a tradition that respected rituals shared between these two men. Had the master given his same response to Genghis Khan or Tamurlane the Great, the other would quite likely have said, "Okay," and lopped off his head (both men had penchants for constructing huge pyramids from their victims' skulls). Likewise, if a typical modern American SWAT team ordered the Zen master to lie face down on the ground—"Don't you realize we're the team who could taser and pepper spray you without blinking an eye! Get the fuck down, motherfucker! Get the fuck down!"—and he refused to follow their instructions, he'd soon find himself lying in his own shit and piss, a sodden mass of muscles that no longer worked. Afterwards he'd find himself facing charges of resisting arrest, quite possibly assaulting a police officer, and worst of all, contempt of cop.[285]

My real breakthrough in understanding this story came when I realized that the Zen master's actions only make sense if at least one of three (unstated of course) premises is in place: either 1) he believes in reincarnation, which means if he dies he's coming back anyway; 2) he believes the material world is not primary, but instead a "hellish illusion" to which the Zen master has been "condemned," which means he won't so much mind leaving; or 3) he's powerless to avert immediate death anyway.

If any of these are accurate, his equanimity makes some sense. And if any of these are accurate for me, then I could consider modeling my own attitudes and behavior on his.

But if his life is precious and meaningful to him—if he is in love not only with his own life but with at least some of the humans in his community, and also with the swirling of fog in the tops of trees, and the way the fog fades in the morning sun, and in love with the way baby bears shimmy up trees when frightened, and with the chattering of squirrels teasing dogs, with the squabbling of songbirds over seeds, with the slow majesty of newts, salamanders, and turtles—and if he has the opportunity through any action to stop the general and his troops from sacking the village, from destroying his own life and the lives of those he loves (*Seven Samurai* comes to mind), then this Zen master's equanimity becomes nothing but a mask for cowardice, stupidity, and an appalling lack of creativity. And surely you can see that if he has the power to somehow stop the shogun's general but does not simply because he believes that the world is not primary, his beliefs would directly serve those who wish to exploit and destroy. Surely then you can also see how these beliefs would be promulgated—pushed very hard—both by those in power and by those who believe themselves powerless, those whose cowardice makes them wish, unconsciously of course, that they actually do have no power.

And why would they wish that? Because then they need not take responsibility for the actions—the sacking of the village, for example—they take no steps to prevent.

❰ ❰ ❰

There is much that is beautiful about Buddhism. I have heard some very wise Buddhists argue that the world is *not* illusion, that the problem is that because of our enculturation and our ego, we do not see the world in all its pain and beauty. Simply because much of what we see is illusion, they say, does not mean that nothing exists: it just means we do not see clearly. I love that. As my friend George Draffan says, "Meditation methods are ways to help us see more clearly, to dismantle our emotional and perceptual projections, to become more sensitive to what's actually going on. . . . Meditation itself is an ingenious collection of tools, spiritual technologies for dismantling habituated patterns and projections." To dismantle habituated patterns would be more than welcome. I could argue against none of this.

But Christians, too, can point to a theoretical Christianity that does not attempt to express "dominion" over the earth and its inhabitants,[286] that does not give other humans the choice of Christianity or death, that does not cause the hatred of women, children, life. Capitalists, too, can express fantasies of how some ideal capitalism can bring peace, justice, and happiness to all (humans). And scientists have their own technotopiae that they, too, use to urge us all onward.

But we have to ask ourselves how these religions are expressed *on the ground, in the real world*—I mean both of these literally—how they play out in the lives of living breathing human beings and others. What have been the effects of Christianity on the health of landbases? Has biodiversity thrived on the arrival of the cross? How has the arrival of Christianity affected the status of women? How has it affected the indigenous peoples it has encountered? We can and should ask the same questions of Buddhism, science, capitalism, and every other aspect of our or any other culture. Not how they play out theoretically, not how their rhetoric plays out, not how we wish they would play out, not how they *could* play out under some imaginary ideal circumstances, but how they *have* played out.

Just as Christianity has so very often been on one hand a tool of empire—as when Emperor Constantine went forth to conquer under the sign of the cross, and as when George W. Bush went forth to conquer because "God told me to"—and on the other hand a tool of subservience to power and escapism by the powerless (or those who believe themselves powerless), as I've described these last few pages, so too Buddhism often becomes yet another means for the traumatized to rationalize escaping the physical world. I cannot tell you how many

Buddhists have said to me—attempting to sound serene, but instead with an odd combination of smugness and brittleness in their voices—that salmon and other creatures are just shifting patterns of energy and are therefore not "real," meaning concern about their fate is not only folly but a barrier to enlightenment. A longtime pacifist activist said to me during an interview, "There are no salmon on one level of existence. There is only the movement of God's eyebrows. I've had the experience of transcending all duality. There's only this kind of rush of consciousness, and a part of that consciousness becomes salmon, and a part of that consciousness becomes time. And the salmon thrive for millions of years, and they go extinct. There's all this momentary burst of consciousness." Because, the story goes, these creatures are nothing but a part of this illusory earth—a "movement of God's eyebrows"—it doesn't matter so much if these creatures are driven extinct. In fact, I've been told, there can be no extinction because the salmon don't exist in the first place, or if there *is* extinction, then it is God's will, God's dream. Further, there is clearly something wrong with me because I remain attached to these creatures. This would be a good opportunity, they say, for me to practice detachment. How can I ever achieve enlightenment if I remain attached to this world?

I have many times experienced my own version of what this interview subject called non-duality, and what I call an uninterrupted state of grace. It has sometimes gone on for months. But for me it does not involve detachment from the world, or perceiving the world as a "rush of consciousness." Instead it is the opposite, a falling deeply into the world, an immersion, until I can feel how trees, insects, rain, soil, humans, the body of the earth, and my own body work together and in opposition, and my response is to say, "Oh, the beauty. The beauty." It is to experience and comprehend complete and joyous participation in the dance that is this extraordinarily wonderful world.

At one recent talk a Buddhist objected to my discussion of violence, saying, and I've heard this one a lot, that there can never be any reason for any form of violence. I did not ask whether she eats. I asked instead what she would do if she saw someone standing in front of her, beating a child.

She said, "I would bear witness to the child's suffering."

"You wouldn't intervene?"

"While using violence to stop the perpetrator might seem helpful in the short term, it would simply throw more violence into the universe—make the universe a more violent place—and in the long run would lead to more violence. I would not intervene."

I responded, "That's all theoretical. If after the show I happen to be walking by an alley, and happen to see that someone is beating you to death with a two-by-four, I strongly suspect all your fancy spirituality will rightly fall by the wayside as you beg me to not simply stand by and bear silent witness to your suffering and to your murder."

She shook her head. "No."

"I don't believe you."

She talked over me, "And it's the same with salmon. In the long run they'll go extinct anyway, and in the end the sun will burn up the earth, so it doesn't really matter . . ."

"Just because everyone in this room will someday die," I responded, "doesn't mean it's okay to torture them to death now. That's absurd. If this is where your spirituality leads you, I want no part of it."

Still other Buddhists tell me I must never act from anger, and must act only from a place of compassion and lovingkindness™ toward oppressors and abusers. I get this shit *all the time.* Just a couple of days ago I received an email from a stranger attempting to point out errors in my thinking. "As a writer there is only so far you can go with hostility and still be effective. In your upcoming radio interview, why don't you talk about you, how are you dealing with your health problems, what did you see or feel recently that inspired you (rather than what made you angry)?" This was a woman, which was sort of odd: usually intrusive men try to tell me what's wrong with my work while intrusive women try to fix my life. But this woman also wrote, "How is your sexuality/sensuality being affected by your increasing mental aggression against forces over which you have little control [sic]. How does the anger affect personal relationships. Are you still hugging trees or do you now have a human in bed with you?"

My first thought was to respond that whether my anger at the dominant culture's destruction of the planet affects my sex life is a question to which she will never know the answer.

One of the main problems with her questions (apart from the fact that my personal life is none of her goddamn business) is the premise that because I'm angry at the culture I'm angry at my friends. That's just plain silly. My anger is not a shotgun. I'm angry at the things that make me angry, and I'm not angry at the things that do not. What a concept.

But, and this is very important, from her perspective it's probably not silly at all. And that's the problem. The central point of R. D. Laing's great book *The Politics of Experience* was, so far as I'm concerned, that people act according to the way they experience the world. If you can understand their experience, you can understand their behavior. This is as true for the criminally insane as it is for capitalists. But once again I repeat myself.

He cites a description of a pathetic lunatic, given by the German psychiatrist Emil Kraepelin: "Gentlemen, the cases that I have to place before you today are peculiar. First of all, you see a servant girl, aged twenty-four, upon whose features and frame traces of great emaciation can be plainly seen. In spite of this, the patient is in continual movement, going a few steps forward, then back again; she plaits her hair, only to unloose it the next minute. *On attempting to stop her movement,* we meet with unexpectedly strong resistance; *if I place myself in front of her with my arms spread out* in order to stop her, if she cannot push me on one side, she suddenly turns and slips through under my arms, so as to continue her way. *If one takes firm hold* of her, she distorts her usually rigid, expressionless features with deplorable weeping, that only ceases so soon as one lets her have her own way. We notice besides that she holds a crushed piece of bread spasmodically clasped in the fingers of her left hand, which she absolutely *will not allow to be forced from her.* The patient does not trouble in the least about her surroundings so long as you leave her alone. *If you prick her in the forehead with a needle,* she scarcely winces or turns away, and leaves the needle quietly sticking there without letting it disturb her restless, bird-of-prey-like wandering backwards and forwards. *To questions* she answers almost nothing, at the most shaking her head. But from time to time she wails: 'O dear God! O dear God! O dear mother!,' always repeating uniformly the same phrases."[287]

Laing says, "If we see the situation purely in terms of Kraepelin's point of view, it all immediately falls into place. He is sane, she is insane; he is rational, she is irrational. This entails looking at the patient's actions out of the context of the situation as she experienced it. But if we take Kraepelin's actions (in italics)—he tries to stop her movements, stands in front of her with arms outspread, tries to force a piece of bread out of her hand, sticks a needle in her forehead, and so on—out of the context of the situation as he experienced it and defined by him, how extraordinary *they* are."[288]

From within the context of industrial capitalism as those enculturated into industrial capitalism experience and define it, destroying one's landbase (and then everyone else's) to increase the size of one's bank account makes sense. From within the context of civilization, as experienced and defined by the civilized— those who consider themselves in the most "advanced state of human society"—the destruction of all other cultures makes perfect sense. When you are bombarded from birth on with images and stories that teach you to perceive women as sexual objects, it should come as no surprise when you treat them as such. Likewise, when you are raised in an abusive household or an abusive culture where relations are based on power, and where those in power routinely use violence to terrorize those they wish to subjugate—when that is your experience of the world, when that is how the world has been defined for you—it may make sense to you to try to gain power over everyone you can. Or, and this brings us back to our discussion, anger may unduly frighten you—when those in power became angry, you suffered.

To be clear: All of this stepping away from anger—the presumption, for example, that anger toward the culture would lead to displacing that anger toward your friends—makes sense if you are afraid of your own emotions (or if you yourself displace your anger), if you are afraid of anger because you have been abused—made powerless in the face of "forces over which you have little control"—and realize in your body that the anger you feel only highlights your own impotence.

The point, it seems painfully (and beautifully) clear to me, is to not eradicate anger, but to try to be clear about when and why and at whom I am angry, and to be mindful of my anger. When appropriate, to let anger inform and even possess me so long as it does not consume me, as I can, when appropriate, let love or fear or joy inform and possess me so long as they too do not consume me. To aim my anger, not displace it, just as I would hope to aim and not displace my love, fear, or joy. I do not mind when someone expresses anger at me for something I have done to him or her. I do, however, mind when someone expresses anger toward me I do not deserve. The same can be said, obviously, for love and other emotions.

My dogs sometimes fight over their food dish, even though there is another a few feet away and even though they love each other even more than they love me. Every time they fight, minutes later they're once again cozying up to each other. This may seem odd, but I like it when I see this process, because each time it reminds me again that anger is just anger—I learn the same lesson each time I hear songbirds scold each other, or see bees tussle, or I snap at my mom or she snaps at me—and I'm reminded that outside the context of an abusive relationship, anger is nothing to be frightened of. Anger is just anger.

Attempts to "transcend" anger emerge from this fear, and also from the same old body-hating traditions that want to rid us of all of our "flawed" animal nature: transcendent spirit (cosmic consciousness, God's eyebrows, and so on), good; animal nature/emotion, bad.

Outside of this abusive context, of course, none of it makes any sense at all.

LOVE DOES NOT IMPLY PACIFISM

At the risk of seeming ridiculous, let me say that the true revolutionary is guided by a great feeling of love. It is impossible to think of a genuine revolutionary lacking this quality.

Ernesto Che Guevara [289]

ANOTHER PROBLEM I HAVE WITH BUDDHISM IS THAT BUDDHISM, like other "great" religions of civilization (including science, and including capitalism), isn't land-based. It's been transposed over space, which means by definition it is disconnected from the land, and also means it values, by definition, abstraction over the particularity of place. A religion is, I think, supposed to teach us how to live (which, if we're to live sustainably, must also mean that it teaches us how to live in a certain place). Also a religion is supposed to teach us how to connect to the divine. But people will live differently in different places, which means religions must be different in different places, and must emerge from specific places themselves, and not be abstracted from these places. Thus a religion that emerged from the Near East a couple thousand years ago may or may not have been helpful then and there, but quite probably will not apply so well to where I live right now. It is insane—literally, in terms of being disconnected from physical reality—to believe that a religion that tells someone how to live in, say, the desert of the American Southwest would be applicable (or even particularly helpful) to someone living in the redwood rainforests of the homeland of the Tolowa. It is similarly insane—and disrespectful of the divinity inherent in any particular place—to believe that a religion that helps experience the divine in the desert will particularly help me experience the divine at the ocean's edge. The places are different. So will be the experience of the divine.

☾ ☾ ☾

Even as I was writing the previous ten or so pages, I could hear in my mind the howl of outraged Buddhist pacifists (mainly white Buddhist pacifists: my Asian Buddhist friends aren't nearly so defensive about Buddhism as are many of the American Buddhists I've encountered, and in fact they often share the same criticisms, both of Buddhism and of American Buddhists). It's all very strange and interesting. I've found that there are many things I can bash with no one raising even an eyebrow, much less a fist. I can bash the unholy trinity of capitalism, Christianity, and corporations. I can bash schooling, wage jobs, civilization. I can bash environmentalists. I can even bash writers who bash civilization. Few seem to mind. But at the slightest hint of criticizing Buddhism (or science, which is another unholy cow that evokes the same response as Buddhism, as does, at least occasionally, pornography) I can see many of the faces in the audience harden and can feel their guts churn, their sphincters start to quiver.

❦ ❦ ❦

During a talk a couple of days ago, I amplified my analysis of Buddhism. I was surprised and pleased that the audience interrupted me with applause when I discussed the possibility that equanimity in the face of the culture's destructiveness can mask "cowardice, stupidity, and an appalling lack of creativity," and can be an avoidance of responsibility for acting to halt the atrocities. But I received an email the next morning that typifies the magical thinking of so many pacifists. The letter read in part: "While I would agree with every word you spoke about our civilization, I wouldn't agree that morality is always situational—there are certain acts that are soul-destroying, and advocating violence is one of them. Little word-games about Buddhist monks or innocent children being harmed are just cheap. I too used to hold the nine-inch nails philosophy—that was before I lived 50 years and had three children,[290] and love. The destruction trope is just another example of our society's harmful philosophy coming in by the back door. You're being co-opted by the need to control things. I hate to see your soul co-opted by the forces of destruction.

"The Great Mother will heal Her body, if she has to do it with cockroaches and finches (look at Galapagos). It is only human survival we are talking about here. We are doomed if we don't change, yes, but the earth will surely endure. So we must first put this argument in the proper Selfish context—i.e., saving our own asses. It is presumptuous and sacrilegious [sic] to speak of saving the earth.

"You must not suggest to these damaged and wounded humans, searching so desperately for meaning and peace, that they start breaking things. The ones that [sic] come to your talks are harmed and frightened. You have some power—there is a dark side and a light side—we all know this in our hearts. Please stay on the side of the light."

I'm sure by now you can parse out the unfounded and unstated premises in this note. The first premise is that morality is abstracted from circumstance, meaning in this case that (direct) violence is always—*under each and every circumstance*—wrong, even when it might be necessary to stop even more violence, implying as well that one has no moral responsibility to halt monstrous acts that happen even on one's own doorstep if stopping those acts would require muddying one's spiritual hands. This is the way of the Good German.

It is the way of the Good American. It's certainly the way of the good dogmatic pacifist.

Next, any attempts to even discuss these possibilities must be dismissed as "word games," "cheap," an example of the culture's "harmful philosophy coming in by the back door," and a need to control. This is all exactly what I meant early on in this book by the "Gandhi shield" pacifists often use to not only keep evil thoughts at bay but to make sure no one else thinks them either.

I don't want to go to the same well too many times, but a discussion by R. D. Laing applies. He wrote: "If Jack succeeds in forgetting something [such as the fact that we have the responsibility—the *obligation*—to stop the horrors of civilization, and the ability to do so, *if we choose to*], this is of little use if Jill continues to remind him of it. He must induce her not to do so. The safest way would be not just to make her keep quiet about it, but to induce her to forget it also.

"Jack may act upon Jill in many ways. He may make her feel guilty for keeping on 'bringing it up.' He may invalidate her experience. This can be done more or less radically. He can indicate merely that it is unimportant or trivial, whereas it is important and significant to her. Going further, he can shift the modality of her experience from memory to imagination: 'It's all in your imagination.' Further still, he can invalidate the content: 'It never happened that way.' Finally, he can invalidate not only the significance, modality, and content, but her very capacity to remember at all, and make her feel guilty for doing so in the bargain."[291]

"This is not unusual. People are doing such things to each other all the time. In order for such transpersonal invalidation to work, however, it is advisable to overlay it with a thick patina of mystification. For instance, by denying that this is what one is doing, and further invalidating any perception that it is being done by ascriptions of 'How can you think such a thing?' 'You must be paranoid.' And so on."[292]

The next unstated premise—and I'm going into such great detail because this woman's letter and the perspective it represents is not unusual, but instead is insanely common—is that a desire to stop atrocities such as the extirpation of species is a manifestation of a "need to control."

I used to have this fear, too, that to affect another's behavior—even when that other is hurting me directly—is to be "controlling." But to believe this is to internalize the rhetoric and worldview of the abuser.

Years ago, if you recall, I was in a couple of emotionally abusive relationships, where the women would call me names, harangue me for days, and so on. When I'd ask them to stop they'd say I was trying to censor or control them.

Finally, a friend asked me, "What will it take for you to say 'Fuck you' to this woman and walk away?"

"I can't do that."

"Why not?"

"That would be rude."

"She's not being rude to you?"

"I don't want to put myself on the same level. I don't want to cross some sort of middle line between us. I can talk about things on my half . . ."

"Ah, you've been to counseling! You can say, 'When you call me names, it makes me feel bad,' but you can't say, 'Cut this shit out!' then hang up the phone . . ."

"Hanging up on someone is unacceptable."

"So it's okay for her to perpetrate unacceptable behavior on you, but you aren't allowed to call her on it, nor even to absent yourself? That's crazy."

I opened my mouth to say something, then shut it, then opened it again, then clamped it shut.

That very night the woman called and began haranguing me. I said "Fuck you!" and hung up the phone. (Unfortunately, and this reveals how stupid denial makes us, it took me quite a while longer to figure out that after hanging up on her I didn't have to answer when she called back! It didn't take much longer than that, though, for me to realize that not only did I not need to answer the phone, I could simply not allow *anyone* to harangue me. If they do, I kick them out of my life. What a concept!)

There is an idea, no, a wish cherished by many, that love implies pacifism. If we love we cannot ever consider violence, even to protect those we love. I'm not sure that mother grizzly bears would agree, nor mother moose (I've heard it said that the most dangerous creature in the forest, apart, of course, from civilized humans, is a moose when you're between her and her child), nor many other mothers I've known. I've been attacked by mother horses, cows, mice, chickens, geese, eagles, hawks, and hummingbirds who thought I was threatening their children. I have known many human mothers who would kill anyone who was going to harm their little ones. If a mother mouse is willing to put her life on the line by attacking someone eight thousand times her size, how pathetic it is that we construct religious and spiritual philosophies that tell us that to attack even those who are killing those we most dearly love—or those we pretend we love—is to not love at all. That leads to the fifteenth premise of this book: *Love does not imply pacifism.*

I have a friend, a former prisoner, who is very smart, and who says that dogmatic pacifists are the most selfish people he knows, because they place their moral purity—or to be more precise, their self-conception of moral purity—above stopping injustice.

Years ago I spoke with the wonderful philosopher and writer Kathleen Dean Moore about why calling the earth our mother is not always helpful. I first asked her what were some of the lies we tell ourselves about our relationship to the land.

She responded, "In order of outrageousness: That human beings are separate from—and superior to—the rest of natural creation. That Earth and all its creatures were created to serve human ends. That an act is right if it creates the greatest wealth for the greatest number of people. That a corporation's highest responsibility is to its stockholders. That we can have it all—endlessly mining the land and the sea—and never pay a price. That technology will provide a way to solve every problem, even those created by technology. That it makes sense to barge salmon smolts past dams to the sea, so that grain can move downriver in barges. That a pine plantation is the same as a forest. That you can poison a river without poisoning your children. And the biggest and most dangerous lie of all: That the Earth is endlessly and infinitely resilient."

I asked why that is so dangerous.

She said: "We are doing damage now—to the atmosphere, to the seas, to the climate—that may be beyond the power of healing. When the Earth is whole, it is resilient. But once it is damaged, the power of the Earth to heal itself seeps away. In a weakened world, if we turn against the land, pour chemical fertilizers onto worn-out fields, sanitize wastewater with poisons, dam more rivers, burn more oil, bear more children, and never acknowledge that there may be no chance of healing, never admit what we have done and what we have failed to do—then, who can forgive us?"

I asked, "Why is this so hard for us to understand? We see evidence all around us."

Her answer: "Long-standing ways of thinking, even the way we talk, reinforce the fiction. Think of the metaphor of the Earth as a mother, and the slogan, 'Love your mother.' What does this mean? It might simply acknowledge that humans are created from matter that comes from the Earth. But so are Oldsmobiles, and that doesn't make the Earth the mother of Oldsmobiles.

"I think the whole 'love your mother' metaphor is just wishful thinking. Mothers can usually be counted on to clean up after their children. They are warm-hearted and forgiving: mothers will follow crying children to their rooms and stroke their hair, even if the child's sorrow is shame at his treatment of his mother. It's nice to think the Earth is a mother who will come after us and clean up the mess and protect us from our mistakes, and then forgive us the monstrous betrayal. But even mothers can be worn out and used up. And then what happens to her children?

"There's an ad from an oil company that shows the image of the Earth along with the caption, 'Mother Earth is a tough old gal.'"

I said, "The implication being that the Earth is invulnerable."

She responded, "A dangerous implication. I wrote a letter to the company saying, 'If the Earth really were your mother, she would grab you with one rocky hand and hold you under water until you no longer bubbled.' Cosmic justice."

It should come as no surprise that the great traditions of pacifism emerge from great religions of civilization: Christian, Buddhist, Hindu.

I recently saw an interview with longtime pacifist activist Philip Berrigan— one of the last before he died—in which he stated more or less proudly that spiritual-based pacifism is not meant to change things in the physical world, but relies on a Christian God to fix things. The interviewer had asked, "What do you say to critics of the Plowshares movement who claim that your actions have not produced tangible results?"

Berrigan answered, and especially note his second and third sentences: "Americans want to see results because we're pragmatists. God doesn't require results. God requires *faithfulness*. You try to do an act of social justice, and do it lovingly. You don't threaten anybody or hurt any military personnel during these actions. And you take the heat. You stand by and wait for the arrest."[293]

I can't speak for Berrigan, but I want to see results because the planet is being killed.

In any case, I think Berrigan is wrong. If there is a Christian God, and if several thousand years of history is any indication, He is not, to use the woman's term, on the side of the light. Given all evidence, I'm not sure I want to count on a Christian God to halt environmental destruction.

The Dalai Lama takes a more rounded, intelligent, and useful view on violence. He is, in addition, very aware of his premises, and tries to state them when he can. He has said, "Violence is like a very strong pill. For a certain illness, it may be very useful, but the side effects are enormous. On a practical level it's very complicated, so it's much safer to avoid acts of violence." He then continued, "There is a pertinent point in the Vinaya literature, which explains the disciplinary codes that monks and nuns must observe to retain the purity of their vows. Take the example of a monk or a nun confronting a situation where there are only two alternatives: either to take the life of another person, or to take one's own life. Under such circumstances, taking one's life is justified to avoid taking the life of another human being, which would entail transgressing one of the four cardinal vows." His next sentence reveals the whole point, and brings this discussion home: "Of course, this assumes one accepts the theory of rebirth; otherwise this is very silly."[294]

All of which leads to the sixteenth premise of the book: *The material world is primary. This does not mean that the spirit does not exist, nor that the material world is all there is. It means that spirit mixes with flesh. It means also that real world actions have real world consequences. It means we cannot rely on Jesus, Santa Claus, the Great Mother, or even the Easter Bunny to get us out of this mess. It means this mess really is a mess, and not just the movement of God's eyebrows. It means we have to face this mess ourselves (even if we do get some help from the Easter Bunny and others). It means that for the time we are here on Earth—whether or not we end up somewhere else after we die, and whether we are condemned or privileged to live here—the Earth is the point. It is primary. It is our home. It is everything. It is "very silly" to think or act or be as though this world is not real and primary. It is very silly and pathetic to not live our lives as though our lives are real.*

IT'S TIME TO GET OUT

There's nothing in a man's plight that his vision, if he cared to cultivate it, could not alleviate. The challenge is to see what could be done, and then to have the heart and the resolution to attempt it.

George F. Kennan[295]

IF YOU'VE GOTTEN THIS FAR IN THIS BOOK—OR IF YOU'RE SIMPLY anything other than entirely insensate—we probably agree that civilization is going to crash, whether or not we help bring this about. If you don't agree with this, we probably have nothing to say to each other (How 'bout them Cubbies!). We probably also agree that this crash will be messy. We agree further that since industrial civilization is systematically dismantling the ecological infrastructure of the planet, the sooner civilization comes down (whether or not we help it crash) the more life will remain afterwards to support both humans and nonhumans.

If you agree with all this, and *if* you don't want to dirty your spirituality and conscience with the physical work of helping to bring down civilization, and *if* your primary concern really is for the well-being of those (humans) who will be alive during and immediately after the crash (as opposed to simply raising this issue because you're too scared to talk about the crash or to allow anyone else to do so either), then, given (and I repeat this point to emphasize it) that civilization is going to come down anyway, you need to start preparing people for the crash. Instead of attacking me for stating the obvious, go rip up asphalt in vacant parking lots to convert them to neighborhood gardens, go teach people how to identify local edible plants, even in the city (*especially* in the city) so these people won't starve when the proverbial shit hits the fan and they can no longer head off to Albertson's for groceries. Set up committees to eliminate or, if appropriate, channel the (additional) violence that might break out.

We need it all. We need people to take out dams and we need people to knock out electrical infrastructures. We need people to protest and to chain themselves to trees. We *also* need people working to ensure that as many people as possible are equipped to deal with the fallout when the collapse comes. We need people working to teach others what wild plants to eat, what plants are natural antibiotics. We need people teaching others how to purify water, how to build shelters. All of this can look like supporting traditional, local knowledge, it can look like starting rooftop gardens, it can look like planting local varieties of medicinal herbs, and it can look like teaching people how to sing.

The truth is that although I do not believe that designing groovy eco-villages will help bring down civilization, when the crash comes, I'm sure to be first in line knocking on their doors asking for food.

People taking out dams do not have a responsibility to ensure that people in homes previously powered by hydro know how to cook over a fire. They do however have a responsibility to support the people doing that work.

Similarly, those people growing medicinal plants (in preparation for the end of civilization) do not have a responsibility to take out dams. They do however have a responsibility *at the very least* to not condemn those people who have chosen that work. In fact they have a responsibility to support them. They especially have a responsibility to not report them to the cops.

It's the same old story: the good thing about everything being so fucked up is that no matter where you look, there is great work to be done. Do what you love. Do what you can. Do what best serves your landbase. We need it all.

This doesn't mean that everyone taking out dams and everyone working to cultivate medicinal plants are working toward the same goals. It does mean that if they are, each should see the importance of the other's work.

Further, resistance needs to be global. Acts of resistance are more effective when they're large-scale and coordinated. The infrastructure is monolithic and centralized, so common tools and techniques can be used to dismantle it in many different places, simultaneously if possible.

By contrast, the work of renewal must be local. To be truly effective (and to avoid reproducing the industrial infrastructure) acts of survival and livelihood need to grow from particular landbases where they will thrive. People need to enter into conversation with each piece of earth and all its human and nonhuman inhabitants. This doesn't mean of course that we can't share ideas, or that one water purification technique won't be useful in many different locations. It does mean that people in those places need to decide for themselves what will work. Most important of all, the water in each place needs to be asked and allowed to decide for itself.

I've been thinking a lot again about the cell phone tower behind Safeway, and I see now how these different approaches manifest themselves in this one small place. The cell phone tower needs to come down. It is contiguous on two sides with abandoned parking lots. Those lots need to come up. Gardens can bloom in their place. We can even do our work side by side.[296]

❲ ❲ ❲

When at talks I've mentioned the three premises above—that civilization will crash, that the crash will be messy, and that the crash will be messier the longer we wait—nearly everyone who has thought about these issues at all agrees with the premises immediately. But at a talk I gave yesterday, one man was looking at me dubiously and shaking his head. I asked him what was up.

"I don't think we're going to crash," he said.

Oh Lord, I thought, *a cornucopian*.

But he surprised me. "It's not future tense," he said. "We're already in it."

I told him I agreed.

❨ ❨ ❨

The next of Dear Abby's warnings about abusive relationships was that you should be very wary if the abuser uses threats of violence to control you. A batterer may attempt to convince you that all men threaten partners, but this isn't true. He may also attempt to convince you that you're responsible for his threats: he wouldn't threaten you if you didn't make him do it.

These are actually three related warnings. As far as relating the first—the use of violence to control—to the larger social level, after my most recent show a man said, "You talk a lot about the violence of this culture. I don't feel I'm particularly violent. Where is the violence in my life?"

I asked him where his shirt was made. He said Bangladesh. I told him that wages in clothing factories in Bangladesh start at seven to eight cents per hour, and max out at about eighteen cents per hour. Now, I know we hear all the time from politicians, capitalist journalists, and other apologists for sweatshops that these wages are good because otherwise these people would simply starve to death. But that's only true if you accept the framing conditions that lead to those wages: Once people have been forced off their land—the source of their food, clothing, and shelter—and the land given to transnational corporations, once people have been made dependent on the corporations that are killing them, sure, it might be better not to starve immediately but to slave for seven cents per hour, starving a tad more slowly.

The question becomes, how much violence did it take to force these people off their land? It is violence or the threat of violence that keeps them working for these low wages.

Cheap consumer goods are not the only place the threat of violence controls our lives. I asked the man if he pays rent.

"Yes."

"Why do you do that?"

"Because I don't own my home."

"What would happen if you didn't pay rent?"

"I would be evicted."

"By whom?"

"The sheriff."

"And what if you refused to leave? What if you invited the sheriff in for dinner? And then after dinner you said, 'I've enjoyed your company, but I haven't enjoyed it all that much, and this is my home, so I would like you to leave now.' What would happen then?"

"If I refused to leave, the sheriff would evict me."

"How?"

"By force, if necessary."

I nodded. So did he.

Then I said, "And what if you were really hungry, and so you went to the grocery store. They've got a lot of food there, you know. And if you just started eating food there, and you didn't pay anything, what would happen?"

"They'd call the sheriff."

"It would probably be the same guy. He's a real asshole, isn't he? He'd come with a gun and take you away. Those in power have made it so we have to pay simply to exist on the planet. We have to pay for a place to sleep, and we have to pay for food. If we don't, people with guns come and force us to pay. That's violent."

The reason (part two of Abby's warning) that batterers may attempt to convince victims that all men threaten partners of course is that if you can get victims to disbelieve in the possibility of alternatives—if you can make your violence seem natural and inevitable—there will be no real reason for them to resist. You will, like the owners of sweatshops, have them exactly where you want them: under your control, with no need to even bother beating them anymore. The larger social equivalent is our culture's frantic insistence that all cultures are based on violence, that all cultures destroy their landbase, that men of all cultures rape women, that children of all cultures are beaten, that the poor of all cultures are forced to pay rent to the rich (or even that all cultures have rich and poor!). Perhaps the best example of this culture trying to naturalize its violence is the belief that natural selection is based on competition, that all survival is a violent struggle where only the meanest, most exploitative survive. The fact that this belief is nearly ubiquitous in this culture despite it being demonstrably untrue, logically untenable (recall the one-sentence disproof from early in this book: those creatures who have survived in the long run have survived in the long run, and if you hyperexploit your surroundings you will deplete them and die; the only way to survive in the long run is to give back more than you take), and a complete distortion of Darwin's elegant ideas, to which it is wrongly attributed, reveals the degree to which we have internalized the perspective of the abusers, and done so against the combined weight of history and common sense.

The third part of Abby's warning was that abusers attempt to convince their victims that the victims are responsible for the abusers' threats: the abuser wouldn't threaten you if you didn't make him do it. This has huge implications for activists. I cannot tell you how many activists have insisted to me that we must never use sabotage, violent rhetoric, and certainly never violence, because to do so will call up a strong backlash by those in power.

This insistence reveals an absolute lack of understanding of how repression works. Abusers will use any excuse to ratchet up repression, and if no excuses are forthcoming, excuses will be fabricated. Recall my discussion of the planned "outbursts" of CIA agents. Recall the Japanese knot-tying art of *hojojutsu*, where every movement tightens the ropes around your throat. Those in power will repress us no matter what we do or don't do. And if we do *anything* they will ratchet it up.

What is our solution? Probably the most commonly chosen solution, which is no solution at all, is to never upset those in power, that is, to use only tactics deemed acceptable to those in power. The main advantage of pursuing this non-option is that you get to feel good about yourself for "fighting the good fight" against the system of exploitation while not actually putting at risk the benefits you gain from this same system. (Have you ever wondered, by the way, why so many more people in the United States support third world rebel groups than participate in similarly open revolt here?)

Well, let's try this on for a solution. What if we prepare ourselves so that each time they ratchet up their repression towards us, we ratchet up our response? If they make us afraid of acting decisively to stop them from exploiting and destroying us and those we love—to stop them from killing (what remains of) the oceans, (what remains of) the forests, (what remains of) the soil—what would it take for us to make them fear to continue this exploitation, this destruction?

Everyone who has ever in any way been associated with perpetrators of abuse will probably agree with this analysis by psychologist and writer Arno Gruen of why abusers *must* continue to ratchet up their exploitation: "[C]atharsis does not work for those people whose anger and rage are fueled by self-hatred, for if it is projected onto an external object, self-hatred is only intensified and is aggravated by actions that are unconsciously perceived deep within as further forms of self-betrayal. Thus, with every additional act of destruction, destructive rage raises its stakes."[297]

The Oglala man Red Cloud spoke of this insatiability of abusers another way: "They made us many promises, more than I can remember. But they only kept but one. They promised to take our land and they took it."[298]

And George Orwell described it again: "It is intolerable to us that an erroneous thought should exist anywhere in the world, however secret and powerless it may be. Even in the instant of death we cannot permit any deviation."[299]

Abusers, and abusive cultures, are insatiable. They can ultimately brook no impediment to their control, to their destructiveness. Harry Merlo, former CEO of the Louisiana-Pacific timber corporation, articulated this mania as well as possible. After logging, he said, "There shouldn't be anything left on the ground. We need everything that's out there. We don't log to a ten-inch top or an eight-inch top or even a six-inch top. We log to infinity. Because it's out there and we need it all, now."

The question becomes, do we have the guts—and the heart—to stop them? Do we care enough about our landbases and the lives of those we love? Do we dare to act?

(((

I need to be clear: to blame members of the resistance for the backlash by those in power when resistors do not follow the agreed-upon rules is yet more acceptance of the abusers' logic: If I hit you, it is only because you made me do it.

When Nazis killed a hundred Jews for every Jew who escaped from a death camp, it was not the Jews' fault the Nazis chose to do this. When Nazis chose to kill a hundred innocent bystanders for every Nazi killed by partisans, it was not the partisans' fault. The choice to kill was the Nazis'. The responsibility was their own. Remember, from the perspective of the exploiters it is always best if you can get your victims to "choose" to participate. Proper limiting of their options will save you from having to use quite so much force. If you can get them to internalize responsibility for the violence you do use, so much the better.

If those in power choose to build a dam, that is their choice. I am not responsible for their decision. If I choose to take out this dam, that is my choice. Those in power are not responsible for my decision. If after that dam is gone, those in power decide to arrest everyone with brown hair, that is their choice. I would not be responsible for their decision.

We all have choices. I have choices. Those in power have choices. You have choices. Even if we choose to not act, we are still making choices.

《 《 《

The next to last characteristic on Abby's list was that the abuser may break or strike objects. There are two variants of this behavior: one is the destruction of beloved objects as punishment. The other is for him to violently strike or throw things to scare you.

To translate the first variant to the larger cultural level we need only consider the logic routinely used by mainstream environmental activists to keep more radical activists in line: "We must be reasonable, or the feds and corporations will cut *all* the forests." The punishment for not being "reasonable" is the destruction of ever more of what we love. Even more to the point, we know what happens as punishment to traditional indigenous people who do not give up their landbase: they will be killed, their landbase destroyed. And extirpation of species can be seen as a form of punishment, too: if the plant or animal (or culture) cannot adapt (conform) to the requirements of civilization, it will—it must—be destroyed.

Who among us has not witnessed the destruction of wild places or creatures we have loved? That this destruction is not always explicitly labeled as punishment seems secondary—exploiters lie as well as exploit—especially when the threat of further damage hangs always over our heads.

To translate the second variant into larger social terms, all we need to do is invoke a phrase used often these days by the U.S. military and politicians: Shock and Awe. This phrase is a euphemism for bombing the hell out of a people in order to terrorize them into doing what you want. Shock and Awe is merely the most recent name for this. George Washington earned the nickname Town Destroyer among the Indians by doing what the name suggests. He did this to punish those who resisted. A bit further back we find Catholic priests and missionaries cutting down the sacred groves of pagans as punishment for their recalcitrance and to preempt any return to the worship of their nonhuman neighbors. Before that, the Israelites clearcut the groves of all who did not bow before their god. They also clearcut the people.

Dear Abby's last characteristic of abusive relationships is the use of any force during an argument: holding you down, physically restraining you from leaving the room, pushing you, shoving you, forcing you to listen. Should we talk about Christianity or death? Should we talk about prisons? How about compulsory attendance at schools? Maybe we should talk about the fact that at protests cops are armed while protesters are not (I wonder who will win arguments between those two groups?). Why don't we cut to the chase and simply remark on the "social contract" imposed upon us by those in power, that those in power grant themselves a monopoly on force (then force us to attend schools where we are taught that the state—a primary instrument of those in power—has, you guessed it, a monopoly on force).

Within this culture there really is one central rule: Might makes right. I can think of no more abusive way to live.

❆ ❆ ❆

A truism of political science seems to be that part of the deal we sign as civilized human beings is that we allow the state to have a monopoly on violence. About a hundred years ago the German sociologist Max Weber *defined* the modern state as maintaining the monopoly on violence, with the exercise of force being authorized or permitted by the state, which means by law. *The monopoly on violence is what a state is.* Maintaining the monopoly on violence is what a state does. Weber states that "the use of force is regarded as legitimate only so far as it is either permitted by the state or prescribed by it. Thus the right of a father to discipline his children is recognized—a survival of the former independent authority of the head of a household, which in the right to use force has sometimes extended to a power of life and death over children and slaves. The claim of the modern state to monopolize the use of force is as essential to it as its character of compulsory jurisdiction and of continuous organization."[300]

Chibli Mallat made the implications clear: "Judicial power wields, through the rule of law, the most sophisticated manifestation of state coercion. There is no rule of law without the state's monopoly of violence."[301]

My friend George Draffan brings it all home: "The modern state rests on the monopoly of legitimate violence and, consequently, on the monopoly of taxation. Moreover, the group that effectively controls means of organized violence also acquires the monopoly over the enforcement of rules of economic and civic life. A weak state, then, is one which has lost the ability to effectively maintain these key monopolies. In late- and post-communist Russia, a constellation of factors led, after 1987, to a progressive privatization of the state. The privatization of the state is understood here as the process whereby the function of protecting juridical and economic subjects was taken over by criminal groups, private protection companies, or units of the state police force acting as private entrepreneurs. The consequence of that can also be defined as the covert fragmentation of the state: the emergence, on the territory under the formal jurisdiction of the state, of competing and uncontrolled sources of organized violence and alternative taxation networks."

It's quite a scam, if you can get people to buy into it. Those in power make the rules, and those in power enforce the rules. If those in power decide to toxify the landscape, toxify they will, and part of the bargain we evidently agree to on being part of this society is that they can use violence to enforce their edicts, and we cannot use violence to resist them. When they are killing the planet this quickly becomes absurd.

Recently in Bolivia a group of Aymara Indians kidnapped and killed an extraordinarily corrupt mayor, after legal means of redress failed. Legal means of redress had never stood a chance: the mayor represents the state, and the legal system supports the state and its representatives. As one of the Indians said, "We would have been satisfied if Altamirano [the mayor] admitted he had made mistakes, or if he had proposed a punishment for himself, or if the authorities had fined him. But none of this happened. What else could we do?"[302] Representatives of the state used this killing—which was definitely a fair execution according to Aymara justice, as well as their only real option for stopping the mayor's thuggery—as an excuse to arrest the leader of a land ownership reform movement, although not even the prosecution claimed he was anywhere near the scene of the kidnapping or execution. The prosecution really had no choice but to pursue this case. Far more is at stake than the murder of one corrupt politician. The prosecution stated, "There is only one justice, the justice of the state, of the law, there cannot be another justice."[303]

Of course a representative of the state would say that.

I disagree. There must be another justice, in fact many other justices. What is justice to the state, to the powerful, is not justice to the poor, to the land. What is justice to the CEO of ExxonMobil is not justice to the polar bears being driven to extinction by global warming. So long as we only believe in the justice of the state, of the law—made by those in power, to serve those in power—so long will we continue to be exploited by those in power. The rule of the state is always, hearkening back to the competing laws of Greek tragedies, in conflict with the rule of the people. And in a culture driven mad, the justice of the state will always be in conflict with the justice of the land.

❆ ❆ ❆

Dear Abby's advice to her readers was, in glorious all caps: "IF YOUR PARTNER SHOWS THESE SIGNS, IT'S TIME TO GET OUT." We can say the same about the culture, and if all caps are good enough for Abby, then by all means they're good enough for me: IF YOUR CULTURE SHOWS THESE SIGNS, IT'S TIME TO GET OUT.

It's time to get out.

COURAGE

Desperation is the raw material of drastic change. Only those who can leave behind everything they have ever believed in can hope to escape.

William S. Burroughs

I LEARNED ABOUT E-BOMBS FROM ONE OF MY STUDENTS—CASEY MADDOX, an excellent writer—at the prison. He wrote an extraordinary novel about someone who is kidnapped and put through a twelve-step recovery program for an addiction to Western civilization. The book's title is *The Day Philosophy Died*, and, as we'll get to in a moment, that title is related to E-bombs.[304]

E-bombs are, to my reckoning, one of the few useful inventions of the military-industrial complex. They are kind of the opposite of neutron bombs, which, if you remember, kill living beings but leave nonliving structures such as cities relatively intact: the quintessence of civilization. E-bombs, on the other hand, are explosive devices that do not hurt living beings, but instead destroy all electronics. Casey calls them "time machines," because when you set one off you go back one hundred and fifty years.

At one point in the novel the kidnappers are going to use a small plane to drop an E-bomb over the Bay Area. They carry the bomb on board inside a casket. The main character asks, "Who died?"

"Philosophy," someone says. "When philosophy dies," that person continues, "action begins."

As they prepare to set off the E-bomb, the main character keeps thinking, "There's something wrong with our plan." The thought keeps nagging him as they do their countdown to the celebration. Five, four, three, two, one. And the main character gets it, but too late. The E-bomb explodes. Their plane plummets.

One of the kidnappers clutches his chest, keels over. He's got a pacemaker. Even nonviolent actions can kill people. At this point, any action, including inaction, has lethal consequences. If you are civilized, your hands are more or less permanently stained deep dark red with the blood of countless human and non-human victims.

Long before he finished the book, Casey showed me where he first read about E-bombs. It was in, of all places, *Popular Mechanics*. If you check the September 2001 issue out of the library—which even has rudimentary instructions for how to construct one—make sure you use someone else's library card. Preferably someone you don't like.

The article was titled, "E-bomb: In the Blink of an Eye, Electromagnetic

Bombs Could Throw Civilization Back 200 Years. And Terrorists [*sic*] Can Build Them for $400."

And that's a bad thing?

The author, Jim Wilson, begins: "The next Pearl Harbor will not announce itself with a searing flash of nuclear light or with the plaintive wails of those dying of Ebola or its genetically engineered twin. You will hear a sharp crack in the distance. By the time you mistakenly identify this sound as an innocent clap of thunder, the civilized world will have become unhinged."

So far so good.

He continues, "Fluorescent lights and television sets will glow eerily bright, despite being turned off. The aroma of ozone mixed with smoldering plastic will seep from outlet covers as electric wires arc and telephone lines melt. Your Palm Pilot and MP3 player will feel warm to the touch, their batteries overloaded. Your computer, and every bit of data on it, will be toast."

I know, I know, this all sounds too good to be true. But it gets even better.

Wilson writes, "And then you will notice that the world sounds different too. The background music of civilization, the whirl of internal-combustion engines, will have stopped. Save a few diesels, engines will never start again. You, however, will remain unharmed, as you find yourself thrust backward 200 years, to a time when electricity meant a lightning bolt fracturing the night sky. This is not a hypothetical, son-of-Y2K scenario. It is a realistic assessment of the damage the Pentagon believes could be inflicted by a new generation of weapons—E-bombs."

When I mention all this at my shows, people often interrupt me with cheers.

The core of the E-bomb idea is something called a Flux Compression Generator (FCG), which the article in *Popular Mechanics* calls "an astoundingly simple weapon. It consists of an explosives-packed tube placed inside a slightly larger copper coil, as shown below. [The article even has a diagram!] The instant before the chemical explosive is detonated, the coil is energized by a bank of capacitors, creating a magnetic field. The explosive charge detonates from the rear forward. As the tube flares outward it touches the edge of the coil, thereby creating a moving short circuit. 'The propagating short has the effect of compressing the magnetic field while reducing the inductance of the stator [coil],' says Carlo Kopp [an Australian-based expert on high-tech warfare]. 'The result is that FCGs will produce a ramping current pulse, which breaks before the final disintegration of the device. Published results suggest ramp times of tens of hundreds of microseconds and peak currents of tens of millions of amps.' The pulse that emerges makes a lightning bolt seem like a flashbulb by comparison."

As good as all this may sound (oh, sorry, I forgot that technological progress is good; civilization is good; destroying the planet is good; computers and televisions and telephones and automobiles and fluorescent lights are all good, and certainly more important than a living and livable planet, more important than salmon, swordfish, grizzly bears, and tigers, which means the effects of E-bombs are so horrible that nobody but the U.S. military and its brave and glorious allies should ever have the capacity to set these off, and they should only be set off to support vital U.S. interests such as access to oil, which can be burned to keep the U.S. economy growing, to keep people consuming, to keep the world heating up from global warming, to keep tearing down the last vestiges of wild places from which the world may be able to recover if civilization comes down soon enough), it gets even better (or worse, if you identify more with civilization than your landbase): After an E-bomb is detonated, and destroys local electronics, the pulse piggybacks through the power and telecommunication infrastructure. This, according to the article, "means that terrorists [sic] would not have to drop their homemade E-bombs directly on the targets they wish to destroy. Heavily guarded sites, such as telephone switching centers and electronic funds-transfer exchanges, could be attacked through their electric and telecommunication connections."

The article concludes on this hopeful note: "Knock out electric power, computers and telecommunication and you've destroyed the foundation of modern society. In the age of Third World-sponsored terrorism,[305] the E-bomb is the great equalizer."[306]

❆ ❆ ❆

I go to the post office. Jim, my favorite clerk there, with whom I often chat as he processes the packages I'm mailing, comments on the heat. It's eighty-five or eighty-six, he says, the second or third highest temperature on record here. I know, cry me a fricking river, but I live on the cool coast of northern California.

"It makes you think about global warming," he says.

I nod, then reply, "Nineteen thousand people dead in Europe from the heat, and the damn newspapers don't even mention global warming." I don't mention that this is more than six times the number killed in the attacks on the World Trade Center. Jim likes my politics, but polite discourse generally demands that we ignore many obvious things.

Now it's his turn to nod. He says, "Did you see those pictures of glaciers melting in Europe?"

"The climate is changing, and those in power won't do anything about it."

"The culture has too much momentum," he responds, "and those in charge have too much money and power for us to stop them."

"That's why my next book is about how to take down civilization."

He looks at me for a moment. "You can write a book about it, but you can't make it happen."

"I can help push in the right direction at the right times, and I think that can make a difference."

"It will come down all right, and pretty soon at that. But it won't be your doing. It will be the system collapsing in on itself."

This is the guy at the Post Office! There are many who know this, but few who speak it out loud. I say, "We can hurry it up."

"It's going to be nasty," he responds.

"It already is."

"That nastiness is exactly why I bought a gun. A thirty-eight."

I'm about to say that's also exactly why I bought a gun a few years ago, but he carries my packages to the big bins in back.

When he returns he says, "It's for myself."

I don't know what he means.

He says, "I don't want to live like that."

"I don't want to live like this."

"I don't want to live like an animal."

"I've got news for you, Jim. You already are an animal."

"I need my electricity. I can't live without it."

I don't say anything. I think, *Is it worth it to you?*

He looks me straight in the eyes, and says, "I'm going to retire in January. Don't do this right now. Give me a few years to enjoy my retirement."

❲ ❲ ❲

It's the next day. I'm flying to Pennsylvania to give a talk. I hope my talk does more good than the oil that's burned to get me there.

I've just learned that the largest ice shelf in the Arctic—a solid feature for 3,000 years—has broken up. I've also just learned that a scientist studying this ice shelf—overseeing the destruction, as it were—stated, "I am not comfortable linking it to global warming. It is difficult to tease out what is due to global warming and what is due to regional warming."[307]

And here's something else I've recently learned. Global warming (or is it just regional warming that somehow seems to happen all over the globe?) has caused phytoplankton to decrease 6 percent in the last twenty years.[308] That is very bad. That is unspeakably bad. When the phytoplankton goes, it's all over.

I left before six this morning. I woke up at 4:15. It's now 8:30. I'm on a plane sitting on a runway in Sacramento. This will be my third take-off of the day. I'm tired, and at least pretending to sleep. My two row partners are talking about the weather. One says, "This was the first year since 1888 that we had more than ten days in a month over the century mark. Eighteen days it was."

I hear the other murmur something.

The first says, "That's damn hot, it is. It sure is damn hot."

It's only going to get worse, I think. And then I try to sleep.

❆ ❆ ❆

I had two dreams. In the first, my father came to my home. I did not want him here. He began to throw rocks at me. I tried to evade the rocks and did not throw any back. His daughter in the dream, who was not my sister, approached me. She spoke. She was pregnant, she said. Her father, my father, was the fetus's father. She was unable to bring herself to have an abortion. This would, she said, be an act of violence she could not commit. Nor could she bear to give birth to this product of rape. She could not bear to continue her father's lineage. Her only choice, she said, was to kill herself. She saw that as the only way to stop the horror that her parent had perpetrated upon her, and to stop the product of that horror growing inside of her.

Two thoughts came to me as I slept. First I noticed that it never occurred to her or to me in the dream to kill her father, my father, nor did she abort the baby, kill her father inside of her, and begin to live her life anew, free from him and his rapes. The second was to recognize that this is of course what we as a culture are doing. We so identify with the poisonous processes that have been forcibly implanted inside of us by our ancestors that we see no way to remove them save suicide. To kill the oppressors, and even to kill their influences they've implanted in us would be a violence we must avoid at all costs. And so we kill ourselves and the world with us. Somehow we do not perceive this as violence.

Several years ago I spoke with Luis Rodriguez, who wrote the wonderful book *Always Running: La Vida Loca: Gang Days in L.A.* He is a former gang member who got out through the literature of revolution. One of the things I asked him was why so many gang kids stand on street corners shooting at mirror images of themselves. If they're so angry, I asked, why don't they at least shoot at capitalists?

He said that part of the answer is that cops pit gang kids against each other. Another part is that the kids want to die. Of course they want to die. They are, after all, teenagers, and one of the things teenagers must do before they can become adults is die to their childhood. The child dies so the adult can be born. But no one is telling these kids that the deaths can be spiritual and metaphorical instead of physical. And so they stand on street corners, killing themselves and killing each other.

Luis also said that when he was younger he wanted to kill every CEO and cop he saw, because they were killing those he loved. But he later realized that he wasn't so interested in killing those individual human beings as he was in killing the relationships that allow them to kill kids. That is, he wanted to break their identities as CEOs or cops, and get them instead to identify with their animal humanity.

I've thought about this a lot in terms of tactics for women (and men) who are threatened with rape. Now, first, I need to say that anyone in that situation can do no wrong: no one can ever complain about anything she may or may not think or say or do, nor at any attitude she may or may not assume. Having said that though, I need to say that something that has helped some women, both as they are being threatened or assaulted and then afterwards, has been to redefine the relationship they suddenly find themselves in. The first step in this redefinition is to change her perception of the relationship from one between a rapist and a victim to one between a rapist and a survivor, that is, to begin to perceive herself not as a victim with no choices (although she may recognize that her range of choices may have been at least temporarily diminished because of the circumstances she finds herself in through no fault of her own) but as someone who is going to use any available means she chooses in order to survive this encounter (or not, as she chooses). For some women this choosing to be a survivor may then lead to them submitting to the rapist's physical demands, allowing him to have her body while her soul remains her own. This is one of the points I think Bertholt Brecht was making in his fable about a man who lives alone who one day hears a knock on his door. When he answers, he sees The Tyrant outside, who asks, "Will you submit?" The man says nothing. He steps aside. The Tyrant enters his home. The man serves him for years, until The Tyrant becomes sick from food poisoning and dies. The man wraps the body, takes it outside, returns to his home, closes the door behind him, and firmly answers, "No." For other women this may mean fighting to the death, preferably his. Still others—many others—do not consciously make the choice to move from victim to survivor in that moment of violation—they are too busy simply surviving to think about labeling themselves as survivors—but they make that choice over time, in the months, years, and decades that follow, as they metabolize what was done to them and their responses. And of course yet others choose different approaches: there are as many approaches to this question of reidentifying oneself from victim to survivor as there are potential victims, potential survivors.

The next step that at least some women pursue in this process of changing their circumstances is to attempt to get the man to no longer identify himself as a rapist, but as something else (one hopes not a murderer). An example may help clarify. One morning in the mid-1970s, my sister was reading in bed when suddenly she felt a man's weight on her back and a knife at her throat. The man said he was going to rape her. She said, "You can do that if you'd like, but I have to tell you that my husband and I are being treated for syphilis. I don't know if you want to risk catching it." Our mother had always told her to keep a prescription bottle by her bedside for exactly this contingency. (And what does it say about our culture that mothers need to prepare their daughters for this possibility, or really, given the rates of rape in our culture, this likelihood?) Fortunately, the man didn't look closely at the bottle, or he would have learned that the original prescription was several years old, for medicines designed to alleviate my sister's migraines, and that the bottle was now full of aspirin. He told her that it wasn't worth the risk, and that instead he wanted all of her money. She had twenty dollars in her purse and she gave him five.[309] He left. The point is that my sister had caused the man to no longer identify himself as a rapist, but as a robber, and to act on that identification. She effectively killed the rapist. Sometimes, when men strongly identify as rapists, it is not possible to kill the rapist without killing the man. So be it.

The first part of our task, then, is to attempt to break our own identification as the civilized and remember that we are human animals living in and reliant on our landbases for survival, to begin to care more about the survival of our landbase than the perpetuation of civilization. (What a concept!) Then we must break our identification as victims of this awful and deathly system called civilization and remember that we are survivors, resolve that we *will do what it takes* so that we—*and those we love, including nonhuman members of our landbase*—will survive, outlast, outlive, defeat civilization. That we will in time dance and play and love and live and die among the plants and animals who will someday grow amidst its ruins. Once we have made that shift inside of ourselves, once we no longer see ourselves as victims of civilization but as its survivors, as those who will not let it kill us or those we love, we have freed ourselves to begin to pursue the more or less technical task of actually stopping those who are killing our landbases, killing us. One way to do that might be to get CEOs, cops, and politicians to identify themselves as human animals living in and reliant on their landbases and to break their identities as CEOs, cops, and politicians. The good news is that some few of them may listen to reason. The bad news is that history, sociology, psychology, and direct personal experience suggest that most—nearly all—will not.

In the second dream, I drove on a small road into a place I'd been before, a place that was wild. But my car could not pass between two small trees. I stopped and got out. I could not get into the wild. I was frustrated. There was a reservoir nearby, and as I walked toward it, it filled with warships. Richard Nixon was lashed, à la Admiral Farragut, to the mast (in this case radar tower) of a ship, flashing his trademark two-fingered salute. The beach was soon packed with patriots pushing me this way and that for not enthusing about the military takeover of the reservoir. The patriots began to party. I struggled to get away, and finally was able to walk alone into the wilderness.

Part of the grammar of my dreams is that when I have multiple dreams in the same sleep, they speak to the same questions. This dream, then, was a follow up to the first, with the first revealing our incapacity to face our predicament, to come up with any response more creative than suicide, and the second making clear that we cannot return to the wild and bring our cars and our machines with us. They will not fit. And so where does this leave us? It leaves us near artificial lakes filled with killers and liars who tie themselves to instruments of war. And it leaves us in the midst of crowds of people who perceive all of these death-machines as good things, and who party among their machines of death. It leaves us needing to find a different way to make it back to the wilderness, back to our home.[310]

❨ ❨ ❨

The most common words I ever hear spoken by any environmentalists any-where are, "We're fucked." Most of these environmentalists are fighting des-perately, using whatever tools they have—or rather whatever legal tools they have, which means whatever tools those in power grant them the rights to use, which means whatever tools will be ultimately ineffective—to try to protect some piece of ground, to try to stop the manufacture or release of poisons, to try to stop civilized humans from tormenting some groups of plants or animals. Sometimes they're reduced to trying to protect one tree.

John Osborn, an extraordinary activist and friend who, when I met him, was the heart and soul of the Spokane, Washington environmental community, has often given his reasons for doing the work: "As things become increasingly chaotic, I want to make sure some doors remain open. If grizzly bears are still alive in twenty, thirty, and forty years, they may still be alive in fifty. If they're gone in twenty, they'll be gone forever."

But no matter what we do, our best efforts are insufficient to the dangers we face. We're losing badly, on every front. Those in power are hell-bent on destroying the planet, and most people don't care.

Many of us know we're fucked. But many don't talk about it, especially publicly. We believe we're alone in this feeling. But we're not.

Just today I got this email: "I attended your talk last night, and was deeply surprised. I hadn't read your works (except in the middle of the night last night), and was skeptical of your message, but not as you might normally think. I stopped going to hear Enviro speakers quite a while back, call it estrangement. I work as an environmental regulator, EPA Water Pollution, and have been doing that for 15 years. I've seen a bit of water go beneath the bridge. I know what's in it.

"I'm a little tired of utopian environmental theory. It's hard to hear someone talk about some perfect future society (spirituality, free love, etc.), when I'm trying to figure out what to do with some damaged place, or a slag pile, or the siting of a new chip mill that can eat 10,000 acres of forest per year. It ain't about theory. It is very, very real.

"In the role of regulator I have to live in the world of what has been done, and what is doable. I've had to understand the brutal limitations of physics, history, law, technology, money, politics, and human folly. We've busted some things we can't fix. We're still doing it. I've had to witness more than I care to. I've had some very sweet victories, even been a Champion a few times. But, we are so FUCKED!!! I've never stated that anywhere. It just seemed like too difficult a truth to share. Thank you for saying that out loud. Hope bashing is OK, we can make more."

I've been bashing hope for many years. Frankly, I don't have much of it, and I think that's a good thing. Hope is partly what keeps us chained to the system. First there is the false hope that suddenly somehow the system may inexplicably change. Or technology will save us. Or the Great Mother. Or beings from Alpha Centauri. Or Jesus Christ. Or Santa Claus. All of these false hopes—all of this rendering of our power—leads to inaction, or at least to ineffectiveness: how, for example, would Philip Berrigan have acted had he not believed— hoped—God would help solve things?

One reason my mother stayed with my father was that there were no bat- tered women's shelters in the fifties and sixties, but another was because of the false hope that he would change. False hopes, as I've written elsewhere, bind us to unlivable situations, and blind us to real possibilities. Does anyone really believe that Weyerhaeuser is going to stop deforesting because we ask nicely? Does anyone really believe that Monsanto will stop Monsantoing because we ask nicely? If only we get a Democrat in the White House, this line of thought runs, things will be okay. If only we pass this or that piece of legislation, things will be okay. If only we *defeat* this or that piece of legislation, things will be okay.[311] Bullshit. Things will not be okay. They are already not okay, and they're getting worse.

One of the smartest things Nazis did to Jews was co-opt rationality, co-opt hope. At every step of the way it was in the Jews' rational best interest to not resist: many Jews had the hope—and this hope was cultivated by the Nazis—that if they played along, followed the rules laid down by those in power, that their lives would get no worse, that they would not be murdered. Would you rather get an ID card, or would you rather resist and possibly get killed? Would you rather go to a ghetto (reserve, reservation, whatever) or would you rather resist and possibly get killed? Would you rather get on a cattle car, or would you rather resist and possibly get killed? Would you rather get in the showers, or would you rather resist and possibly get killed?

But I'll tell you something important: the Jews who participated in the War- saw Ghetto uprising, including those who went on what they thought were sui- cide missions, had a higher rate of survival than those who went along peacefully. Never forget that.

HOPE

Hope is the real killer. Hope is harmful. Hope enables us to sit still in the sinking raft instead of doing something about our situation. Forget hope. Honestly and candidly assessing the situation as it truly stands is our only chance. Instead of sitting there and "hoping" our way out of this, perhaps we should recognize that realizing the truth of our situation, even if unpleasant, is positive since it is the required first step toward real change.

Gringo Stars

Hope is the leash of submission.

Raoul Vaneigem

The cure for despair is not hope. It's discovering what we want to do about something we care about

Margaret Wheatley[312]

IT ISN'T MERELY FALSE HOPES THAT KEEP THOSE WHO GO ALONG ENCHAINED. It is hope itself.

Hope, we are told, is our beacon in the dark. It is our light at the end of a long, dark tunnel. It is the beam of light that against all odds makes its way into our prison cells. It is our reason for persevering, our protection against despair (which must at all costs, including the cost of our sanity and the world, be avoided). How can we continue if we do not have hope?

We've all been taught that hope in some better future condition—like hope in some better future heaven—is and must be our refuge in current sorrow. I'm sure you remember the story of Pandora. She was given a tightly sealed box and was told never to open it. But, curious, she did, and out flew plagues, sorrow, and mischief, probably not in that order. Too late she clamped down the lid. Only one thing remained in the box: hope. Hope, the story goes, was "the only good the casket held among the many evils, and it remains to this day mankind's sole comfort in misfortune." No mention here of action being a comfort in misfortune, or of actually *doing something* to alleviate or eliminate one's misfortune. (*Fortune*: from Latin *fortuna*, akin to Latin *fort-*, *fors*, chance, luck: this implies of course that the misfortune that hope is supposed to comfort us in is just damn bad luck, and not dependent on circumstances we can change: in the present case, I don't see how bad luck is involved in the wretched choices we each make daily in allowing civilization to continue to destroy the earth.)

The more I understand hope, the more I realize that instead of hope being a comfort, that all along it deserved to be in the box with the plagues, sorrow, and mischief; that it serves the needs of those in power as surely as a belief in a distant heaven; that hope is really nothing more than a secular version of the same old heaven/nirvana mindfuck.

Hope is, in fact, a curse, a bane.

I say this not only because of the lovely Buddhist saying, "Hope and fear chase each other's tails"—without hope there is no fear—not only because hope leads us away from the present, away from who and where we are right now and toward some imaginary future state. I say this because of *what hope is*.

More or less all of us yammer on more or less endlessly about hope. You wouldn't believe—or maybe you would—how many editors for how many

329

magazines have said they want me to write about the apocalypse, then enjoined me to "make sure you leave readers with a sense of hope." But what, precisely, is hope? At a talk I gave last spring, someone asked me to define it. I couldn't, and so turned the question back on the audience. Here's the definition we all came up with: Hope is a longing for a future condition over which you have no agency. It means you are essentially powerless.

Think about it. I'm not, for example, going to say I hope I eat something tomorrow. I'll just do it. I don't hope I take another breath right now, nor that I finish writing this sentence. I just do them.[313] On the other hand, I hope that the next time I get on a plane, it doesn't crash.[314] To hope for some result means you have no agency concerning it.

So many people say they hope the dominant culture stops destroying the world. By saying that, they've guaranteed at least its short-term continuation, and given it a power it doesn't have. They've also stepped away from their own power.

I do not hope coho salmon survive. I will do what it takes to make sure the dominant culture doesn't drive them extinct. If coho want to leave because they don't like how they're being treated—and who could blame them?—I will say good-bye, and I will miss them, but if they do not want to leave, I will not allow civilization to kill them off. *I will do whatever it takes.*

I do not hope civilization comes down sooner rather than later. I will do what it takes to bring that about.

When we realize the degree of agency we actually do have, we no longer have to "hope" at all. We simply do the work. We make sure salmon survive. We make sure prairie dogs survive. We make sure tigers survive. We do whatever it takes.

❨ ❨ ❨

Casey Maddox wrote that when philosophy dies, action begins. I would say in addition that when we stop hoping for external assistance, when we stop hoping that the awful situation we're in will somehow resolve itself, when we stop hoping the situation will somehow not get worse, then we are finally free—truly free—to honestly start working to thoroughly resolve it. I would say when hope dies, action begins.

❨ ❨ ❨

Hope may be fine—and adaptive—for prisoners, but free men and women don't need it.

Are you a prisoner, or are you free?

❆ ❆ ❆

People sometimes ask me, "If things are so bad, why don't you just kill yourself?"

The answer is that life is really, really good. I am a complex enough being that I can hold in my heart the understanding that we are really, really fucked, and at the same time the understanding that life is really, really good. Not because we're fucked, obviously, nor because of the things that are causing us to be fucked, but despite all that. We are fucked. Life is still good. We are really fucked. Life is still really good. We are *so* fucked. Life is still *so* good.

Many people are afraid to feel despair. They fear that if they allow themselves to perceive how desperate is our situation, they must then be perpetually miserable. They forget it is possible to feel many things at once. I am full of rage, sorrow, joy, love, hate, despair, happiness, satisfaction, dissatisfaction, and a thousand other feelings. They also forget that despair is an entirely appropriate response to a desperate situation. Many people probably also fear that if they allow themselves to perceive how desperate things are that they may be forced to actually *do something* to change their circumstances.

Despair or no, life is good. The other day I was lying by the pond outside my home, looking up through redwood needles made translucent by the sun. I was happy, and I thought, "What more could anyone want?"[315] Life is so good. And that's all the more reason to fight.

Another question people sometimes ask is, "If things are so bad, why don't you just party?"

Well, the first answer is that I don't really like parties. The second is that I'm having great fun. I love my life. I love life. This is true for most activists I know. We are doing what we love, fighting for what and whom we love.

I have no patience for those who use our desperate situation as an excuse for inaction.[316] I've learned that if you deprive most of these people of that particular excuse they just find another, then another, then another. The use of this excuse to justify their inaction—the use of any excuse to justify inaction— reveals nothing more nor less than an incapacity to love.

At one of my recent talks someone stood up during the Q & A and announced that the only reason people ever become activists is to make themselves feel better about themselves. Effectiveness really doesn't matter, he said, and it's egotistical to think it does. He trotted out the old line about how the natural world doesn't need our help. At least he averred that the natural world exists, as opposed to being the movement of some god's eyebrows, but the end result was the same old narcissism.

I told him I disagreed.

He asked, "Doesn't activism make you feel good?"

"Of course, but that's not why I do it. If I only want to feel good, I can just masturbate. But I want to accomplish something in the real world."

"Why?"

"Because I'm in love. With salmon, with trees outside my window, with baby lampreys living in sandy stream bottoms, with slender salamanders crawling through the duff. And if you love, you act to defend your beloved. Of course results matter to you, but they don't matter to whether you make the effort. You don't simply hope your beloved survives and thrives. You do what it takes. If my love doesn't cause me to protect those I love, it's not love. And if I don't act to protect my landbase, I'm not fully human."

A while back I got an email from someone in Spokane, Washington. He said his fifteen-year-old son was wonderfully active in the struggle for ecological and social sanity. But, the father continued, "I want to make sure he stays active, so I feel the need to give him hope. This is a problem, because I don't feel any hope myself, and I don't want to lie to him."

I told him not to lie, and said if he wants his son to stay active, he shouldn't try to give him hope, but instead to give him love. If his son learns how to love, he will stay active.

❨ ❨ ❨

A wonderful thing happens when you give up on hope, which is that you realize you never needed it in the first place.[317] You realize that giving up on hope didn't kill you, nor did it make you less effective. In fact it made you more effective, because you ceased relying on someone or something else to solve your problems—you ceased *hoping* your problems somehow get solved, through the magical assistance of God, the Great Mother, the Sierra Club, valiant tree-sitters, brave salmon, or even the Earth itself—and you just began doing what's necessary to solve your problems yourself.

Because of industrial civilization, human sperm counts have been cut in half over the last fifty years. At the same time, girls have begun to enter puberty earlier: 1 percent of three-year-old girls have begun to develop breasts or pubic hair, and in only the last six years, the percentage of girls under eight with swollen breasts or pubic hair has gone from 1 percent to 6.7 percent for white girls, and 27.2 percent for black girls.[318]

What are you going to do about this? Are you going to hope this problem somehow goes away? Will you hope someone magically solves it? Will you hope someone—anyone—will stop the chemical industry from killing us all?

Or will you do something about it?

When you give up on hope, something even better happens than it not killing you, which is that it kills you. You die. And there's a wonderful thing about being dead, which is that once you're dead they—those in power—cannot really touch you anymore. Not through promises, not through threats, not through violence itself. Once you're dead in this way, you can still sing, you can still dance, you can still make love, you can still fight like hell—you can still *live* because *you are* still alive, in fact more alive than ever before—but those in power no longer have a hold on you. You come to realize that when hope died, the you who died with the hope was not you, but was the you who depended on those who exploit you, the you who believed that those who exploit you will somehow stop on their own, the you who depended on and believed in the mythologies propagated by those who exploit you to facilitate that exploitation. The socially constructed you died. The civilized you died. The manufactured, fabricated, stamped, molded you died. The victim died.

And who is left when that you dies? You are left. Animal you. Naked you. Vulnerable (and invulnerable) you. Mortal you. Survivor you. The you who thinks not what the culture taught you to think, but what you think. The you who feels not what the culture taught you to feel but what you feel. The you who is not who the culture taught you to be but who you are. The you who can say yes, the you who can say no. The you who is a part of the land where you live. The you who will fight (or won't) to defend your family. The you who will fight (or won't) to defend the others you love. The you who will fight (or won't) to defend the land upon which your life and the lives of those you love depend. The you whose morality is not based on what you have been taught by the culture that is killing the planet, killing you,[319] but on your own animal feelings of love and connection to your family, your friends, your landbase. Not to your family as self-identified civilized beings but as animals who require a landbase, animals who are being killed by chemicals, animals who have been formed and deformed to fit the needs of the culture.

When you give up on hope—when you are dead in this way, and by being so are really alive—you make yourself no longer vulnerable to the co-optation of rationality and of fear that Nazis perpetrated on Jews and others, that abusers perpetrate on their victims, that the dominant culture perpetrates on all of us. Or rather it is the case that the exploiters frame physical, social, and emotional circumstances such that victims perceive themselves as having no choice but to perpetrate this co-optation on themselves. But when you give up on hope, this exploiter/victim relationship is broken. You become like those Jews who participated in the Warsaw Ghetto uprising.

When you give up on hope, you lose a lot of fear. And when you quit relying on hope, and instead begin to just protect those you love, you become dangerous indeed to those in power.

In case you're wondering, that's a very good thing.

❨ ❨ ❨

I'm talking to a friend, an ex-con, who says he thinks revolutions only take place when some critical mass of people get to what he calls the "fuck it" point: the point where things are so bad that people are finally ready to just say *fuck it* and do what needs to be done.

I can't say I disagree.

It reminds me of a talk I gave a few months ago. I spoke of how so many of my students at the prison fully recognized civilization's destructiveness and were ready to bring it all down. Afterwards someone from the audience stood and said that he was a public defender, and that his experience with his clients was radically different. They did not, he said, want to bring it all down. They merely wanted a bigger piece of the capitalist pie.

What he said struck me immediately as true. But I did not know how to merge that truth with what my former students had told me. Later that night a friend made it clear: the public defender and I were dealing with people who were at different parts of the process of being eaten by the state. The people he worked with had merely been arrested. Perhaps some still thought the system was fair. Perhaps others thought they could beat the system. Perhaps still others hoped merely that the system would not destroy them. None of them had yet reached the "fuck it" point. My students, on the other hand, were at a maximum security prison, many for the rest of their lives. There was no longer any reason for them to believe in the system. They had nothing left to lose.

❨ ❨ ❨

We know what those in power do to those who threaten that power. Jeffrey Leuers burned three SUVs in an act of symbolic resistance, and was sentenced to more than twenty-two years in prison, a far longer sentence than that given to rapists, to men who beat their wives to death, to chemical company CEOs who give so many of us cancer. If we were to seriously threaten the perceived entitlement of those in power to convert the living world into consumer products to be sold, they would kill us.

I don't particularly want to die. I love living, and I love my life. But I'll tell you something that helped me lose at least some of the fear I have that those in power will kill me if I threaten their perceived entitlement to destroy the planet. I asked myself: What's the worst they can do to me? Effectively, the worst they can do is kill me. Yes, they can torture me—as they do to so many—or they can put me in solitary confinement in a tiny box—as they also do to so many—but I would hope (there's that word) that in those cases I'd be able to kill myself if necessary. Well, so far as I can figure, if they kill me, most probably one of three things will happen. One possibility is that when we die, it's "boom, boom, out go the lights," in which case I'll just be dead, and I won't know anything anyway. Another possibility is that after we die we go "somewhere else," whatever that means, in which case I'll just keep fighting them from there. And a third possibility is that after we die we get reincarnated. If that happens, I'll follow the lead of the eighteen-year-old Indian Kartar Singh (Sardar Kartar Singh Saraba, or sometimes Shaheed Kartar Singh Saraba) who fought to drive the British from his home, and who in 1915 was betrayed and caught. When the magistrate overseeing the case was about to choose whether to hang him or imprison him for life, Kartar Singh stated: "I wish that I may be sentenced to death, and not life imprisonment, so that after re-birth, I may endeavour to get rid of the slavery imposed by the whites. If I am born as a female, I shall bear lion-hearted sons, and engage them in blowing to bits the British rulers."[320]

The court decided he was too dangerous to be allowed to live.

I hope he came back to fight again.

《 《 《

The man from the EPA continued, "I'm glad you're not a pacifist. I'm peace-loving myself, but have long studied martial arts. I don't consider this a contradiction. Sometimes danger is a form of protection. There's a reason that even peaceful wild things are born with thorns and claws. The real questions are: how and when you should 'open the can of whoop ass' (that's redneck talk).

"I'm glad that you're willing to eat meat yet you question how meat is produced. This is a very important distinction. I wrote a discharge permit for one
of the largest slaughterhouses in the world. Five thousand cows per day, plus processing of meat from the equivalent of five thousand cows per day killed in off-
site slaughterhouses. That's a lot of slaughter. Pollution output like a big city. This
is the most economically efficient production of meat the world has ever seen,
but highly polluting and unconscionably cruel. I believe it hurts us as a people
to allow this cruelty to animals, and it hurts our souls to pretend meat is raised
in some peaceful rural barnyard.

"You mentioned that you thought that things might go with a Bang. Since 9/11,
I have been working on security issues, vulnerability assessments, response
plans, etc. I know a bit about these matters and agree that there is a very real possibility of use of "weapons of mass destruction" by the U.S. or others. My pet theory, however, is not a bang, but a whimper. As you said, the gasoline party is
over. We've passed the halfway mark of mineable petroleum supply, and the last
half will be harder to extract economically than the first half. (Old Jed won't
find more bubbling crude without high tech equipment and expensive extraction methods.)

"Meanwhile, world consumption is growing.

"As oil, water, and key minerals go into shorter supply, the slow squeeze will
begin. Power structures, political and otherwise, need power to stay in power.
It's hard to run an Empire on an empty tank, and the political/economic pow-
erhouses could find themselves coughing to a stop in some very bad neighborhoods. That is happening now.

"In the twilight of a civilization, the state of emergency or crisis can last a century. There will be key watershed events within that cycle, but in terms of
human experience, this cycle is evolutionary, punctuated by big scary events.
Sort of like low-level warfare. Actually, it is characterized by low-level warfare.
I believe we're at the point of key events in this cycle. Our collective decisions
are critically important right now. I am saddened that we're so collectively
asleep at the wheel, so enamored with the trivial and our trinkets.

"When I look at key points of crux, I think they focus around energy, water,
and food. Gee that was hard: DUH! . . . The basics. The world industrial complex is geared up for overproduction, just as some key resources become scarce.
When hungry people are overproducing widgets, while rich people go in debt
to overconsume widgets, this will produce unexpected feasts and famines. We
can expect more surprises from the energy sector. Infrastructure can be a very
fragile thing if not actively maintained and sustained. Our dependence on

genetically altered monoculture for food crops and animals sets us up for rapid spread of disease. There is a looming Dust Bowl (overgrazing) in China, which will greatly disrupt domestic food production, and this will spread ripples in the pond. We have rolled our own tit into this wringer. You are right: we are really FUCKED . . .

"I noted that some people were very disturbed by the fact that you consider some form of societal collapse is imminent. ('I'm twenty, I want a life, what do I do?') That one surprised me: I realized that I have considered societal crisis as an ongoing given, while others have not. Again, this as an evolving process, which will have flashpoints and key moments of decision.

"There is a way out, but it requires a certain minimum level of focus and engagement from the larger public. Unfortunately, the Bread and Circuses have paid off for Korporate Amerika. Most people are fairly satiated and numb, and they don't have a place to put that vague gnosis of getting screwed. If something happened and the bulb switched on, we could use our remaining wealth as seed money. I wouldn't mind a little Utopian thinking if it were practical and focused, with a vision of a minimized ecological footprint. If we don't embrace that little downgrade of lifestyle now, we will pay dearly, and not that far down the road. I suspect that the downgrade will be forced upon us by the slow squeeze of economic downturn, etc.

"I think folks missed your message of healing. You managed to cry it through. Wish I did that more. You are right: Life is Wonderful, friends are loving, and there is a group of people who are 'getting it.' I am blessed with an occupation that allows me to push in the right directions, a wonderful son, good friends, a herd of nice old bonsai trees, and a bumper crop of watermelons. Lately, my relationships have been deeper than I thought possible. I'm rethinking things. My deepest wishes are changing. All of these are good reasons to stop the Pollyanna routine and get a little busy . . . Nothing wrong with being a mean old protective Earth Daddy. After all, a real good dog knows who to bite."

THE CIVILIZED WILL SMILE AS THEY TEAR YOU LIMB FROM LIMB

CIVILIZED MAN SAYS: I am Self, I am Master, all the rest is Other—outside, below, underneath, subservient. I own, I use, I explore, I exploit, I control. What I do is what matters. "What I want" is what matter is for. I am that I am, and the rest is women and the wilderness, to be used as I see fit.

Christina M. Kennedy [321]

IN THE LAST 24 HOURS, OVER 200,000 ACRES OF RAINFOREST WERE destroyed. Thirteen million tons of toxic chemicals were released. Forty-five thousand people died of starvation, thirty-eight thousand of them children. More than one hundred plant or animal species went extinct because of civilized humans.

All of this in one day.

❮ ❮ ❮

I don't think most people care, and I don't think most people will ever care. We can trot out whatever polls we want to try to prove most Americans actually do care about the Environment™, Justice™, Sustainability™—that they care about anything beyond being left alone to numb themselves with alcohol, cheap con sumables, and television. We can cite (or make up) some poll saying that all other things being equal, 64 percent of Americans don't want penguins to be driven extinct (unless saving them will even slightly increase the price of gaso-line); or we can cite (or make up) some other poll saying that 22 percent of American males would prefer to live on a habitable planet than to have sex with a supermodel (this number climbs to 45 percent if the men are not allowed to brag about it to their friends).[322] But the truth is that it's just not that important to most people—*it* in this case being the survival of tigers, salmon, traditional indigenous peoples, oceans, rivers, the earth; *it* also being justice, fairness, love, honesty, peace. If it were, "most people" would do something about it.

Sure, most people would rather that they themselves be treated with at least the pretense of justice, fairness, and so on, but so long as those in power aren't aiming their Peacekeepers™ at me, why should I care if brown people living on a sea of oil a half a world away get blown to bits? Likewise, so long as the price of my prescription anti-depressants stays reasonably low and the number of TV channels on my satellite dish stays high, why should I care that some stupid fish can't survive in a dammed river? It's survival of the fittest, damn it all, and I'm one of the fit, so I get to survive.

Another way to talk about people not caring what happens to the world is to talk about rape and child abuse. Most rapes are committed not by burly strangers breaking into women's homes, nor by pasty-faced perverts lurking outside schools and in internet chat rooms, but instead by fathers, brothers, uncles, husbands, lovers, friends, counselors, pastors: those who purport to love the women (or men) they hurt. Similarly, most children are not abused by thugs who kidnap them and force them to act in porn films, but by their caretakers, those, once again, who purport to love them, who are supposed to help them learn how to be human beings. And of course these caretakers are taking care to teach these children how to be civilized human beings: teaching them that the physically powerful exploit and do violence against the less physically powerful; teaching them that exploiters routinely label themselves—and probably believe themselves—caretakers as they destroy those under their care; teaching them that under this awful system that's the *job* of caretakers; teaching them that life has no value (for of course we are all born with the knowledge that life has value, a knowledge that must be beaten, raped, and schooled out of us).

Those doing the raping, beating, schooling, are not only some group of strange "others": "trailer trash," "foreigners," "the poor." They include respected members of this society. Within this culture, they're normal people. Their behavior has been normalized.

If normal people within this culture are raping and beating even those they purport to love, what chance is there that they will not destroy the salmon, the forests, the oceans, the earth?

❆ ❆ ❆

A few years ago I had an agent at a prestigious literary agency. The agency's address, if this gives an indication of how fancy schmancy the organization is, was One Madison Avenue (an entire floor, even!). I sent my agent the first seventy pages of the manuscript for *A Language Older Than Words*. She read them, then told me that if I cut the family stuff and the social criticism, she thought I'd have a book. She also told me that I was too angry. If I would only tone down the book and not frighten fence-sitters, she said, I'd have myself a bestseller.

I was shocked. I was of course familiar with the old artistic/literary line, "The devil comes promising a larger audience," but it never occurred to me I'd have the chance to sell out this early in my career.

I responded that there was an old blues DJ I liked to listen to who often said after spinning a song, "If you're not moving after that one, you're dead from the butt down." Well, I said, if you're not angry and frightened now, after everything this culture has done, you're dead from the heart out.

In retrospect, that might not have been the most relational thing I could have said.

We had this conversation the same day U.S.-backed troops massacred the MRTA members who had taken over the Japanese ambassador's house in Peru. I said to her, "If the MRTA members are going to give their lives, the least I can do is tell the truth. You're fired."

Her request—that I tone things down to not offend fence-sitters—is the non-battle-cry of cowards everywhere: Too scared even to say that they themselves are frightened, they resort to telling others—for their own good, of course—to tone down their words or actions so some mythical third party won't be affronted or frightened. *You must never blow up a dam*, they tell us, *or mainstream Americans will consider all environmentalists terrorists. You will actually hurt the cause of salmon.* Likewise, *You must never demand an end to old-growth logging (or even think about stopping industrial forestry), or you will alienate potential political allies.* And, *You must never speak out against capitalism (industrialism, utilitarianism, Christianity, science, civilization, and so on) or no one will take you seriously.*

It's not always cowards who say such lines. Sometimes it's people who for whatever reason fail to grasp the insatiability and utter implacability of the dominant culture's death urge. There were (and are) Indians—many of them—who pleaded with their relations to not upset the civilized: if only we all go along with this latest of the ever-shifting demands of the civilized, the logic went (and goes), we will finally be left somewhat alone on the remnants of our land. And there were Jews—many of them—who fell into the trap Nazis laid, baited with false hopes. If only we are reasonable, the logic once again goes, they, too, will be reasonable. If only we show ourselves to be good and worthy Germans—in some cases even good and worthy Nazis—the mass of good Germans will speak and act to protect us from harm.

What a load of horseshit.

It's easier to see this sad gullibility in retrospect than in the present, isn't it? It always is.

I think it's just as much a mistake to count on help from the mass of good Americans as it was from the mass of good Germans. Some will certainly help, but I don't think there will ever be a mass awakening, where suddenly the majority, or even significant minority, of people do what is best for their landbase.

When I lived in Spokane, I had a friend with whom I would get together for dinner once a month or so. Sometimes we'd go to the symphony, sometimes to pick up trash by the side of a road. And we'd talk. Given what you know about me from my books you can probably guess that I often found myself itching to talk about taking down civilization. That's not an itch I generally leave unscratched. But I was delicate, because nice as this person was, and as dedicated to cleaning up roadside trash, he was definitely what my former agent would have called a fence-sitter. When I'd get too explicit about the need to take down civilization he'd too-quickly make a joke, or get distracted, or suddenly remember something important he had to tell me on some other subject—any other subject—or he would get angry at me about something that didn't actually make him angry. So I learned to keep it light, to only hint, to make smaller and smaller talk while the world burned.

Fast-forward a decade to my last week before I left Spokane. He called me on the telephone. I could tell he was both excited and agitated.

He said, "I did it. I made the plunge."

"What did you do?" I thought maybe he was getting married, though so far as I knew he wasn't dating anyone.

He said, "I wrote a twenty dollar check to a local environmental organization."

I told him, sincerely, that I was happy for him.[323]

The seventeenth premise of this book—and this is sort of a combination of the second premise, that this culture will not undergo a voluntary transformation, and the tenth, that most members of this culture are insane—is that *it's a mistake (or more likely, denial) to base our decisions on whether our actions will or won't frighten fence-sitters or the mass of Americans.*

Sure, we can let the potential response of these people be one more piece of information that helps to influence our choices, but we must always remember that we are only responsible for our own actions. Just as we are not responsible for the choices—retributive or otherwise—made by those in power as putative response to any action we may take, so, too, we are not responsible for the response or non-response of the mass of Americans (or Czechs, Liberians, or Indonesians, for that matter).

Here's another way to put the seventeenth premise: The mass of civilized people will never be on our side.[324] I'm not saying by this that we should give up on educating or informing people (I am, after all, a writer: educating and informing is what I *do*). I'm saying, first, that we need to try to be aware of where our identification lies—with whom or what we identify—and we need to ask ourselves: If what the mass of Americans want is in opposition to what your own particular landbase needs, which do you choose to support? If it comes down to stark choices—which of course it already does—on which side will you take your stand (recognizing also that refusing to choose is just another way of choosing the default)?[325]

Second, I'm saying that we need to be aware that we have a finite amount of time each day and a finite amount of time in our lives, so if we actually hope to accomplish something tangible we need to choose wisely how we spend that time. Some people may feel it's the best use of their time to inch fence-sitters closer to falling to the side of the living, and by all means they should do that. I don't think most fence-sitters are effectively reachable, and so I do not write for them. I write for people who already know how horrible civilization is, and who want to do something about it. I want to encourage them to be more radical, more militant, just as others have encouraged me.

Further, we need to recognize that educating people will only go so far toward saving salmon, sturgeon, marlins, prairie dogs, forests, rivers, glaciers, oceans, skies, the planet. At some point we have to actually *do* something.

The problem is not and has never been that the mass of people do not have enough information, such that if we just present them with enough facts they will strive for justice, for sanity, for what is best for their landbase. Think again about rape. Rape is not caused by a lack of information. Similarly, it doesn't take a genius to figure out that dams kill salmon, or that deforestation kills creatures who live in forests. Would it have merely required information to get the whites who slaughtered Indians (or who took their land after the soldiers had done the slaughtering) to stand with these Indians against members of their own culture? Would it require that today, as traditional indigenous people continue to be put in reserves, concentration camps, prisons, and graves, and as their land continues to be stolen? When cancer kills those we love—our grandparents, brothers, sisters, children, friends, lovers—when chemicals cause little girls to develop breasts and pubic hair, when pesticides make children stupid and sickly, the problem is not education. The problem has never been education. To believe that it is, is to buy into yet one more lie that keeps us from acting to protect ourselves.

Or maybe it's not one more lie, but the same old lie, the same old faith-based excuse for inaction, except that this time instead of it being some mythical god or great mother who will save us if only we act in good enough faith—if only we are nice enough, kind enough, loving enough (using the culture's self-serving and toothless definition of love) to our exploiters—it is some just-as-mythical mass of Americans who will somehow save the day if only—if *only*—we are innocuous enough to not frighten them off (and not coincidentally, if only we do not upset those in power).

Even moreso than most people not being on our side, if we were to truly act in defense of our landbases, of our bodies, we would quickly find ourselves hated by the exploiters (of course), the fence-sitters, mainstream Americans, mainstream liberal activists. (My goodness, if mainstream social justice activists assault people, hold them for cops to arrest, and chant complaints about having their demos ruined just because some people break a few windows, imagine what these same activists would do if people began to strike more than symbolic blows against this death culture?) We would find ourselves hated by everyone who identifies more closely with civilization than with their landbase.

In *The Culture of Make Believe* I was attempting among other things to understand the relationship between exploitation, contempt, a sense of entitlement, threats to that entitlement, and hatred. I had learned that after the American Civil War the number of lynchings in the American South increased by at least a couple orders of magnitude. I wanted to know why. My understanding came when I happened across a line by Nietzsche, "One does not hate when one can despise." I suddenly understood that perceived entitlement is key to nearly all atrocities, and that any threat to perceived entitlement will provoke hatred.

Here's what I wrote:

"Europeans felt that they were (and are) entitled to the land of North and South America. Slave owners clearly felt they were entitled to the labor (and the lives) of their slaves, not only in partial payment for protecting slaves from their own idleness, but also simply as a return on their capital investment. Owners of nonhuman capital today feel they, too, are entitled to the 'surplus return on labor,' as economists put it, as part of their reward for furnishing jobs, and to provide a return on *their* investment in capital. Rapists act on the belief that they are entitled to their victims' bodies. Americans act as though we are entitled to consume the majority of the world's resources, and to change the world's climate. All industrialized humans act like we're entitled to anything we want on this planet."[326]

I then wrote:

"From the perspective of those who are entitled, the problems begin when those they despise do not go along with—and have the power and wherewithal to not go along with—the perceived entitlement. That's where Nietzsche's statement comes in, and that's where hatred of the sort I'm trying to get at in this book becomes manifest. Several times in this book I have commented that hatred felt long and deeply enough no longer feels like hatred, but more like tradition, economics, religion, what have you. It is when those traditions are challenged, when the entitlement is threatened, when the masks of religion, economics, and so on are pulled away that hate transforms from its more seemingly sophisticated, 'normal,' chronic state—where those exploited are looked down upon, or despised—to a more acute and obvious manifestation. Hate becomes more perceptible when it is no longer normalized. Another way to say all of this is that if the rhetoric of superiority works to maintain the entitlement, hatred and direct physical force remain underground. But when that rhetoric begins to fail, force—and hatred—waits in the wings, ready to explode."[327]

The point as it relates to the current book is that if you think the exploiters responded with fury and great violence when capitalists were merely disallowed from owning human beings[328]—when that particular perceived entitlement was thwarted—just imagine the backlash when civilized humans are stopped from perpetrating the routine exploitation that characterizes, makes possible, forms the basis of, and is the essence of their way of life.

The next few pages of *The Culture of Make Believe* continue to elaborate on this idea and I'd like to quote them now at length:

"Pretend that you were raised to believe that blacks—niggers would be more precise in this formulation—really are like children, but strong. And pretend that niggers working for whites is simply part of the day-to-day experience of living. You do not question it any more than you question breathing, eating, or sleeping. It is simply a fact of life: whites own niggers, niggers work for whites.

"Now pretend that someone from the outside begins to tell you that what you are doing is wrong. This outsider knows nothing of the life you live and that your father and his father lived. To your knowledge this outsider has never walked the fields and actually watched the slaves work, has never gone over the figures to see that your farm wouldn't be viable without these slaves, and doesn't know the slaves well enough to know that they, too, could not survive without the things you provide for them. Pretend that your slaves listen to this outsider, and because of this, your relationship with them begins to deteriorate, even to the point that you begin to lose money.

"If it were me—had I been raised under these circumstances and with those beliefs—I think it possible that once I got over my initial shock at the temerity of this outsider meddling in something that is none of his or her business, I would have become angry, and perhaps felt eventually outrage towards this interloper who was threatening to ruin my way of life. Raised in those circumstances, it would have taken more courage than most of us have, I think, to admit that one's way of life is based on exploitation, and to gracefully begin to live a different way.

"It's easy enough at this remove to simply say that slaveholders were immoral, and that members of the KKK and other hate groups were a bunch of stupid bigots with whom we have nothing in common.

"But are you sure?

"Try this. What if instead of owning people, we're talking about owning land. Someone tells you that no matter how much you paid to purchase title to some piece of land, the land itself does not belong to you. No longer may you do whatever you wish with it. You may not cut the trees on it. You may not build on it. You may not run a bulldozer over it to put in a driveway. All of those activities are immoral, because they're based on your exploitation of a living thing: in this case the land. Did you ask the land if it wants you to build on it? Do you care what the land thinks? But the land can't think, you say. Ah, but that's just what you think. It is how you were taught to think. Let's say further that your livelihood and your way of life are based on working this land—the outsiders call it exploiting—and that if the outsiders have their way you'll be out of business. Again and again they tell you that you are a bad person, a stupid bigot, because you refuse to see that your way of life is based on the exploitation of something you don't perceive as having any rights—or sentience—to begin with.

"Angry yet?

"Then how about this? Outsiders take away your computer because the process of manufacturing the hard drive killed women in Thailand. They take your clothes because they were made in sweatshops, your meat because it was factory farmed, your cheap vegetables because the agricorporations that provided them drove family farmers out of business (or maybe because lettuce doesn't like to be factory farmed: 'lettuce prefers diversity,' say the outsiders), and your coffee because its production destroys rainforests, decimates migratory songbird populations, and drives African, Asian, and South and Central American subsistence farmers off their land. They take your car because of global warming, and your wedding ring because mining exploits workers and destroys landscapes and communities. They take your TV, microwave, and refrigerator because, hell, they take the whole damn electrical grid because the generation of electricity is, they say, so environmentally expensive (dams kill salmon, coal plants strip the tops off mountains and generate acid rain, wind generators kill birds, and let's not even talk about nukes). Imagine if outsiders wanted to take away all these things—without your consent—because they had determined, without your input, that all of these things are exploitative and immoral. Imagine that these outsiders actually began to succeed in taking away these parts of your life you see as so fundamental. I'd imagine you'd be pretty pissed. Maybe you'd start to hate the assholes doing this to you, and maybe if enough other people who were pissed off had already formed an organization to fight back against these people who were trying to destroy your life—I could easily see you asking, 'What do these people have against me anyway?'—maybe you'd even put on white robes and funny hats, and maybe you'd even get a little rough with a few of them, if that was what it took to stop them from destroying your way of life."[329]

This is the typical response of the civilized to any threat to their perceived right to exploit. Recall once again Thomas Jefferson's explanation of what would happen to those Indians who fought back: "In war, they will kill some of us; we shall destroy all of them."[330] Unfortunately, Indians and their allies have not yet been able to stop the grinding of this machine-culture. Yet they have still received that fury for even trying, and often for merely existing and showing to their exploiters that other ways of being are possible (and desirable).

You really wanna see some hatred? You wanna see some violence? Thwart the civilized. Shut them down. Stop them from destroying the planet.

The civilized will smile as they tear you limb from limb.

THEIR INSANITY WAS PERMANENT

Now, were Columbus and his fellow European exploiters simply "greedy" men whose "ethics" were such as to allow for mass slaughter and genocide? I shall argue that Columbus was a *wétiko*, that he was mentally ill or insane, the carrier of a terribly contagious psychological disease, the *wétiko* psychosis. The Native people he described were, on the other hand, sane people with a healthy state of mind. Sanity or healthy normality among humans and other living creatures involves a respect for other forms of life and other individuals, as I have described earlier. I believe that is the way people have lived (and should live).

The *wétiko* psychosis, and the problems it creates, have inspired many resistance movements and efforts at reform or revolution. Unfortunately, most of these efforts have failed because they have never diagnosed the *wétiko* as an insane person whose disease is extremely contagious.

Jack D. Forbes [331]

WHY CIVILIZATION IS KILLING THE WORLD, TAKE EIGHTEEN. AN ACTUAL conversation that took place in an exercise center near Seattle. Men and women walked on treadmills as they stared at televisions, read books, or looked in mirrors.

One woman said, "I can't handle my neighbors' trees. I wish she'd cut them down when the crane comes through. After the last storm a branch came right through my deck."

Another woman responded, "I know what you mean. Last year no one in my neighborhood wanted to cut their trees. Luckily, when I had the crane come, everyone on our cul-de-sac changed their minds, and we were able to get rid of sixteen of those trees."

The first: "I still have two trees left. I'm ordering the crane this year. I don't want one to fall on my house."

A third woman, an environmentalist, said, "An arborist could thin the branches so the wind will go through them and the tree won't fall."

The first: "If anyone comes out, the tree goes!"

The third: "Wow. I was just thinking about all the things trees do for us. They exchange carbon dioxide for oxygen. They provide homes for animals, who are fun to watch. They —"

A fourth woman interrupted: "Trees are a mess. You know, the manager here had fourteen taken out of her yard when she moved in so she could have some light. And my neighbor has this stupid 150-year-old tree that just has to go. Its roots are pushing up our three-thousand-dollar shed. No tree is worth that."

First woman: "I'll replant anyway. Just not with some ugly evergreen. Maybe a dwarf tree."

☾ ☾ ☾

I have to admit it discourages me that at this late date we still have to fend off this argument that we must not tell the truth for fear we will frighten or anger the mass of people. Certainly an examination of history shows a greater willingness of the mass of people to participate in the atrocities of the culture than to oppose them. How is it that a sure-fire way for a president to increase his standing in the polls is to invade yet another defenseless country? Or, compare how many Germans were in the Wehrmacht in World War II—or how many were just good Germans—to how many were part of the resistance. One of the reasons members of the resistance knew they had to kill Hitler was because he was so magnificently popular among the majority of people: if Hitler were allowed to speak, they knew the people would listen.

❨ ❨ ❨

WHY CIVILIZATION IS KILLING THE WORLD, TAKE NINETEEN. Two words: Detroit Tigers. No, not because the Tigers are so terrible that they threaten life as we know it—although they are bad, historically bad, bad enough that if there were a hypothetical contest between the 2003 Tigers and the legendary 1899 Cleveland Spiders (20 wins, 134 losses: .130 winning percentage), the only reason the 2003 Tigers would win is because everyone who played for the Spiders is long-since dead[332]—but because more people care about Detroit Tigers than real ones.

I've commented elsewhere how deeply it saddens me that hundreds of thousands of Americans attend sporting events each night, and millions more watch on TV, yet if we try to get a rally together to do something—*anything*—to save salmon, we're lucky to get fifteen people, and they're the same ones who showed up last week to protest the circus, and the week before to hold signs decrying increases in the military budget.[333] You could argue that the difference is advertising—if smooth-voiced announcers constantly exhorted us to blow up dams, and if newspapers daily devoted a dozen pages to the travails of endangered species, then more people would care.

Maybe.

I doubt it.

There's a deeper point to be made here, which is that what people want can to some degree be told—more or less tautologically—by what they do. If more people go to see the Detroit Tigers every summer night than do *anything* to save real tigers from extinction, it's probably because that's what they want to do.[334]

❨ ❨ ❨

Or maybe that's what they think they want.

Or maybe that's what they've been taught to want. Or maybe that's what everything in the culture has led them to want. Or maybe that's what everything in the culture has traumatized them into wanting. (And yes, it's pretty traumatizing to watch the Tigers, but I'm talking about something deeper here.) Or maybe these wants are toxic mimics of real wants. Or maybe they're what people have become addicted to.

My students at the prison who'd been addicted to crank often said they started taking drugs because the drugs felt so good (especially in contrast to the often-not-so-quiet desperation of their lives), yet soon found themselves taking the drugs no longer to feel good but to keep from feeling bad.

I have known women who were sexually abused as children who as adults loathed and feared sex, and who at the same time became extremely promiscuous. They were disallowed from saying no as children, and were trained well in the ways of subsuming themselves in order to please men. Now, these women went to the bars voluntarily, yes? Doesn't that mean they wanted to? They picked up the men for sex. No one put a gun to their head, yes? Doesn't that mean they wanted this?

But it didn't make them feel good. They told me this later. Many hated it. Or did they? They thought they loved it. They thought it validated them. They thought it was what they wanted. But did they? What did they really want?

And the men. What did they want? If all they wanted was to get off, they could have grabbed some hand lotion and saved themselves the trouble of dressing up. If all they wanted was an ego boost, they could have got that through conversation. What did they really want?

Let's go further. What did my father want when he beat or raped us? On one level he obviously wanted to do what he did, or he would not have done it. He had choices, didn't he?

Or did he?

After high school I attended the Colorado School of Mines, a well-thought-of engineering school. I did this because I got an academic scholarship and because I'd been told—I'd internalized—that anyone who got through calculus in high school would be an idiot to pass up an opportunity like this: After college I would be sure to get a high paying job, and isn't that the point of life? Never mind that I didn't like my high school math and science classes. But I still wanted to go to that school, didn't I? Or I certainly wouldn't have gone. Or would I?

These questions go to the heart of everything I'm writing about in this book, and go to the heart of how we'll get out of this mess we're in. We'll talk about how in a while, but first I want to bring in another piece of this puzzle. I receive a lot of letters commenting on the books I've written, and many letters specifically about *A Language Older Than Words,* but no one has ever mentioned what

I've always thought of as one of the most important sections of that book. This is the section where I describe how scientists set out to intentionally drive monkeys insane, to turn them into, to use their words, "monster mothers." Now, part of the reason I put in that section is to ask the implicit question: What sort of evil people would set out to drive some group of others insane? (The answer, of course, is that normally we call these people advertisers, corporate journalists, drill sergeants, prison guards, teachers, or quite often parents.) But the real point is that the treatment these monkeys received from those who were already themselves psychopathological turned the monkeys irredeemably and irrevocably insane. Their insanity was permanent.[335] They were forever unreachable. They were incapable of normal social relations, including normal sexual relations, and had to be impregnated by use of what the human psychopaths called "rape racks." (We can ask, once again, what sort of twisted psyches could conceive such a device.[336]) Nothing other monkeys or humans could do would ever reach these violent and pathetic creatures.[337] Their only relief from this pain of being who they'd been made into—and the only relief for those who then had the misfortune to come in contact with these insane monkeys—came through their own eventual deaths.

Recall the central point of R. D. Laing's *The Politics of Experience*: People act according to the way they experience the world. If you can understand their experience, you can understand their behavior. So, a woman is taught at five years old that she will receive what she thinks is love when she is violated by her caretaker (she may also receive financial rewards). She is also taught that if she resists she will suffer violence and abandonment. How does this affect her later experience, her later behavior? I'm speaking not just of her sexual behavior, but other aspects, too. We can ask similar questions about my father. How did the abuse he suffered affect how he perceives the world, how he *is* in the world, how he treats the world around him? And how did my own abuse affect my perceptions, my being, how I treat the world around me? How did my schooling affect the decisions I made or didn't make about going to a school to study a subject I did not enjoy? Did I want to go there? Who was the I who did?

I need to be clear. I'm not saying that every woman who was sexually abused hates sex, or is unreachable, or must follow some self-destructive path. I'm not saying that everyone who is abused ends up abusing others. I'm not saying that education is never helpful. I'm not saying there is no reason to write. I'm not saying that no one changes. I'm not saying that no one is reachable. I am saying that there are those who are not reachable. There are those who will never be reachable. There are those who have been driven permanently insane, and especially when they have the full power of social (including financial, police, military, and public opinion) support behind them, they will never change, never stop their destructive behavior. The only relief from this pain of being who they've been made into—and the only relief for those who then have the misfortune to come in contact with these insane monkeys, or to be more precise, insane apes—will be through their own eventual deaths. In many ways this is merely a psychological restating of Planck's observation on scientific revolution: "[A] new scientific truth does not triumph by convincing its opponents and making them see the light, but rather because its opponents eventually die, and a new generation grows up that is familiar with it."[338] Only this time we're not talking about something so superficial as scientific beliefs, but the emotional, perceptual, psychological, spiritual foundations of people's (and society's) personalities and worldview. It's as a friend wrote me recently: "Un-metabolized childhood patterns almost always trump adult-onset intellectualizations. Sure, one of the reasons we don't resist more effectively than we do—or less ineffectively—is because the cops will kill us if we do. But I think even more important are the internalized forms of oppression, the transparent mental shackles that continue to curtail our movement without us even being aware of them."

((((((

The same person wrote: "It feels right to say one of the fundamental reasons people don't resist even when it's obvious that those in charge are destroying us is that so many of us just never psychologically grew up."[339]

((((((

So we understand each other: We *need* a healthy landbase. That is the most important thing in the world.[340] While a healthy landbase is not the only thing that matters, it is undeniably true that without a healthy landbase, nothing else matters.

((((((

It should be obvious that what is true on the personal level is even more true on the social level. One reason I have recovered from my childhood to the degree that I have is that I have worked very hard at it, and have had the loving support of my friends, my mother, and my sisters. If I've had to work this hard to make a life after only a formative decade of violence when I was young (as well as coercive schooling, ubiquitous advertising, and the other ways our psyches are routinely—almost mechanically—hammered into, or rather, out of, shape); and when there are so many people who have for whatever reasons not had the opportunity or ability to work toward a recovery, and so who are passing on their pain to those others who have the misfortune of coming into contact with them (and we should acknowledge that those suffering this misfortune include at this point more or less every human and nonhuman on the planet); and when this culture rewards anti-social behavior (meaning behavior that destroys human and non-human communities); how much more difficult it is and must be for an entire culture to change.

More clarity: When I say that most people don't care, I mean this in the most popular sense of the word *care*, as in, "If people just cared enough about the salmon, they would act to protect them from those who are killing them." Obviously they don't care, or they would do what it takes to save them: We're not *that* stupid, and these tasks are not cognitively challenging, once you drop the impossible framing conditions of civilization's perpetual growth and perceived divorce from the natural world (and its perceived divorce from consequence).

There is a deeper sense, however, in which having been inculcated into this death cult(ure), we do care about salmon and rivers and the earth (and our own bodies): we hate them all and want to destroy them. Otherwise why else would we do it, or at least allow it to happen?

Fortunately, there is an even deeper sense in which we do care. Our bodies know what is right, if only we listen to them. Beneath the enculturation, beneath the addiction, beneath the psychopathology, our bodies remember that we are meant for something better than this, that we are not apart from our human and nonhuman communities, but a part of them, that what we allow to be done to our landbase (or our body) we allow to be done to ourselves. Our bodies remember a way of being not based on slavery—our own and others'—but on mutual responsibility. Our bodies remember freedom. Our bodies remember that our intelligence is meant for something better than building monuments to death, that our intelligence is meant to help us connect to the rest of the world, to understand, communicate, relate. Our intelligence is meant, as are the particular intelligences of rivers and manatees and panthers and spiders and salmon and bumblebees, to help us realize and participate—play our part—in the beautiful and awesome symphony that is life.

There are many who will never be able to reach these memories, to accept them in a way that leads them away from their addiction to slavery, their addiction to civilization. That is a tragedy: personal, communal, biological, geological.

But there are others—many of them—who can and do remember the knowledge of bodies, and who are willing to do what is necessary to protect their bodies, their landbases, to stand in solidarity with salmon, grizzlies, redwoods, voles, owls, to work with these others—as humans have done forever outside the iron shackles of civilization—for the benefit of the larger community. And that is a beautiful and powerful and moral thing.

It's also really fun.

You should try it sometime.

《 《 《

If those in power really aren't reachable, and if the majority of people probably will never act to defend their—and our—landbases and bodies, and if the culture is in fact enacting a death urge that will lead to planetary annihilation unless it is stopped, and if you care about your body, your landbase, what are you going to do? What are the right actions to take?

ROMANTIC NIHILISM

One needs something to believe in, something for which one can have whole-hearted enthusiasm. One needs to feel that one's life has meaning, that one is needed in this world.

Hannah Senesh[341]

DURING THE CONVERSATION IN WHICH MY FORMER AGENT TOLD ME that if I ever wanted to reach an audience, I'd have to tone down my work, she also told me that I was a nihilist.

I felt vaguely insulted. I didn't know what a nihilist was, but I knew from her tone that it must be a bad thing. I pictured an angry teenager leaning against a building, wearing black slacks, turtleneck, and beret, scowling and chain-smoking.

But that's not me, so I looked up *nihilist* in the dictionary.

The first definition—that life is meaningless and that there are no grounds for any moral truths—clearly doesn't fit me. Nor is it true that I do not believe in truth, beauty, or love.[342] The second definition— that the current social order is so destructive and irredeemable that it needs to be taken down to its core, and to have its core removed—fits me like a glove, I suppose the kind you'd put on to not leave fingerprints.

I've had a lot of conversations with Casey about nihilism, and about how the whole black turtleneck thing really doesn't work for me. And how I rarely scowl. Emma Goldman is famously (and incorrectly) quoted as saying, "If I can't dance, I don't want to be part of your revolution."[343] Well, I don't like to dance, but if I can't laugh, then you can start the revolution without me.

One day Casey said, "I've got you figured out."

I raised my eyebrows.

"You," he said, "are a romantic nihilist." And then he laughed.

So did I. I laughed and laughed. Yes, I thought, a revolution of romantic nihilists. I would be down for that. Count me in.

❨ ❨ ❨

I did a talk in Portland the other day. I heard that afterwards something of a firestorm erupted on a local discussion website, as some pacifists attacked me for not adhering to the One True Way of Social Change™, and then non-pacifists responded, pacifists re-responded, and so on. A friend told me not to bother going to read the whole thing ("There's nothing useful. Lots of heads in the sand.") but did send me one post that seems to me to capture the essence of what I'm trying to get at (in four short paragraphs instead of hundreds of pages). Here it is:

"Himalayan blackberries are not native to Oregon. Their hideous thorny brambles have taken over huge tracts of land here. They kill native species. They hurt like hell when you step on one or fall into a clump of them. If you try to hack them down they'll grow back (they are tough suckers). If you try to pull them out by the roots their thorns bury themselves in your thumb and fester. The best thing to do for a big field full of blackberries would be to burn it, then bulldoze the hell out of it. Get them out of there down to every last root.

"The social, political and psychological state that we find ourselves in is the cultural equivalent of blackberries. Our culture is invasive, destructive, painful, and should never have been planted in the first place. We are a part of it (whether we want to be or not).

"Derrick Jensen wants to burn it all down.

"I want to drive the bulldozer."

❆ ❆ ❆

A few months ago the editors of *The Ecologist* started a new feature in their magazine: Each issue they ask an environmentalist or writer a series of questions about the books that have most deeply influenced them, and what books they would like to recommend to others. Many of the books are those we might expect, *Small is Beautiful, When Corporations Rule the World, The Lorax.* One writer evidently decided to forego modesty, and recommended his own books.

They asked me. I guess I must have been in a black turtleneck mood, because I let fly with a response that could charitably be described as scowling, if such is possible in writing.

Question one: Which book first made you realise that something was wrong (with the planet/political system/economic system, etc)?

My answer: It wasn't a book. It was the destruction of place after place that I loved. And it was the complete insanity of a culture where so many people work at jobs they hate: What does it mean when the vast majority of people spend the vast majority of their waking hours doing things they'd rather not do? The culture itself convinced me something was wrong, by being so extraordinarily destructive of human happiness and, far more importantly, the world itself.

That said, Neil Evernden's *The Natural Alien* was the first book I read that let me know I was not insane: that the culture is insane. It was the first book I read that did not take the dominant culture's utilitarian worldview as a given.

Question two: Which one book would you give to every politician?

Answer: One that explodes.

Before you freak out, let's change the question and see what you think: Which one book would you give to Hitler, Goering, Himmler, and Goebbels?

Let's ask this another way: Would a book have changed Hitler? I don't think so. Unless it exploded.

And before you freak out at the comparison of modern politicians to Hitler and his gang, try to look at it from the perspective of wild salmon, grizzly bears, bluefin tuna, or any of the (fiscally) poor or indigenous human beings. Those in power now are more destructive than anyone has ever been. And they are for the most part psychologically unreachable. And if someone does reach some politician, that politician will no longer be in power.

I recently shared a stage with Ward Churchill. He said the primary difference between the U.S. and the Nazis is that the U.S. didn't lose.

I responded with one word: "Yet."

Question three: What book would you give to every CEO?

Answer: See above.

Question four: What book would you give to every child?

Answer: I wouldn't give them a book. Books are part of the problem: this strange belief that a tree has nothing to say until it is murdered, its flesh pulped, and then (human) people stain this flesh with words. I would take children outside and put them face to face with chipmunks, dragonflies, tadpoles, hummingbirds, stones, rivers, trees, crawdads.

That said, if you're going to force me to give them a book, it would be *The Wind in the Willows*, which I hope would remind them to go outside.

Question five: It's 2050. The ice caps are melting, sea levels are rising. You're only allowed one book on the Ark. What is it?

Answer: I wouldn't take a book, and I wouldn't get on the ark. I would kill myself (and take a dam out with me). I do not want to live without a living landbase. Without a living landbase I would already be dead. No book would even remotely compensate. Not a million books. Not a million computers. Not a million people would compensate.

NECK DEEP IN DENIAL

We do not err because truth is difficult to see. It is visible at a glance. We err because this is more comfortable.

Alexander Solzhenitsyn[344]

ANYBODY NOT NECK-DEEP IN DENIAL MUST BY NOW UNDERSTAND
that the global economy is utterly incompatible with life. That much is clear. But
why is that the case? Understanding that took me years, even though, when you
get to the bottom of it, it's pretty damn obvious. Here it is: A global economy
effectively creates infinite demand. There you have it. That's a problem, because
no natural community—not even one so fecund as the salmon used to be, or
passenger pigeons, or cod, and so on *ad absurdum*—can support infinite
demand, especially when nothing beneficial is given back. All natural communities survive and thrive on reciprocity and cycles: salmon give to forests who
give to salmon who give to oceans who give to salmon. A global economy is
extractive. It doesn't give back, but follows the pattern of the machines that
characterize it, converting raw materials to power.[345] Combine an extractive
(machine) economy with infinite demand, and you've got the death of pretty
much everything it touches. Duh. I first gained this understanding from an
email someone sent me. She lives in Canada and wrote that until a few years previous her valley had been full of grizzly and black bears. She used to see maybe
a dozen bears on an average spring, summer, or fall day. Now she was lucky to
see one a week, and it was usually the same bear. The difference, she said, was
that hunters had discovered the Chinese market for bear gall bladders. The market would consume as many gall bladders as the hunters could take. So they
took them all. It was immediately clear to me that the local human community
could have killed basically as many bears as they wanted for gall bladders,
because I'm sure the market is pretty small there. And besides, if they kill all the
bears, how will they get more gall bladders tomorrow? But as soon as you open
up the market to the entire world, not only do you lose the face-to-face feedback
of seeing your future supplies dwindle on the altar of today's profits, but the
demand for something even as esoteric as gall bladders becomes more or less infinite. No population can support that. That is exactly what happened to great
auks, passenger pigeons, Eskimo curlews, cod, salmon, sperm whales, right
whales, blue whales, humpback whales, roughy, sharks, white pine, redwood.
Everything. No population can support infinite demand. No population can
survive a global economy. The problem is inherent, not soluble by any amount
of tinkering.[346]

The same argument reveals, by the way, how it is that within this culture every technological innovation is turned to evil. Let's say I live in a human-sized community, less than a hundred and fifty people or so.[347] I invent something. Within that functioning community—one in which we know we'll be living on this land we love forever, and so we have to get along not only with each other but with all our nonhuman neighbors—we will then have ways to make decisions how to use (or not use) this technological innovation. I've been told, for example, that the Okanagans of what is now British Columbia divide their community for decision-making purposes into four groups by proclivity and expertise. One group is the youth, which doesn't necessarily mean the young, although they often are. These people have tremendous creative energy, and yearn for change that will bring a better future. They're creative, and theoretical, and they tend to move and think quickly. The next group are the elders, who are concerned with protecting traditions. They move slowly. They're interested in the sacred and in deeper awareness. The next group, the fathers, are more action-oriented, and are concerned with security, sustenance, and shelter. Members of the final group, the mothers, the nurturers, want to make sure everyone is taken care of. They process a lot and ask, "How's *everyone* going to be affected by this?"[348] Members of these four groups will formulate their opinions on the innovation, and facilitators will help the community and its leaders come to an eventual decision. So, let's say I invent something with both beneficial and harmful uses, depending on who's using it. We as a community decide whether we want it, whether it will enhance our lives and the lives of our nonhuman neighbors, and how we will use it, if we use it at all.

Now, let's say I invent this thing, and let's say it does have serious harmful uses, and let's say the community tells me not to use it. Let's say I ignore them. This would be exceedingly strange in the first place. Picture a healthy family that has decided as a unit that they do not want any of their members (or anyone else) to put poisons on their food, nor to toxify the water they drink, nor to toxify the air they breathe. What reasonable member of this family community would be so horrid as to proceed anyway? But let's say I do. I don't know why I do. Maybe I'm a capitalist. Or maybe I'm a sociopath. Maybe the former is a subcategory of the latter. Within a healthy functioning community I would be dissuaded from acting such, or if that didn't work I would be disallowed, or if that didn't work I would be exiled or killed. I would not be allowed to harm the community in this way.

But, and here's the point, when your invention moves beyond the local community, when, as Mumford stated approvingly as a purpose of civilization, you "make available to all men [*sic*] the discoveries and inventions and creations, the works of art and thought, the values and purposes that any single group has discovered,"[349] you move beyond face-to-face accountability, which means there are no longer those immediate and vital checks on harmful uses. Further, and even worse, let's say I invent something we in our community perceive as having only beneficial uses. Our community says it's great to go. But just as if you have a big enough economy *someone* is going to come up with a way to make money off bear gall bladders, thus guaranteeing the bears' demise half a world away, if you have a big enough pool of people with access to the original invention, *someone* is going to figure out a way to use for ill almost anything you make. Remove accountability, create mass communication, and voila! Suddenly everyone's harmfully using this previously beneficial invention. And if everyone else is doing it, wouldn't I be a damn fool to do otherwise?[350]

((((

Okay, so maybe I'm wrong. Not about civilization killing the planet. That's obvious. But about the whole Earth Mother/Benevolent God/Santa Claus/Easter Bunny thing. Maybe the Great Spirit is watching over us, and is going to help us out of this mess.

The last few days I've been thinking about a parable I heard when I was young. A Christian is walking down a road in India. Suddenly a throng of people comes running the opposite direction. When they get close enough, he hears them cry that an elephant has gone crazy (or maybe sane, depending on your perspective), and is trampling people up ahead. The Christian says, "I am not worried. God will take care of me."

He keeps walking.

Another crowd rushes by delivering the same message. He responds, "I am not worried. God will take care of me."

He keeps walking.

Yet another crowd. Same message. Same response: "I am not worried. God will take care of me."

He keeps walking.

He sees the elephant.

He's a little worried.

But God will take care of him.

The elephant sees him. The elephant rushes him. The elephant stomps him flat.

As he is dying, he turns against God, curses, and moans, "God, why didn't you take care of me?"

Then he hears the voice of God, clear and strong, "You idiot! Why do you think I sent all those people to warn you?"

I have no doubt that when the people who are relying on the Great Mother to clean up their toxic messes die, the Great Mother will say to them something similar: "You idiot! Why do you think I sent all of those catastrophes to warn you? What do you think was the message behind global warming, behind little girls getting pubic hair, behind mass extinctions, behind the epidemic of cancer?"

☾ ☾ ☾

A series of dreams. In the first, I'm in a canyon. Like the Grand Canyon, it's huge. Also like the Grand Canyon, it's on the Colorado River. But it's near the ocean. I can hear the waves. Like the Colorado, the river no longer reaches the sea, but dies in sand and dirt, its water—its blood—sucked away by cities, by the civilized, held back by dams no one in this dream dreams of removing. In this dream, hydrologists and geologists and environmentalists and all sorts of other -ists dig little trenches in the sand where they place little fishes one by one in the hopes that water will magically rise up from the soil to keep the fishes alive. The ocean roars in the distance, the fish flop and die on the dry and sandy soil, the -ists stroke their chins in consternation, standing in the shadow of the dams, and do what pathetically little is available to them to save the river that they themselves are helping to kill by their stupidity and blindness.

This is what we do.

Later that night, I dreamt I was fighting a lich: a user of magic who had been not living, not dead for several thousand years. In this dream I had magic, too, but of a different sort, and each time he tried to freeze me in place, or suck away my life—as he had done to so many others, and as he must do if he is to continue to not-live, not-die—I struck him back twice as hard as he tried to strike me. He began to fear me, and then he began to weaken. Soon it became clear he was going to die. He kept fighting—because that was what he had done for so long—but suddenly I understood that not only was it my task to not let him kill me, and not only was it my task to kill him, but even more it was my task to release him from his undead state, to grant him the release that all undead[351] secretly (even to themselves) desire. It was my task to teach him a lesson known to every tadpole, every raindrop, every sea anemone, every mountain, every elephant, every uncivilized human being: how to die. It was my task to finally and completely kill him.

This is what we must do.

But the dreams did not end there. Still later in the night, I was given a box of puppies, which I carried through a city. Although all of the puppies were from the same litter, many were tiny, smaller than the smallest runts I've ever seen. I had to hurry to return them to their mother. I searched and searched for a way out of the city, and at last reached a forest. There, their mother waited for them. I gave her back her children. Some, I knew, would live. And I knew that some would die.

This, I knew in the dream, is true as well for all of us—human and nonhuman alike—who are boxed up and separated from our source of sustenance, who are being killed by the fumes and emptiness of everything our cities represent and are. Some will live, and some will die. And I knew in the dream also that this is just as true for those of us who fight the system, those of us who fight the lich, those of us who do not merely dig tiny trenches in the barren sand below the big dams that need to be taken down: some of us will live and some of us will die.

MAKING IT HAPPEN

Every individual who wants to save his humanity—and, indeed his skin—had better begin thinking dangerous thoughts about sabotage, resistance, rebellion, and the fraternity of all men [and women] everywhere. The mental attitude known as "negativism" is a good start.

Dwight MacDonald[352]

Individuals have international duties which transcend the national obligations of obedience.

U.S. et al v. Goering et al[353]

Anyone with knowledge of illegal activity and an opportunity to do something about it is a potential criminal under international law unless the person takes affirmative measures to prevent the commission of crimes.

Tokyo War Crimes Tribunal[354]

AT THIS REMOVE, WHO ARE THE REAL HEROES OF THE THIRD REICH? Hitler, Bormann, Himmler, Goering? I think not: I don't imagine many parents gift their male children with the first name Adolf anymore. You may as well call your kid Caligula. Those who ran the Third Reich are rightly reviled. So are their lieutenants, people like Frank, Eichmann, and Kaltenbrunner, people who were deservedly hanged for carrying out the evil (if the word *evil* is to mean anything at all, it must apply here) plans of their leaders. The same is true for the loyal generals, like Keitel and Jodl, both of whom were hanged for planning and waging aggressive war (U.S. generals would be well-served to read *Justice at Nuremberg* and other texts describing their own fate should justice ever befall them—that is, if they can stand how itchy this reading might make their own necks). And the same is true for their propagandists, like Goebbels and Streicher (to save capitalist journalists from actually having to venture into the unknown territory of performing independent research, I'll just say right out that Goebbels killed himself, and Streicher was hanged for the effect his lies had on the furthering of atrocity).

No, the real heroes of the Third Reich are not the now-dead equivalents of Bush, Cheney, Rumsfeld, Powell, and company. Nor are they the also-dead equivalents of Gates, Hurwitz, Trump, and others whom we're taught to admire and emulate. Nor are they the equivalents of Peter Jennings, Tom Brokaw, Dan Rather, Barbara Walters, and others who lie to us, distract us, while the world is murdered.

The real heroes of the Third Reich are those like Axel Freiherr von dem Bussche and Ewald Heinrich von Kleist, those who took it upon themselves to try through any means necessary to stop the evil in which they found themselves immersed. It is Count Claus Von Stauffenberg (killed by the Nazis July 20, 1944), who had lost an eye and an arm fighting for Germany, and still managed to plot and plan for years, and to plant the bomb that on July 20, 1944, nearly succeeded in killing Hitler. It is Ludwig Beck (killed by the Nazis July 20, 1944), who in 1938 resigned as Chief of the German General Staff rather than lead his country into war, after that becoming the spiritual leader of the native resistance. It is Admiral Wilhelm Canaris (tortured,[355] then killed by the Nazis April 9, 1945), leader of German military intelligence (the Abwehr), who made

sure his organization passed on full information to the Allies, and who did
everything he could to take down the Nazis. It is the brilliant general (and
field marshal) Erwin Rommell (forced by the Nazis to commit suicide October 14, 1944) who used his position of privilege to the advantage of the resistance. It is Hans Von Donhanyi (tortured, then killed by the Nazis April 9,
1945), an Abwehr agent who successfully led a group of Jews disguised as
Abwehr agents to Switzerland. It is Hans Oster (tortured, then killed by the
Nazis April 9, 1945), who used his position to provide explosives to the resistance. It is Jesuit Priest Alfred Delp (tortured, then killed by the Nazis February 2, 1945), who recognized the role the Christian church played in Hitler's
popular support,[356] and did everything he could to counter that, including
advocating Hitler's assassination. It is General Henning von Treskow, who
had long worked toward the assassination of Hitler, and who blew himself up
with a hand grenade after the failure of the July 20, 1944 plot, leaving us his final
words, which we may wish to modernize and take to heart: "Now the whole
world will attack us and abuse us. But I am still absolutely convinced that we
have acted rightly. I believe Hitler to be not only the arch-enemy of Germany,
but also the arch-enemy of the world. When in a few hours' time I appear
before the judgement seat of God to give an account of my deeds and omis-
sions, I believe I shall be able to answer with a good conscience for what I
have done in the struggle against Hitler. Just as God once promised Abraham
that he would spare Sodom, if only ten just men could be found in it, so I
hope God will not destroy Germany because of us. None of us can complain
about his death. Whoever joined us, put on the shirt of Nessus. A man's moral
worth begins only when he is ready to give his life for his convictions."[357]

                                    (((

Six or seven years ago I gave testimony before several panels of the National
Marine Fisheries Service, the Northwest Power Planning Council, the United
States Fish and Wildlife Service, and other agencies overseeing the murder of the
salmon. The ostensible purpose of these panels was for citizens to give representatives of government and industry input concerning the fact that dams on the
Columbia and other rivers kill salmon. The real purpose was for all of us—
myself included—to make ourselves feel good by pretending to do something
useful while we stood by and watched salmon rapidly slide to extinction.

Here's the testimony I gave at one such panel:

"In 1839 Elkanah Walker wrote in his diary, 'It is astonishing the number of salmon which ascend the Columbia yearly and the quantity taken by the Indians.' He continued, 'It is an interesting sight to see them pass a rapid. The number was so great that there were hundreds constantly out of the water.' In 1930 the *Coeur d'Alene Press* wrote, 'Millions of chinook salmon today lashed into whiteness the waters of northwest streams as they battled thru the rapids.' The article went on to say that 'the scene is the same in every northwest river.' *The Spokesman-Review* noted that at Kettle Falls, 'the silver horde was attacking the falls at a rate of from 400 to 600 an hour.'

"And now? In order to serve commerce this culture dammed the rivers of the Columbia watershed. Local groups and individuals—including those who knew the salmon most intimately, the Indians—fought against the federal government and the river industries, but dams were built, and now most runs of salmon in the Northwest and California are extinct or on the verge.

"The destruction of the salmon is not unique. It is the story of this culture. After a leak of poisonous gas from Union Carbide's plant in Bhopal, India, killed up to fifteen thousand human beings and injured up to five hundred thousand, an anguished doctor made the common-sense proclamation that the company 'shouldn't be permitted to make poison for which there is no antidote.' That's what dams have been since the beginning: 'a poison for which there is no antidote.'

"In order to make the cultural pattern perfectly clear, here are more poisons this culture has created without creating antidotes: It created the toxic mess at the Hanford Nuclear Reservation with no consideration for how to clean it up; before the first atomic bomb's detonation, scientists feared the explosion would create a chain reaction destroying the atmosphere, yet they proceeded; this culture has clearcut its way across this continent—indeed across the planet—with no thought to an inability to restore those forests; politicians do their damnedest to allow pollution of aquifers with no clue how to clean them up; global warming, the ozone hole, acid rain, and other results of technological 'progress' are examples of poisons for which there are no antidotes.

"Why does this culture do this? One reason is that within this culture knowledge and technological 'progress' are driven by fiscal profitability. This fiscal profitability inevitably involves forcing others to pay for the economic activities of the producers. The Downwinders—and all humans and nonhumans who will live in eastern Washington for the next 250,000 years—pay for Hanford with their health; those who drink from Spokane's aquifer pay with their health for the economic activities of those who pollute it; the salmon and those of us who would have eaten them—or merely watched them climb

Kettle Falls—pay for the profits of the industries that have turned the rivers into a series of lakes.

"Recently, Senator Slade Gorton commented on salmon: 'There is a cost beyond which you just have to say very regrettably we have to let species or subspecies go extinct.' I would turn that statement around: There is a cost beyond which you just have to let destructive pieces of technology go extinct. There is a cost beyond which you have to let a treasonous collaboration between government and industry go extinct. There is a cost beyond which you have to let destructive worldviews go extinct. There is a cost beyond which you have to let civilization go extinct. The extinction of the salmon is not a price I'm willing to pay to support the irrigators, barging industry, aluminum industry, and producers of electricity, each of which is fighting desperately to cause salmon to go extinct.

"It may be incorrect to say outright that dams are 'a poison for which there is no antidote.' There is a realistic way to save salmon. I'm not speaking, of course, of the runs already extinct. The culture will forever carry that crime on our collective conscience. But other runs can be saved by a simple expedient. Remove dams that kill salmon. Blow them up. Even from a strictly economic perspective (in other words, from a perspective that ignores life), the dams aren't necessary: Randy Hardy, Bonneville Power Administration Head, admits there is a 'glut of power on the market at rates lower than' that of the dams. Yet instead of removing dams the Administration's response is to approach state and federal governments to request further subsidies. The public pays to kill the salmon. Corporate interests obstruct the removal of dams just as dams stand in the way of salmon on their way to spawn. For years politicians have studied the salmon to death, with each study revealing what we already know, that dams kill salmon. We've known this forever: laws were passed during the reigns of both Richard the Lionheart and Robert I (Robert the Bruce) in the twelfth and fourteenth centuries forbidding the erection of fixtures that would impede the passage of salmon on rivers and streams.[358]

"Steve Clark of the Bureau of Reclamation gave us the real reason for the studies, when he said that he wished that salmon would go extinct so that we can 'get on with living.'

"Industry representatives at this and other panels have repeatedly stressed the need for proven solutions. I will give them a proven solution: blow the dams and allow the Columbia to once again be a wild river. It is time for us to stop studies that have been a mere stalling tactic on the part of politicians and the business interests they represent. It is time to find a way to remove the dams—dams that are killing salmon—so that we, and the salmon, can get on with living."

I received a standing ovation for that testimony (from the audience, obviously, not from the panelists), and throughout the rest of the evening, many of the speakers said simply, "I support the Jensen alternative. Blow the dams."

Weeks later, I gave the following testimony at another panel:[359]

"Every morning when I wake up I ask myself whether I should write or blow up a dam. Every day I tell myself I should continue to write. Yet I'm not always convinced I'm making the right decision. I've written books, good ones, and people have read them. At the same time I know it's not a lack of words that's killing salmon, but rather the presence of dams.

"Anyone who lives in this region and who knows anything about salmon knows the dams must go. And anyone who knows anything about politics knows the dams will stay, at least for now. Scientists study, politicians and businesspeople lie and delay, bureaucrats hold sham public-input meetings, activists write letters and press releases, I write books and articles, and still the salmon die. It's a cozy relationship for all of us but the salmon.

"In the 1930s, prior to building the dams, the United States government knew the dams would kill salmon, and proceeded anyway. One reason they proceeded, and they were very explicit about this, is that salmon are central to many of the region's indigenous cultures, and much as killing buffalo helped bring Plains Indians to terms, the government knew killing salmon would break the collective cultural back of the region's Indians. This is all a matter of public record. I repeat, one explicit reason dams were built was to destroy salmon stocks, and thus destroy native cultures. This is genocidal behavior under the law. It is a Crime Against Humanity, and anyone who participates in it, to this day, is guilty of a Crime Against Humanity.

"Make no mistake. The dams are instruments of genocide, just as surely, explicitly, and intentionally as the gas chambers at Treblinka, Birkenau, and Auschwitz. Lest you think this connection is spurious, no less an authority than Adolf Hitler said he based his genocidal lebensraum policy on the 'Nordics,' as he called them—that's us—of North America who'd had 'the strength of will,' in his words, to exterminate an 'inferior' people.[360] Just as Hitler's genocide was only able to take place through the witting or unwitting assistance of hundreds of thousands of bureaucrats, technicians, scientists, businesspeople, and politicians who were merely 'doing their jobs,' so, too, with this ongoing genocidal and ecocidal project.

"From the inside, it's possible to rationalize any horror. First person account after first-person account of genocide and ecocide reveals that the psyches of nearly all high-level perpetrators are surrounded by an almost impenetrable wall of denial and abstract justification. Nazis never killed Jews; they used 'scientific treatments' to improve the health and vitality of the German nation. By the same token, members of this culture have never killed Indians or destroyed their cultures; it's merely Manifest Destiny to 'overspread the continent.' Likewise, none of you on this panel are killing salmon, you're producing electricity and helping irrigators.

"Any of you who represent Kaiser Aluminum, Bonneville Power, or other corporate, commercial, or governmental interests—which are fundamentally the same—and who fail to see how you are lending your talents to a genocidal project—why mince words, how you are committing genocide—are in famous company. While on trial for his life in 1961, part of Adolph Eichmann's defense was that no one ever told him what he was doing was wrong. Eichmann was merely running a railroad, efficiently transporting human cargos east and cargos of clothing, hair, and gold fillings west. His hands were clean. He killed not a single Jew. Yet by lending his talents to the project he was responsible—and was ultimately held responsible—for the deaths of millions. The same holds true today for each of you. You are merely trying to improve corporate profits, or make the region's economy run more smoothly, or otherwise just 'doing your job.' But 'doing your job' in this case means committing ecocide and genocide.

"I say this to every bureaucrat here, to every representative of Kaiser Aluminum, the Bonneville Power Administration, Senator Slade Gorton's office, Senator Larry Craig's office, Senator Jim McClure's office, to all members of the Northwest Power Planning Council: I will not allow you Eichmann's excuse. What you are doing is wrong. It is genocidal conduct under the 1948 United Nations Convention on Punishment and Prevention of the Crime of Genocide, to which the United States *is* a signatory. I plan on someday seeing each and every one of you brought to justice and accountability for your crimes, and I want for it to be a matter of public record that you have been told that what you are doing is wrong.

"As for the rest of us, those of us who care about salmon, we must learn the difference between real and false hopes. Sea minks, great auks, passenger pigeons, Eskimo curlews, Carolina parakeets, great runs of salmon. You would think by now we would have learned that this economic and political structure is antithetical to life on this planet.

"We keep hoping that somehow corporations like Kaiser Aluminum will do the right thing.[361] By now we should have learned. To expect corporations to function differently than they do is to engage in magical thinking. The specific and explicit function of for-profit corporations is to amass wealth. The function is *not* to save salmon, nor to respect the autonomy or existence of indigenous peoples, nor to protect the vocational or personal integrity of workers, nor to support life on this planet. Nor is the function to serve communities. It never has been and never will be. To expect corporations to do anything other than the purpose for which they are expressly and explicitly designed, that is, to amass wealth at the expense of human and nonhuman communities, is at the very least poor judgment, and more accurately delusional.

"Similarly, after Hanford, Rocky Flats, the Salvage Rider, dams, governmental *inaction* in the face of the Bhopal, the ozone hole, global warming, the greatest mass extinction in the history of the planet, surely by now there can be few here who still believe the purpose of government is to protect us from the destructive activities of corporations. At last most of us must understand that the opposite is true: that the primary purpose of government is to protect those who run the economy from the outrage of injured citizens.

"The responsibility for protecting our landbases thus falls to each of us. This means that all of us who care about salmon need to force accountability—*force accountability*—onto those causing their extinction; we must learn to be accountable to salmon rather than loyal to political and economic institutions that do not serve us well. If salmon are to be saved, we must give BPA and Kaiser Aluminum a reason to save them. We must tell these institutions that if they cause salmon to go extinct, we will cause these institutions to go extinct. And we must mean it. We must then say the same to every other destructive institution and to those who run them, and we must act on our words; we must do whatever is necessary to protect our homes and our landbases from those who are destroying them. Only then will salmon be saved. Only then will the genocide stop.

"Saving salmon from extinction means taking out dams. Everyone knows this. Even the Corps of Engineers now acknowledges this. But there is a vast difference between acknowledgement and action. So we must tell the government that if it will not help us in this, if it will not back up our resolve to save salmon, to stop the committing of genocide, to save our communities, if it will not remove the dams, then it must be us who do so. Again, we must mean it.

"When dams were erected on the Columbia, salmon battered themselves against the concrete, trying to return home. I expect no less from us. We too must hurl ourselves against and through the literal and metaphorical concrete that keeps us imprisoned within an economic and political system that does not blanche at committing genocide and ecocide.

"I've been told that before making important decisions, members of many native cultures would ask, 'Who speaks for wolf? Who speaks for salmon?' I ask that here. If salmon were able to take on human manifestation, to assume your body, or yours, or yours, or yours, what would they do?

"And why aren't you doing it?"

The response by members of the panel? They called security on me.

The response by the rest of us, myself included? The dams still stand. The salmon still slide toward extinction.

So much for discourse.

❨ ❨ ❨

It's one thing, as my friend Jim at the Post Office pointed out, to talk or write about taking out dams, to talk or write about taking down civilization, to talk or write about protecting the landbases where we live, and it's quite another thing to make it all happen.

❨ ❨ ❨

I'm riding in a car with my friend Carolyn Raffensperger. It's late, and we're making good time across northern Iowa, in part because everyone else drives so fast. If I drive 85, everyone passes me. Driving 75 in, say, Oregon, makes me the fastest driver on the road. Carolyn asks what I hope to accomplish with my work.

I say, "I would like to change discourse so that we start talking honestly and deeply about bringing down civilization."

She responds immediately: "That's not what you want."

"You're right," I say. "That's not what I want. I want to bring it all down."

"Yes," she says."

❨ ❨ ❨

I need to be explicit. While I think it's pretty easy (and necessary) to make a moral and tactical case for the assassination of Hitler, I'm not attempting to make a moral or tactical case for assassinating Bush, or for that matter, any other American political figure.

In the early days of the resistance to the Nazis, many still believed it possible to overthrow the regime without killing Hitler.[362] But, as Peter Hoffmann notes in his crucial book *The History of the German Resistance 1933–1945*, "As the war went on influential opposition circles came to realize that the removal of the dictator in person, his murder in other words, was an essential prerequisite to the success of any attempted *coup*. A sacred oath had been sworn to him; in strict legal terms and in the minds of the unthinking citizenry and soldiery, the majority in fact, he was the legally established warlord and Supreme Commander. Unless, therefore, its Supreme Commander were first removed, the Army could not be counted upon; yet it was the sole instrument with which a *coup* could be carried out."[363] Pacifists can complain all they want about this statement, but those in the resistance knew more about this than the pacifists or I ever will. And Hitler said much the same thing in his own inimitable way, "There will never be anyone in the future with as much authority as I have. My continued existence is therefore a major factor of value. I can, however, be removed at any time by some criminal or idiot."[364]

I don't believe that's the case in the United States. I'm sorry to break the news to you, George, but I don't think you're as central to the continuation of the U.S. corporate (or, following Mussolini, fascist) state as Hitler was to his fascist (or, chasing Mussolini back the way he came, corporate) state. If you were assassinated by, say, an extremely dedicated pretzel, I'm certain that literally hundreds of millions of people worldwide would feel a certain sense of relief (but those people of course don't count, since most of them are poor, I mean, terrorists), yet the sad truth is that the United States economy would trundle on, destroying the lives of countless humans and nonhumans the world over.[365]

The question becomes, what do we want to accomplish? The honest answer to that question will point us toward some probable courses of action. (Similarly, examination of our actions and inactions will probably make clear what we really want.)

If we want to bring down the Nazis, we probably have to kill Hitler (among many other tasks). The question becomes a technical one: how do we do it? Similarly, if we want to save salmon, we face six relatively straightforward technical tasks: 1) remove dams, 2) stop deforestation, 3) stop commercial fishing, 4) stop the murder of the oceans, 5) stop industrial agriculture (which destroys waterways by erosion and pollution run-off), and 6) stop global warming, which means stopping the oil economy. With the exception of global warming, which may soon enter a runaway phase, these are very doable, in fact should be reasonably easy for a species and a people who pride ourselves on our problem-solving abilities. The problems only seem insoluble when we refuse—like the Nazi doctors—to look outside the confines of this extractive, exploitative social structure, and outside of a mythology that causes many to pretend that one can kill the planet and live on it. We can't have dams and salmon. We can't have deforestation and salmon. We can't have commercial fishing and salmon. We can't have global warming and salmon. If we want salmon, we have to stop each of these.

What would we do, I ask myself again and again, if we fully internalized the understanding that the government is a government of occupation, and the culture is a culture of occupation? What would we do if space aliens (or commie pinko Russkies, or Islamofascists, or ChiComs, or whoever is the enemy of the day) were erecting and maintaining dams on rivers we love and rely on, if they were cutting down forests we love and rely on, vacuuming oceans we love and rely on, changing the climate? Wouldn't stopping them become a series of straightforward if perhaps daunting tasks? Isn't that what happens when we cease to identify with the culture that is killing the planet, and remember to identify with our own landbases?

Several pages ago I outlined some possible courses of action for those who don't want to personally participate in bringing down civilization but who agree that: a) civilization will crash, b) the crash will be messy, and c) because civilization is systematically destroying the planet, the longer civilization lasts the worse things will be. Now, however, I want to ask the other half of that question: if we agree with each of those premises, and if we do want to bring it all down, how do we do that?

FULCRUMS

So many objections may be made to everything, that nothing
can overcome them but the necessity of doing something.

Samuel Butler[366]

IF WE'RE GOING TO TALK ABOUT BRINGING DOWN CIVILIZATION, WE NEED TO talk about fulcrums.[367]

If you recall, Archimedes said something to the effect of "Give me a long enough lever and a place to stand and I can move the world." Well, he was being concise; by emphasizing the length of the lever and the place to stand he left off the lever's other crucial component: the fulcrum. Archimedes could have the longest and strongest board in the universe, and the most solid place to stand, and he still wouldn't have been able to leverage his strength without that pivot point.

The purpose of a lever is to transmit or modify (often magnify) power or motion. I can bend metal with a crowbar I couldn't budge using muscles alone. I can crack nuts easily with a nutcracker, and moving heavy weights is a piece of cake with a wheelbarrow.

What does this have to do with taking down civilization?

Everything.

So long as the dominant culture is still dominant—by which I mean so long as its exploitative mindset holds sway over what's left of the hearts and minds of the people who run this culture—there will always be a disproportionate number of people willing to kill to perpetuate it (to gain or maintain the power, or the promise of power, associated with being an exploiter[368]) compared to the number who are willing to fight to protect life. It's Jefferson's line all over again: "In war they shall kill some of us; we shall destroy all of them." And those who are willing, ready, and oftentimes eager to destroy those who threaten the hegemony of those in power often include their hired guns: Those in power worldwide have about 20 million soldiers and 5 million cops at their command. In the U.S. alone, these numbers are about 1.4 million soldiers and 1.4 million cops (one-third of whom are prison guards), the primary function of whom is to use violence or its threat to serve those in power. Far worse, nearly all of us have allowed ourselves to become convinced of the righteousness of Premise Four of this book: that violence flows only one direction, that it is right and just for servants of power to kill in that service (yet it is proper for their leaders to inevitably declaim on the regrettability of these inevitable murders), and it is blasphemy for the rest of us to fight back.[369] This latter is as true for mountain lions who fight

389

back against those who wish to destroy their habitat as it is for humans who fight back against those who wish to destroy their habitat.

All of this is a roundabout way of saying that those in power have the luxury of using that power inelegantly. They can and often do simply overwhelm us with sheer force. ("Shock and awe" is one of the currently preferred terms.) Those of us fighting for life, on the other hand, need to learn how and where to find appropriate fulcrums to amplify our efforts.

《 《 《

From the perspective of members of the German resistance in World War II, Hitler was certainly one such fulcrum. Killing just this one man would have multiplied their efforts to the tune of saving millions of lives. Had someone killed him before the war started—and some tried—the effects of their efforts would have been multiplied by tens of millions.

《 《 《

One man acting alone very nearly curtailed World War II. No, I'm not talking about Neville Chamberlain, Winston Churchill, Josef Stalin, or even Ludwig Beck. I'm talking about Georg Elser.

Who?

Georg Elser was a German who hated what Hitler was doing to his country, and especially what he was doing to workers. Further, he understood that the Nazis were driving his country to war, and thought that by murdering Hitler he would be doing something both great and good.[370] In other words, he was a good German, if we just this once use that term sincerely.

He knew that every year on November 8 Hitler gave a speech at the Löwen-bräu restaurant[371] in Munich in honor of his failed 1924 putsch against the Weimar Republic. In 1938 Elser attended the speech to reconnoiter the hall. Realizing almost immediately there was no way he could get close enough to shoot Hitler with a pistol, he determined to build a bomb.

He got a job at a quarry for the express purpose of stealing 120 pounds of explosives. ("The entrance to the explosives storage bunker was sealed by a door, to which Vollmer [the quarry's director, who later was sentenced to two years in prison merely for having hired Elser] held the key. Elser secured three different keys of the approximate size and returned to the quarry late one night to try them all. Two would not penetrate the keyhole; the third went in easily enough, but would not turn. Elser patiently filed down the key until it turned and the tumblers in the lock slipped out of place. The door swung open to reveal a treasure trove of explosives. It was as easy as that."[372] Four or five times each night Elser snuck into the bunker to steal very small amounts of explosives, until he had as much as he needed.) He was also able to obtain a 75 mm military artillery shell,[373] as well as other necessary tools: planes, hammers, squares, tin shears, saws, rulers, pliers, clocks, a battery, and so on.

Late the next summer, Elser entered phase two of his plan. He moved to Munich, rented a room, and told his landlady he would be gone each night working at a laboratory on a super-secret invention. His invention was the bomb.

Each night he went to the Löwenbräu for dinner and stayed till near closing. He then moved to a deserted gallery and waited unmoving till the place closed and everyone left. After that came his real work. Here's how one historian described it: "Working by the weak beam of a flashlight shrouded with a blue handkerchief, Elser carefully prised away the molding that surrounded a rectangular section of the column [just behind where Hitler would speak]. Then he carefully drilled a small hole in one upper corner of the veneer panel and inserted the tip of a special cabinetmaker's saw. With exquisite care, Elser began cutting away the panel. He worked three or four hours, then cleaned up evidence of his work before falling asleep in a chair. The painstaking sawing a few millimeters at a time, the replacing of the molding, the picking up of each grain of sawdust after each stint of work— none of this tried the craftsman's patience. He spent three nights just removing the panel. No trace of his tampering could be detected.   He chipped out a cavity, bits and pieces at a time, using hammers and steel hand drills of various diameters. Each tap reverberated through the empty hall, sounding to Elser like pistol shots. When some obstruction required heavier blows than usual, he waited for noises from the street to cover the sounds. Since he worked during the pre-dawn hours, he often had to wait a long time between hammer blows."[374]

When finally he completed carving out the cavity he attached a sheet of steel beneath the wood so a security guard's tapping would reveal no hollow space. He inserted the explosives and a five-day timer he'd made from alarm clocks.

Because he knew that each year Hitler's speech ran from 8:30 to 10:00 PM, Elser set the timer for 9:20.[375] Unfortunately for Elser, and for the world, Hitler's plans changed at the last moment, and he spoke from 8:00 until 9:12. The bomb went off seven minutes later, and killed those who were standing right where Hitler had been.

Elser was arrested one hundred yards from the Swiss border. In his pockets he had a postcard from the Löwenbräu, technical drawings of shells and detonators, and so on. He spent the next several years in a concentration camp, and was killed by the SS two weeks before Germany surrendered.[376]

☾ ☾ ☾

If we're going to talk about fulcrums, we need to also talk about bottlenecks. Anyone who has ever driven on a freeway knows precisely what a bottleneck is. You're driving along fine at 69 miles per hour on a six-lane highway. You top a hill and hit your brakes because the person in front of you hit his brakes, because the person in front of him hit her brakes. Traffic slows to a crawl. People begin frantically changing lanes, trying to find one that will get them through this mess three minutes sooner. Eventually people realize they need to get into the center lane (you realize this about ten seconds after you got into the left lane, and just as three semis creep by you in the center). At long last you come across the problem: a car broken down in the left lane and a cop parked on the right. Moments later, you're zooming again at precisely four miles over the speed limit, but for that forty-five minutes of traffic jam, you had the full bottleneck experience so beloved of motorists everywhere. Or one more example. Take a hose (or a pipeline). Kink it (or disable a pumping station). It doesn't matter how smoothly the water (or oil) flows through the rest of the hose (or pipeline). If there is a kink (or a disabled pumping station) in even one place, the water (or oil) will not flow. Bottleneck!

Now, how does this apply on a larger scale?

Albert Speer, Minister of Armaments for the Third Reich, later commented that the Allied bombing efforts could have been more effective had they more often targeted bottlenecks. One small example of this was that when the Allies bombed tractor factories, the Germans were no longer able to manufacture engines for tanks and airplanes there until the factories had been rebuilt, but when the Allies bombed ball bearing plants, the Germans were hindered from rebuilding factories. You need ball bearings to manufacture manufacturing plants. Ball bearing plants were bottlenecks in the process.

Here's an example of the Allies not hitting bottlenecks: the firebombing of Hamburg, which killed tens of thousands of people and destroyed much of the city, cost less than two months of productivity.[377] As a result of not targeting bottlenecks, Allied bombing reduced total German production by only 9 percent in 1943, and by building new factories, overworking undamaged factories, and diverting consumer production towards military ends, the Germans still met their production targets.[378]

But it ends up that ball bearing plants were trivial bottlenecks compared to others. Transportation networks, for example, were an even larger bottleneck. Eventually the Allies were able to destroy about two-thirds of the German rolling stock.[379] A United States military analysis later determined that the difficulties this caused the Germans in moving raw materials and finished goods made the attacks on railroads "the most important single cause of Germany's ultimate economic collapse."[380]

We all know (and Hitler knew this too) that oil was another bottleneck.[381] You can have the most powerful tanks in the world, and without oil they're just big hunks of steel. Without oil you have no modern army. Heck, without oil, you have no modern civilization. Keep that in mind. Hitler's understanding of these basic facts was one reason for his ultimately fatal choice to try to take the oil fields of the Caucasus instead of just pushing toward Stalingrad. Further, once the Allies started pounding the German synthetic oil industry—hitting the selected targets again and again and again—they were able to reduce monthly oil production from 316,000 to 17,000 tons.[382] These shortages obviously crippled the German war economy.

Just so we're clear that there are lots of bottlenecks, and that a little creativity can discover them, here's another bottleneck from World War II: industrial diamonds. Industrial grinding and drilling is almost impossible without diamonds. Both the Nazis, who had on hand only an eight-month supply, and DeBeers, which controlled the world's diamond supply, knew this. The Nazis smuggled several million carats into Germany. DeBeers could have acted to stop them—and thus effectively stopped wartime production, which means effectively stopped the war—but did not.

The new questions become: What are some of civilization's bottlenecks? What are some of civilization's limiting factors? Like transportation networks, oil, and industrial diamonds for the Nazis, what are some of the objects or processes that, if interdicted, could cause civilization to grind to a halt?

Similarly, where can we find fulcrums, pivot points, to magnify our efforts? Where do we put the levers, what do we use for fulcrums, how and when and how hard do we push to help topple this culture of death?

Are these fulcrums psychological? I hear all the time that it would do no good to take out dams, for example, because that would leave intact the mindset that leads to their erection in the first place. We need to change hearts and minds, I am told, and once these hearts and minds have been changed everything else will fall into place. Civilization will disappear because people are no longer insane enough to want it.

But maybe that question is too vague. Whose hearts and minds are we trying to reach? Where do we place our efforts in changing hearts and minds to achieve the most effect? Is it among the politically and economically powerful? Is it among the "mass of Americans"? Is it among the disaffected? Is it among the poor? Is it among the so-called criminal classes? Is it among the cops and the military? These latter, after all, have a lot of guns.[383] Where will we achieve the most good?

Are the fulcrums spiritual? People value what they consider sacred. They sacralize what they value. Perhaps we should attempt to desacralize power for power's sake. Perhaps we should attempt to break down the divine right of science, the divine right of corporations, the divine right of production, the divine right of nation-states. Perhaps we should attempt to help people to remember that spiders who live in their bathrooms are sacred, as are salmon who spawn in rivers outside their homes, plants who push up through sidewalks, salamanders who live high in the hollows of ancient redwoods, their own bodies, their own experiences, their own sexuality, their own flesh free from industrial carcinogens. Where do we place the levers, the fulcrums, to help people remember that they are humans living in a landbase, that they are animals?

Are these fulcrums personal, such that, like Hitler, the "removal" of this or that person will make a tangible difference? Would it help the redwoods and workers of northern California to make sure Charles Hurwitz, CEO of MAXXAM, does not damage them from his high-rise home in Houston, Texas? If so, where and how and when do we act in this way?

In cases where it's not the individual CEO, but the position—where social framing conditions make it so that most people who would take up that position share the same deadened worldview that would cause them to commit the same atrocities—where then do we place the levers and fulcrums? Do we go CEO to CEO, "removing" them one by one? We always hear that the machine-like characteristics of corporations mean that CEOs are simply cogs—albeit large ones—in these community-destroying institutions, and so it would do no good to remove them. It's an odd argument to make, even when I make it myself (as I did a few pages ago).[384] There are few who suggest that simply because arresting or killing one rapist does not stop other men from raping, that this means we should not stop whatever rapists we can through any means necessary. Yet when it comes to CEOs the argument seems to hold: Someone else will just take this one's place, so we must not stop this one personally. In fact, we must allow him to continue to be rewarded with millions of dollars per year in salaries and stock options. Where are the fulcrums to stop these people, these institutions? Where are the bottlenecks?

Or perhaps the fulcrums are social. Perhaps instead of (or in addition to) removing individual CEOs, we need to change the social institutions that themselves amplify the destructive efforts of these individuals. Charles Hurwitz does not kill redwoods by cutting them down. He kills them by ordering them cut down, or even more abstractly, by ordering someone to maximize profits. Are there counterlevers we can use to pry away his levers of power? Are there social means by which we can do that?

Or perhaps, as was also true of the Nazis, some of the fulcrums are infrastructural. John Muir is famously noted as saying, "God has cared for these trees, saved them from drought, disease, avalanches, and a thousand tempests and floods. But he cannot save them from fools." The thing is, a fool couldn't cut down trees by him or herself. I used to think that we were fighting an incredibly difficult battle in part because it takes a thousand years of living to make an ancient tree, while any fool can come along with a chainsaw and cut it down in an hour or two. I've since realized that's all wrong. The truth is that thriving on a living planet is easy   the whole forest, for example, conspires to grow that tree and every other, and *we* don't have to do anything special except leave it alone—while cutting down a tree is actually a very difficult process involving the entire global economy. I wouldn't care how many ancient redwoods Charles Hurwitz cut down, if he did it all by himself, scratching pathetically with bloodied nails at bark, gnawing with bloody teeth at heartwood, sometimes picking up rocks to make stone axes. To cut down a big tree you need the entire mining infrastructure for the metals necessary for chainsaws (or a hundred years ago, whipsaws), the entire oil infrastructure for gas to run the chainsaws, and for trucks to transport the dead trees to market where they will be sold and shipped to some distant place (once Charles had downed the tree by himself, I would wish him luck transporting it without the assistance of the global economy); and so on. It takes a whole lot of fools to cut down a tree, and if we break the infrastructural chain at any point, they won't be able to do it.

The same is true, of course, for the rest of this culture's destructive activities, from vivisection to factory farming to vacuuming the oceans to paving the grasslands to irradiating the planet: every one of this economy's destructive activities requires immense amounts of energy and worldwide economic, infrastructural, military, and police support to accomplish. If any one of these supports fail—I want to emphasize, if *any one of these supports fail*—the destructive activities will be curtailed. Where do we place our levers?

Or maybe the fulcrums are all of the above. Maybe changing people's hearts has a place. Maybe so do all the others, and maybe we should pursue them all, according to our gifts, proclivities, and opportunities.

The bottom line so far as fulcrums and bottlenecks: What will it take to stop this culture of death before it kills the planet?[385]

VIOLENCE

I believe there will ultimately be a clash between the oppressed and those who do the oppressing. I believe there will be a clash between those who want freedom, justice and equality for everyone, and those who want to continue the system of exploitation. I believe there will be that kind of a clash, but I don't think it will be based on the color of the skin.

Malcolm X[386]

I'M SURE BY NOW WE'VE ALL HEARD THE CLICHÉ ABOUT HOW ESKIMOS have something like ninety-seven words for snow. It ends up that's kind of bullshit. First, they're not Eskimos, but Inuits. Second, the translations for their words for snow aren't all that exciting, kind of like "fluffy snow," "hard snow," "cold snow," and so on. The reason they have so many words for snow is that they don't have adjectival forms the way that English has.

Along these lines, though, I do think we need more words in English for violence. It's absurd that the same word is used to describe someone raping, torturing, mutilating, and killing a child; and someone stopping that perpetrator by shooting him in the head. The same word used to describe a mountain lion killing a deer by one quick bite to the spinal column is used to describe a civilized human playing smackyface with a suspect's child, or vaporizing a family with a daisy cutter. The same word often used to describe breaking a window is used to describe killing a CEO and used to describe that CEO producing toxins that give people cancer the world over. Check that: the latter isn't called violence, it's called production.

Sometimes people say to me they're against all forms of violence. A few weeks ago, I got a call from a pacifist activist who said, "Violence never accomplishes anything, and besides, it's really stupid."

I asked, "What types of violence are you against?"

"All types."

"How do you eat? And do you defecate? From the perspective of carrots and intestinal flora, respectively, those actions are very violent."

"Don't be absurd," he said. "You know what I mean."

Actually I didn't. The definitions of violence we normally use are impossibly squishy, especially for such an emotionally laden, morally charged, existentially vital, and politically important word. This squishiness makes our discourse surrounding violence even more meaningless than it would otherwise be, which is saying a lot.

The conversation with the pacifist really got me thinking, first about definitions of violence, and second about categories. So far as the former, there are those who point out, rightly, the relationship between the words violence and violate, and say that because a mountain lion isn't violating a deer but simply killing the deer to eat, that this would not actually be violence. Similarly a human who killed a deer would not be committing an act of violence, so long as the predator, in this case the human, did not violate the fundamental predator/prey relationship: in other words, so long as the predator then assumed responsibility for the continuation of the other's community. The violation, and thus violence, would come only with the breaking of that bond. I like that definition a lot.[387]

Here's another definition I like, for different reasons: "An act of violence would be any act that inflicts physical or psychological harm on another."[388] I like this one because its inclusiveness reminds us of the ubiquity of violence, and thus I think demystifies violence a bit. So, you say you oppose violence? Well, in that case you oppose life. You oppose all change. The important question becomes: What types of violence do you oppose?

Which of course leads to the other thing I've been thinking about: categories of violence. If we don't mind being a bit *ad hoc*, we can pretty easily break violence into different types. There is, for example, the distinction between unintentional and intentional violence: the difference between accidentally stepping on a snail and doing so on purpose. Then there would be the category of unintentional but fully expected violence: whenever I drive a car I can fully expect to smash insects on the windshield (to kill this or that particular moth is an accident, but the deaths of some moths are inevitable considering what I'm doing). There would be the distinction between direct violence, that I do myself, and violence that I order done. Presumably, George W. Bush hasn't personally throttled any Iraqi children, but he has ordered their deaths by ordering an invasion of their country (the death of this or that Iraqi child may be an accident, but the deaths of some children are inevitable considering what he is ordering to be done). Another kind of violence would be systematic, and therefore often hidden: I've long known that the manufacture of the hard drive on my computer is an extremely toxic process, and gives cancer to women in Thailand and elsewhere who assemble them, but until today I didn't know that the manufacture of the average computer takes about two tons of raw materials (520 pounds of fossil fuels, 48 pounds of chemicals, and 3,600 pounds of water; 4 pounds of fossil fuels and chemicals and 70 pounds of water are used to make just a single two gram memory chip).[389] My purchase of the computer carries with it those hidden forms of violence.

There is also violence by omission: By not following the example of Georg Elser and attempting to remove Hitler, good Germans were culpable for the effects Hitler had on the world. By not removing dams I am culpable for their effects on my landbase.

There is violence by silence. I will tell you something I did, or rather didn't do, that causes me more shame than almost anything I have ever done or not done in my life. I was walking one night several years ago out of a grocery store. A man who was clearly homeless and just as clearly alcoholic (and inebriated) approached me and asked for money. I told him, honestly, that I had no change. He respectfully thanked me anyway, and wished me a good evening. I walked on. I heard the man say something to whomever was behind me. Then I heard another man's voice say, "Get the fuck away from me!" followed by the thud of fist striking flesh. Turning back, I saw a youngish man with slick-backed black hair and wearing a business suit pummeling the homeless man's face. I took a step toward them. And then? I did nothing. I watched the businessman strike twice more, wipe the back of his hand on his pants, then walk away, shoulders squared, to his car. I took another step toward the homeless man. He turned to face me. His eyes showed he felt nothing. I didn't say a word. I went home.

If I had to do it again, I would not have committed this violence by inaction and by silence. I would have stepped between, and I would have said to the man perpetrating the direct violence, "If you want to hit someone, at least hit someone who will hit you back."

There is violence by lying. A few pages ago I mentioned that journalist Julius Streicher was hanged at Nuremberg for his role in fomenting the Nazi Holocaust. Here is what one of the prosecutors said about him: "It may be that this defendant is less directly involved in the physical commission of crimes against Jews. The submission of the prosecution is that his crime is no less the worse for that reason. No government in the world . . . could have embarked upon and put into effect a policy of mass extermination without having a people who would back them and support them. It was to the task of educating people, producing murderers, educating and poisoning them with hate, that Streicher set himself. In the early days he was preaching persecution. As persecution took place he preached extermination and annihilation. . . . [T]hese crimes . . . could never have happened had it not been for him and for those like him. Without him, the Kaltenbrunners, the Himmlers . . . would have had nobody to carry out their orders."[390] The same is true of course today for the role of the corporate press in atrocities committed by governments and corporations, insofar as there is a meaningful difference.

For years I've been asking myself (and my readers) whether these propagandists—commonly called corporate or capitalist journalists—are evil or stupid. I vacillate day by day. Most often I think both. But today I'm thinking evil. Here's why. You may have heard of John Stossel. He's a long-term analyst, now anchor, on a television program called *20/20*, and is most famous for his segment called "Give Me A Break," in which, to use his language, he debunks commonly held myths. Most of the rest of us would call what he does "lying to serve corporations." For example, in one of his segments, he claimed that "buying organic [vegetables] could kill you." He stated that specially commissioned studies had found no pesticide residues on either organically grown or pesticide-grown fruits and vegetables, and had found further that organic foods are covered with dangerous strains of E. coli. But the researchers Stossel cited later stated he misrepresented their research. The reason they didn't find any pesticides is because they never tested for them (they were never asked to). Further, they said Stossel misrepresented the tests on E. coli. Stossel refused to issue a retraction. Worse, the network aired the piece two more times. And still worse, it came out later that *20/20*'s executive director Victor Neufeld knew about the test results and knew that Stossel was lying a full three months before the original broadcast.[391] This is not unusual for Stossel and company.[392] I recently spoke with one environmentalist/teacher who was interviewed by him who said, "It was nothing but a hit piece. He sliced and diced the interviews with me and the grade schools students to make it seem as though we'd said things we hadn't, and as though we hadn't been able to answer questions that we had. He edited the piece to make the children look stupid." Another called him "the worst motherfucker on the planet," which is saying quite a lot. And now I've got another Stossel story to add to the evidence when he joins the ghost of Streicher in the docket. I got a call a while ago from one of *20/20*'s reporters, who wanted to talk to me about deforestation. The next "myth" Stossel is going to debunk, she said, is that this continent is being deforested. After all, as the timber industry says, there are more trees on this continent today than there were seventy years ago. She wanted a response from an environmentalist. I told her that 95 percent of this continent's native forests are gone, and that the creatures who live in these forests are gone or going. She reiterated the timber industry claim, and said that Stossel was going to use that as the basis for saying, "Give me a break! Deforestation isn't happening!" I said the timber industry's statement has two unstated premises, and reminded her of the first rule of propaganda: if you can slide your premises by people, you've got them. The first premise is the insane presumption that a ten-inch seedling is the same as a two-thousand-year-old tree. Sure, there may be more seedlings today, but there are a hell of a lot fewer ancient trees. And many big timber corporations cut trees on a fifty-year rotation, meaning that the trees will never even enter adolescence so long as civilization stands. The second is the equally insane presumption that a monocrop of Douglas firs (on a fifty-year rotation!)[393] is the same as a healthy forest, that a forest is just a bunch of the same kind of trees growing on a hillside instead of what it really is, a web of relationships shimmering amongst, for example, salmon, voles, fungi, salamanders, murrelets, trees, ferns, and so on all working and living together. Pretty basic stuff. But, she asked, aren't there more of some types of wildlife today than ever before? I responded by telling her that one of the classic lies told by the Forest Service and the timber industry is that because there are more white-tailed deer now than before, that means forests must be in better shape. The problem is that white-tailed deer like the edges between forest and non-forest, so more white-tailed deer doesn't mean more forests: it means more edges, which really means more clearcuts. To say, I continued, that more white-tailed deer means more forests is simply another lie. I talked to her for more than an hour, and by the end she seemed to really understand these points. I made clear that the only way you can make Stossel's leap—from saying that there are more trees today than there were seventy years ago, to saying that deforestation isn't happening—is if you're either ignorant of these premises or you're lying. As George Draffan and I wrote in *Strangely Like War*, "To even imply that a tree farm on a fifty-year rotation remotely resembles a living forest is either extraordinarily and willfully ignorant, or intentionally deceitful. Either way, those who make such statements are unfit to make forestry decisions."[394] She understood that. We sent her a copy of the book. She said they might have me on the program. They didn't, which is fine. But here's the point. Stossel went ahead with the program anyway. Further, he explicitly said that an indicator that deforestation isn't happening is that white-tailed deer are increasing. He had been made fully aware that his statements are untrue. He was made fully aware of the facts. These facts—that seedlings are different than ancient trees, that monocrops of trees are different than forests, and that increasing numbers of white-tailed deer are not an indicator that forests are increasing—are neither controversial nor cognitively challenging. They are not opinions. They are facts as clear as water is wet and fire is hot and ancient trees are ancient. This means he no longer had the first excuse, ignorance.[395] Like Streicher, he is committing violence by lying: violating the truth, violating what is sacred in words and discourse, violating our psyches, and paving the way for further violation of the forests.

☾ ☾ ☾

All writers are propagandists. That doesn't mean we're all liars. Some are liars. Some are not.

℃ ℃ ℃

I probably shouldn't pick on Stossel. He's not the only liar. The entire culture is based on lies, from the most intimate and personal to the most global. The smartest lines I ever wrote were in *A Language Older Than Words*: "In order for us to maintain our way of living, we must tell lies to each other, and especially to ourselves. It's not necessary that the lies be particularly believable, but merely that they be erected as barriers to truth. These barriers to truth are necessary because without them many deplorable acts would become impossibilities. Truth must at all costs be avoided."[396] Members of abusive families lie to each other and to themselves in order to protect the violent perpetrators (they convince themselves—and are convinced by the perpetrators and by the entire family structure—that they are protecting themselves), and to keep their violent social structures intact. Members of this abusive culture lie to each other and to themselves in order to protect this culture's violent perpetrators, and to keep this culture's violent social structures intact. We tell ourselves we can destroy the planet—or rather, for those of us who care, allow it to be destroyed—and live on it. We tell ourselves we can perpetually use more energy than comes in from the sun every year. We tell ourselves that a 90 percent decline in large fish in the oceans may not be unreasonable. We tell ourselves that if we are peaceful enough that those in power will stop the killing. We tell ourselves that civilization is the most desirable form of social order, or really the only one. We tell ourselves things are going to be okay.

Stossel is not the only liar.

SPENDING OUR WAY TO SUSTAINABILITY

The whole individualist what-you-can-do-to-save-the-earth guilt trip is a myth. We, as individuals, are not creating the crises, and we can't solve them. Take our crazy energy consumption. For the past 15 years the story has been the same every year: individual consumption—residential, by private car, and so on—is never more than about a quarter of all consumption; the vast majority is commercial, industrial, corporate, by agribusiness and government.[397] So, even if we all took up cycling and wood stoves it would have a negligible impact on energy use, global warming and atmospheric pollution. I mean, sure, go ahead and live a responsible environmental life; recycle, compost, ride a push-bike; but do it because it is the right, moral thing to do—not because it's going to save the planet.

If we really want to understand why this happened we have to ask ourselves another question: "Why is it that we seem willing to live with the threat of apocalypse rather than trying to seriously alter a world where consumption, of anything, is seen as unrelieved virtue, production, of anything, is regarded as a social and economic necessity, and more, of anything (like children or cars or chemicals or PhDs or golf courses or recycling centres), is unquestioningly accepted?"

Kirkpatrick Sale[398]

IT IS ABSURD FOR SOMEONE TO SAY HE OR SHE DOESN'T BELIEVE in violence. That's like saying you don't believe in death. Certainly one can say that one doesn't want to participate in certain forms of violence, just like one can say that one doesn't want to cause certain forms of death. But violence, like death, is simply a part of life, no larger nor smaller, no more nor less important than any other. In fact it's inseparable from the others. We all participate in violence daily. The only questions are our degree of awareness, and what we do with that awareness.[399]

《 《 《

Tonight I tried to save a wasp. I failed. I was standing in line on a Jetway. The flight had been delayed. Lots of people were cranky. For whatever reason, I wasn't.

The wasp was beautiful. She was a small hunting wasp, with delicate translucent wings and a body the color of peaches. I saw her long before I got to her. Each person in front of me in line looked at her. I prayed no one would smash her. No one did. I got there. I wanted to reach up and grab her to carry her to the small space between the Jetway and the airplane wall to release her to the outside. But I hesitated, mainly because I didn't want to be noticed "doing something odd" in line. I wasn't sure what to do.

I made up my mind when I heard the woman behind me ask her boyfriend, "Can I borrow your shoe?"

I reached up, cupped the wasp in my hand, closed my fingers gently, brought my hand to my chest. People behind me in line gasped. The wasp got out from between my fingers, flew to the ceiling lights. I reached as high as I could, standing on my toes, and missed her again and again. Each time I almost had her, she flew a few inches away.

Finally I had her. Because she was so high that I could not cup her, I had to hold her gently between my thumb and all four fingers. I brought her to my chest.

She stung me. The stings of hunting wasps barely hurt. The venom is meant instead to paralyze spiders or caterpillars, depending on the species of wasp, so she can lay her eggs inside her intended prey. They only sting in defense when all other options are gone, and when they're terrified.

The sting startled me, and I accidentally let go my grip. She flew back up to the ceiling. The line moved on. I should have stayed back and tried again, but I didn't. I got on the plane, and hoped she made it out on her own.

☾ ☾ ☾

In many ways the story of the wasp highlights a distinction between two forms of violence, one of which I evidently didn't like, and one of which I evidently didn't care to think about. The former is direct and by omission. It seemed clear to me that if I didn't do something, this wasp would die, either by being smashed for no good reason by someone wielding a shoe, or by eventually starving or being poisoned in the sterile airport environment. I knew that if I could help her outside she would at least have the chance to make it somewhere away from the concrete and kerosene fumes of the runways, where she might find whole fields of caterpillars or spiders, and where she might find a male wasp eagerly awaiting her attentions. I did not want to stand by and let her die this unnatural death.

The latter—the type of violence I evidently didn't care to think about—was that I was getting on a plane. If whenever I drive I smash moths against my windshield, I think it's safe to presume airplanes do the same to wasps (as well as moths, spiders, birds, and everything else that cannot get out of the way of this big metal bullet pushing through the air at several hundred feet per second). And far greater than this is the habitat damage wreaked by the airline, oil, aluminum, electricity, and other industries all necessary to get this thing in the air. I'm sure many fine fields of fat caterpillars and spiders are systematically sacrificed on the sacred altar of air travel. But it's perhaps better if we don't speak of that kind of violence, don't you think?

☾ ☾ ☾

I don't want to take this logic too far, however, and suggest that because I boarded this plane that I'm responsible for all the creatures killed by the airline industry. The truth is that had I not flown, the airplane would still have killed those wasps, and the industry would still have destroyed those fields. Sure, I would have cost the airline money, and United's gross income for the year would have been $400 less than $38 billion, which I suppose makes me responsible for about 1/95,000,000th of the damage caused by this one airline.

I don't have a lot of patience for those who blame "all of us consumers" for damage caused by the economic and social system, those who say, "We're all in this together,"[400] and who point out, "If we didn't buy tickets, the airline industry would go broke." Well, first, if we didn't buy airline tickets, the feds would bail them out. All major industries rely on massive subsidies of public moneys to stay afloat. Second, if we're going to throw out a fantasy about the mass of Americans rising up to not buy airline tickets, why dream so low? Why not dream big and have this same fantastic mass of people start taking out dams? Why don't we have them storm vivisection labs and factory farms to liberate tormented animals? Why not have them dismantle the entire infrastructure? (Oh, because that might lead to real change, and we don't even want to *dream* about that.) The same people who tell me I can make a difference by not buying an airline ticket quite often tell me I shouldn't try to take out a dam because taking out one lone dam wouldn't accomplish anything. And not buying one lone airline ticket will?

The point, once again and as always, is leverage.[401] Sure, I support individuals and sometimes even industries I believe are headed the right direction through spending my hard-earned dollars in places and ways that are less destructive,[402] and similarly, insofar as possible, I don't support through my spending individuals and industries that are especially destructive, but I also recognize that far more needs to be done than this. I am not *merely* a consumer, much as those in power would like for me to define myself as such. The tools of consumerism are but one set available to me. The trick is to know when and how to use that set, and when and how to use others. The trick, to put it another way, is to leverage my efforts, to make my own small force have larger effects. The questions: What do I want to move?[403] What do I use for levers? Where do I place the fulcrums? How hard and when do I push?

❆ ❆ ❆

There are other problems with attempting to spend or boycott our way to sustainability. The first is that it simply won't work. Spending won't work because within an industrial economy nearly all economic transactions are destructive. Because the industrial economy—indeed a civilized economy—is systematically, inherently, functionally, and inescapably destructive, even buying "good things" isn't really doing something good for the planet so much as it is doing something not quite so bad. Let's say I purchase organic lettuce at the grocery store. That's a good thing, right? Well, not particularly. The problem is that the mass cultivation of lettuce—organic or not—still destroys soils, and its transportation to market still requires the use of oil. I suppose if I purchased lettuce grown in small-scale permaculture beds from my next door neighbor, I'd be doing something even less bad, but this is rare enough to be the exception that makes the rule crystal clear.[404] For an act to be sustainable, it must benefit the landbase, which means the soil, the critters who live in the soil, the plants who live on the soil, the animals who eat the plants, the animals who eat the animals, the insects and others who turn the dead back into soil. Producing, marketing, or purchasing organic lettuce doesn't do that. Rare indeed within our culture is the economic activity that improves the landbase (and that doesn't pay taxes, to boot, since more than 50 percent of the discretionary federal budget goes to pay for war). And don't throw up your hands in despair and give me the old saw about how *all* human activities damage landbases: noncivilized people have lived on landbases for a very long time without destroying them, in fact enhancing their landbases according to the needs of the landbases.

The problem is not our humanity. The problem is this culture—this *entire* culture—and slight changes in spending habits won't significantly stop the destruction.

That's not to say we shouldn't enact whatever changes we can to make whatever difference we can—remember, we do need it all—and buying organic lettuce is better than buying pesticide lettuce, on any number of levels. It's just to say that when I spoke earlier of this culture being a culture of occupation, of the government being a government of occupation, of the economy being an economy of occupation, I wasn't speaking metaphorically or hyperbolically. I was speaking sincerely, literally, physically, in all seriousness and truth. If we were Russians living under the German occupation in 1943, would we believe we could stop the Nazis by buying products made by German companies we like a little more and not buying them from I.G. Farben and other companies we don't like?

The same is true for boycotts. We can't boycott our way to sustainability any more than we can spend our way to it. The industrial economy, as is true for any economy of occupation (which means any civilized economy), is fundamentally a command economy (defined as "an economy that is planned and controlled by a central administration"). I know, I know, we've all been fed the line that "our" economy is based on some mythical thing called the free market, and that whatever it produces is by definition what we want. But I don't want depleted uranium any more than I want depleted oceans. Do you? So how did we get them? If the economy really were free, why are armed military and police necessary to secure producers' access to resources? And even if it were a "free market," that wouldn't help our landbases, since these markets do not value those parts of our landbases not perceived as productive (in other words, not obviously amenable to exploitation). And as mentioned before, in a global economy, free market or not, any wild thing that is vulnerable to exploitation (in other words, is valuable) will either be domesticated—enslaved—or exploited to extinction. But it's worse than this. It's not a free market anyway. Remember the words of Dwayne Andreas: "There's not one grain of anything in the world that is sold in the free market. The only place you see a free market is in the speeches of politicians."[405] Economist Brad DeLong puts this another way: "As producers and employees many of us live in an economy that is better thought of as a *corporate* economy: an economy in which patterns of economic activity are organized by the hands of bosses and managers, rather than one in which the pattern of activity emerges unplanned by any other than the market's invisible hand."[406] Yet another way to say all this is to note that, as alluded to above, all sectors of the economy, in fact the economy as a whole, would collapse almost immediately without huge subsidies. If every person in the country suddenly decided to somehow boycott, for example, the oil industry—which of course won't happen, for any number of obvious reasons—the U.S. and other governments would merely increase the subsidies to that sector of the economy, and probably for good measure arrest the boycott organizers on racketeering charges.[407]

Another reason we can't spend our way to sustainability is that we will *always* be outspent by those who are actively destroying the world. Destroying the world is how they make their money. It is always how they have made their money: through production, through the conversion of the living to the dead, through forcing others (the natural world, human communities) to pay the price for their activities. If you don't produce—that is, destroy—you won't make money. That still isn't to say that there aren't degrees of destructiveness: the damage caused by a permaculture farmer hand-delivering his lettuce leaves to his neighbors would be trivial compared to the damage caused by a full-on industrio-chemical lettuce agricorporation, but, and this is the point, so would his profits. That's why those who profit from this destructiveness will *always* have more money than we do, and will always be able to outspend us. An example should make this clear. Let's say I make a boatload of money writing and selling books. Oops, scratch that, since the manufacture of books—even on recycled paper using soy-based inks—requires lots of water, energy (ghost slaves), and raw materials. In other words, it's very destructive. Okay, so let's say instead I make a boatload of money making a boatload of money (in other words, I haul out my trusty printing press, and I just *make* the damn stuff). Oops, I can't do that, since the counterfeiting of money requires high-quality papers and lots of presumably toxic inks, lots of energy, and so forth. In other words, that's very destructive too. So okay, darnitall, let's say instead I just walk to a bank (wearing only used clothing taken from the dumpster behind Goodwill), and I *take* a boatload of money. I do this at night, because I don't want to threaten or scare any of the tellers, or perform any other action that might be construed as violent. Even better, I don't go to a bank, but go at night to Wal-Mart, and sneak in through an open door. I don't want to break a window, because there are those who would consider this an act of violence. I don't blow the safe because there are those who would consider *this* an act of violence. But let's say the safe is open. I take a boatload of money. Or if the safe isn't open, I take a bunch of consumer items, fabricate some receipts (okay, so this takes paper, but we'll just ignore that) and return them over the next days and weeks and months for a boatload of money. Wal-Mart, with its $258.6 billion in revenues, isn't going to miss it.[408] The point is that I somehow find a way to acquire a boatload of money that a) didn't cause me to "produce"—in other words, destroy—anything, and b) didn't cause me to pay taxes—in other words, to pay the government so it can destroy things. The question becomes, what am I going to do with this cash? Let's say I do what I actually would do if I acquired a boatload of cash: I buy some land and set it aside. Let's ignore the fact that in so doing I'm reinforcing the extremely damaging idea that land can be bought and sold. I buy an entire small creek drainage, and I set to work to improve habitat in that drainage for salmon, Port Orford cedars, mountain lions, Pacific lampreys, red-legged frogs, and so on. I create a sanctuary, a place where salamanders, newts, tree frogs, towhees, phoebes, and spotted owls can thrive and live as they did before the arrival of our awful culture. I've done a good and great thing, maybe even as good and great as what Elser tried to do. But now I find I want to protect more land, because these creatures need more habitat. What do I have to do? Because I pulled this land out of production, and thus am not "making any money" off of it, I have to write more books, print more money, make more trips to Wal-Mart, and unless I've figured out non-destructive ways to acquire cash—like the nocturnal trips to Wal-Mart—then I'm basically creating sacrifice zones elsewhere that I do not see so that the land I do see can be protected. I have to do this every time I want to protect more land.

Now, let's contrast that with someone who purchases this entire watershed not to create a sanctuary but to cut the trees. That person will "make money" off the land by harming it, and can use that money to purchase more land, where that person can cut more trees and make more money, and use that money to buy more land, and so on until there's nothing left. See, for example, Weyerhaeuser, or any other timber (or other) corporation.

Because the civilized economy is extractive, because it rewards those who exploit humans and nonhumans, that is, because it rewards those who do not give back to the landbase what it needs, that is, because it rewards people for disconnecting themselves from the reciprocal relationships that characterize (human or nonhuman) sustainable economies (and relationships), those who value the accumulation of money or power over life will always have more money or power than those who value life over money or power.

❰ ❰ ❰

After a talk I gave last year in Portland, Oregon, several of us anarchists wanted to grab a bite to eat. One said he knew of a place that served great organic food and paid workers a livable wage.

"Sounds perfect," I said.

"One problem," he responded. "None of us can afford to eat there."

❰ ❰ ❰

Heck, what does it say about this culture and its economics that people must pay for food? And what does it say about this culture and its economics that a very few very large corporations control a very large majority of the food supply?

Worse yet, if people are going to be forced to pay for food, what does it say about this culture and its economics that we face a two-tier system of paying, where it's cheaper to buy food that has been raised using poisons than it is to buy food that has been raised without using poisons, which means where the rich have enough money to buy organic, and the poor do not? How strange is it that you have to pay extra to be exposed to fewer poisons? It is for this reason, by the way, that I am opposed to labeling genetically modified foods.[409] It's not good enough for me to simply make it possible for the rich to pay extra to not ingest these artificial mutations. That is morally wrong. And because the government has not stopped and will not stop those who can make a buck by releasing these organisms (and pesticides) into the world, and into our bodies, it falls upon us to stop them. How are we going to do it?

❰ ❰ ❰

Sure, it's a good thing to try to do good with your money. And sure, because this strange and destructive economic system based on ownership and exploitation has pretty much overrun the globe it is extremely difficult to avoid participating in it (which means, among other things, that we shouldn't beat ourselves up too much for purchasing the vehicle we need to carry explosives to dams [or kids to soccer practice], nor should we beat ourselves up if we buy some pesticide-laden, genetically modified pseudofoods at the grocery store [smothered in monosodium glutamate they taste so very yummy, don't you think?]). But we must never forget that if we attempt to economically go head to head with those who are destroying the planet, we will always be at a severe, systematic, inescapable, and functional disadvantage. Not buying an airline ticket won't do squat. But all is not lost. The question, yet again: Where are the fulcrums? How do we magnify our power?

                              (((

Here's the problem. Two people walk through a forest. One considers how extraordinarily beautiful the forest is, and how wonderful it is to be alive. The other notices how much of this forest could be turned into immediate fiscal profit, and thinks about how that could be done. Question: Which of these people will probably make more money off the forest? Question: Within this culture, which of these people is more likely to end up in a position of power, making decisions that affect the human and nonhuman communities in and around the forest? Question: How do those of us who care stop them from destroying the forests?

EMPATHY AND ITS OTHER

All places and all beings of the earth are sacred. It is dangerous to designate some places sacred when all are sacred. Such
compromises imply that there is a hierarchy of value, with
some places and some living beings not as important as others. No part of the earth is expendable; the earth is a whole
that cannot be fragmented, as it has been by the destroyers'
mentality of the industrial age. The greedy destroyers of life and
bringers of suffering demand that sacred land be sacrificed so
that a few designated sacred places may survive; but once any
part is deemed expendable, others can easily be redefined to
fit the category of expendable. As Ruth Rudner points out in
her article "Sacred Land," what spiritual replenishment is possible if one must travel through ghastly fumes and ravaged
lands to reach the little island or ocean or mountain that has
been preserved by the label sacred land?

There can be no compromise with these serial killers of life
on earth because they are so sick they can't stop themselves.
They would like the rest of us to embrace death as they have,
to say, "Well, all this is dead already, what will it matter if they
are permitted to kill a little more?" Even among the conservation groups there is an unfortunate value system in place that
writes off or sacrifices some locations because they are no
longer '"virgin." Those who claim to love and protect the
Mother Earth have to love all of her, even the places that are
no longer pristine. *Ma ah shra true ee*, the giant serpent messenger, chose the edge of the uranium mining tailings at Jack-
pile Mine for his reappearance; he was making this point
when he chose that unlikely location. The land has not been
desecrated; human beings desecrate only themselves.

Leslie Marmon Silko[410]

WHY CIVILIZATION IS KILLING THE WORLD, TAKE TWENTY. In the Q & A after a recent talk, a woman said that part of the problem is that most of the people she knows who care about the health and well-being of oppressed humans and salmon and trees and rivers and the earth—life—do so because, by definition, they care about others. They empathize. They feel connections with these others. They identify with these others.[411] Those who don't care about the health and well-being of oppressed humans and salmon and trees and rivers and the earth—life—don't care because, also by definition, they don't care. They don't empathize.[412] They don't feel connections with these others. That's a problem, she said.

She's right. That's a big problem. Those of us who value life over things and control value life over things and control. Those who value things and control over life value things and control over life. Sure, many environmentalists are jerks, and I'm sure some CEOs are very nice people. Robert Jay Lifton made the point that many of the Nazi guards at concentration camps and even many of the upper level SS officers were good family men,[413] and many people have pointed out that there are many torturers who "do it for a living," and who when they go home are not horrid people. Lifton called this split in one's psyche *doubling*, which he defines as the formation of a second self-structure morally at odds with one's prior self-structure.[414] It's a defense mechanism that allows people to continue to perpetrate violent behavior, he says, whether that behavior is more direct, as in murdering Jews face to face, or less direct, as in designing or building nuclear bombs or running a corporation. I have tremendous respect for Lifton, and have been deeply influenced by his important work, but within this extremely violent culture I'm not sure doubling is quite so prominent as we would normally think.

I would instead see this as a manifestation of typical abusive behavior. Abusers, as is true for most all of us who live in this abusive society, are exquisitely sensitive to power structures, knowing on whom they can project their unmetabolized rage and to whom they must bend their knee. In other words, they are intimately acquainted with Premise Four of this book, and know the precise circumstances under which it will not only be acceptable but fully expected for them to perpetrate violence on those beneath them and suck up to those above: Unfortunately, too few SS guards fragged their officers. Further, Susan Griffin has written extensively about what constituted "normal" family structures within that particular German culture, and the relationship between familial abuse and the larger violence of the Nazis.[415] We could make the same argument today: a normal family within this larger culture is pretty damn violent. This doubling then is not quite so dramatic as it may have seemed.

I'd see the problem instead as the numbing that is a normalized and necessarily chronic state within this culture, an inculcation into the rigid world of Premise Four, where people's empathies are numbed by the routine violence done to them, then trumped by ideology and what Lifton calls "claims to virtue"—Lifton makes clear that before people can commit any mass atrocity they must have a "claim to virtue," that is, they must consider what they're doing not in fact an atrocity but something good[416]—such that they can feel good about themselves (or rather seem to feel good about themselves) as they oppress others to maintain their lifestyle, then go home and dandle their babies on their knees. This is how many Nazis could maintain semblances of lives as they did not kill Jews but rather purified the Aryan race. This is how Americans could maintain façades of happiness as they didn't kill Indians but fulfilled their Manifest Destiny. This is how the civilized can pretend to be emotionally healthy as they do not commit genocide and destroy landbases, but instead take what they need to develop their "advanced state of human society." This is how we can all pretend we are sane as no one kills the planet, but as people maximize profits and develop natural resources.

As well as asking myself each day whether I want to write or blow up a dam, each day I ask myself whether all my talk of saving salmon or old growth or migratory songbirds is just another claim to virtue. I mean, don't those at the center of empire always say they're only perpetrating (defensive) violence against those who want to destroy their[417] lifestyle? And aren't I saying that I'm considering (defensive) violence to maintain a lifestyle that I want? One wants consumer goods, the other wants wild salmon. What's the difference? Maybe my desire to liberate rivers is just a mask to cover an urge to destroy dams, or more broadly just an urge to destroy. I don't *feel* I have a generic urge to destroy but presumably neither do CEOs. That's the wonderful thing about denial: you generally don't know you're in it. But that's one reason I'm trying to lay out my premises so explicitly. I don't want to lie to myself, and I don't want to lie to you.

Each day when I ask whether my work is just an elaborate claim to virtue, I keep coming back to the same answer: clean water. We need clean water to survive. We need a living landbase to survive. We do not need cheap consumables.

We do not need a "purified Aryan race." We do not need to fulfill a Manifest Destiny to overflow the continent or world. We do not need an "advanced state of human society" (even if that *were* an accurate definition of civilization). We do not need to maximize profits or "develop natural resources." We do not need oil, computers, cell phone towers, dams, automobiles, pavement, industrial farming, industrial education, industrial medicine, industrial production, industry. We do not need civilization. We—human beings, human animals living in healthy, functioning communities—existed perfectly fine without civilization for the overwhelming majority of our existence. However, we do need a living landbase. This is not a claim to virtue. This is just true.

Each day I remember that I am not wrong because I come back to understanding that every stream in the United States is now contaminated with carcinogens. I come back to the fact that wild salmon, who survived tens of millions of years of ice ages, volcanoes, the Missoula Flood, for crying out loud,[418] are not surviving one hundred years of this culture. I come back to knowing there is now dioxin in every mother's breast milk. I come back to the knowledge that tigers, great apes, and amphibians are being exterminated. Now. This is all real. This is the real world.

Each day I understand anew the simplemindedness that would cause someone to think that just because claims to virtue are sometimes used to justify violence that all reasons for violence are artificial justifications. I fall into this trap myself all too often. Too many people within this culture do that. But this trap is just that, a trap: the mother mouse made this clear to me, as have all those mothers and others who care enough for the health and well-being of those they love to fight for them. There are some things worth fighting for, worth dying for, worth killing for.

Now, I understand that inculcation into civilization's insane ideology has caused many people in this culture to believe that the others whom this culture is killing are not actually alive: after all, a river doesn't feel, does it? Nor do animals in zoos or in factory farms, nor certainly do plants in factory farms, nor stones in quarries.

But does someone's prior indoctrination mean they need not be stopped?

This I know: Indigenous peoples have entirely different relationships to each other and to the land, based on perceiving "nature" as consisting of beings (including humans) to enter into relationship with, not objects to be exploited. This I know, too: those working to protect land they love are working to protect land they love, and those destroying the land must not love it, or surely they would not destroy it.

Part of what I'm getting at is that those who value things and control more than life can be more likely to kill to gain things or control than if these values were reversed. Obviously: they value things and control more than they do life. As we see. On the other hand, if we value life over control or things, we're less likely to kill even to defend life. As we also see. When groups holding these different values come into conflict this functional difference makes for a grotesquely uneven contest, or if you will allow me the language, battle.[419]

This was true of the plots against Hitler. Many plotters argued over whether to kill Hitler as he blithely caused millions to die. Even during the July 20, 1944, coup attempt the plotters merely arrested Hitler's henchmen. When the coup failed that night these same henchmen didn't hesitate to kill the plotters, or at least the lucky ones: others they tortured before killing.

We've seen this same disparity time and again in interactions between the civilized and the indigenous. We can read account after account of the indigenous welcoming the civilized as guests, showering them with gifts, giving them food, keeping them alive, and we can read account after account of the civilized killing, dispossessing, enslaving the indigenous. Years ago I heard an account of the Indian writer Sherman Alexie saying he wished he would have been alive five hundred years ago to greet Christopher Columbus. Alexie described what he would have done to Columbus with a bow and arrow, or hatchet, or axe, or gun, or chainsaw, then concluded by saying, "No, I wouldn't have done that. I would have invited him in and fed him dinner, because that's what my people do."

This is what many Indians did. Some in time learned that their generosity and kindness was not only misplaced, but in this case suicidal. Some Indians of course have fought back. And when they do? "In war they shall kill some of us; we shall destroy all of them."

We see this same thing today, every moment of every day. Those who run governments and corporations routinely lie, steal, cheat, murder, imprison, torture, dispossess, cause people to disappear. They make and use no end of weapons. We, on the other hand, make really cool papier-mâché masks and pithy signs. Some of us even write really big books.[420] We try to act honorably.

There is of course nothing wrong with acting honorably, and with having empathy. Those are both good and important things. These qualities are supposed to guide our lives. But what do we do when faced with people who are themselves not honorable, and who lack empathy?

Part of the problem is that in general abusers know what they want and know what they'll do to get it. They want to control everything they can and destroy what they can't. They'll do *anything* to achieve that. We, on the other hand, for the most part don't even know what we really want, and in any case we're not sure what we're willing to do to accomplish it.

€ ((

I know what I want. I want to live in a world with more wild salmon every year than the year before, a world with more migratory songbirds every year than the year before, a world with more ancient forests every year than the year before, a world with less dioxin in each mother's breast milk every year than the year before, a world with wild tigers and grizzly bears and great apes and marlins and swordfish. I want to live on a livable planet.

€ ((

Richard Slotkin wrote an excellent book called *Gunfighter Nation: The Myth of the Frontier in Twentieth-Century America*. It's part of a trilogy, the other two components of which are *Regeneration Through Violence* and *The Fatal Environment*. Slotkin examines, among other things, the portrayal in popular fiction of conflicts between those at the center of the American empire and their enemies—for the most part those whose land they want to steal. Because writers of popular fiction are, like other writers, propagandists, he's interested in their role as boosters of empire and articulators of the means by which acts of aggression are rationalized. A pattern Slotkin makes clear is that in book after book (and in real life) the agents of empire *always* want to fight fair—to fight "by civilized rules"—but *every time* they're prevented from doing so because the other side fights dirty. Whites want to deal with Indians fairly, but because Indians are savages (or as the Declaration of Independence puts it: "merciless Indian Savages, whose known rule of warfare, is an undistinguished destruction of all ages, sexes and conditions"), if we are to combat them, well, we have to fight like they do (or rather like we pretend they do) and slaughter them all (or as Jefferson put it, "destroy them all") as we take their land (of course taking their land only in defensive warfare: as Jan Van Riebeeck commented on the similar conquest of South Africa by the Dutch, that the land had been "justly won by the sword in defensive warfare, and that it was now our intention to retain it"[421]). It was the same story in the Philippines, where the United States wouldn't have had to exterminate the natives (one American military officer stated: "We exterminated the American Indians, and I guess most of us are proud of it, or, at least, believe the end justified the means; and we must have no scruples about exterminating this other race standing in the way of progress and enlightenment, if it is necessary"[422]), that is, we wouldn't have had to destroy them all (General Jacob H. Smith: "I want no prisoners. I wish you to kill and burn; the more you kill and burn the better you will please me"[423]), if the nasty Filipinos hadn't fought unfair first (which, as in Premise Four of this book, means fighting back at all). It was the same in the Korean War, where the Americans would have fought fairly if only the damn commies would have played by the rules. And in Vietnam, where we wouldn't have had to napalm the country and massacre literally millions of noncombatants if only they, too, would not have fought dirty. The same is true today where we have to break the rules to fight the terrorists, an enemy who, according to the President of the United States, "hides in shadows and has no regard for human life. This is an enemy that [sic] preys on innocent and unsuspecting people and then runs for cover."[424] If only terrorists would play by the rules, then we would too. But they don't, so, regrettably, we must just this once fight dirty.

If you've ever seen a cop movie, I'm sure you've seen this same plot. Dirty Harry would and could be clean if only the bad guys were not so terribly dirty. And it's not just Harry who is dirty: the same is true for cop after cop in movie after movie. It's a genre convention.

It wasn't really possible for me to see cop or war movies the same after reading Slotkin's work. Nor was it possible for me to see civilized wars the same.

I received confirmation of this pattern yet again just today, as I read the justification by a U.S. soldier for the torture of Iraqi noncombatant prisoners, which includes rape, sodomy, taking pictures of them while forcing them to masturbate, taking pictures of them while forcing them to simulate sex, sensory deprivation, water deprivation, forcing them to kneel or stand for hours, attaching electrical wires to their genitals, forcing them to stand on boxes holding electrical wires and telling them that if they step off the box they will die, putting a saddle on at least one woman in her seventies and riding her around while telling her that she is a donkey, and of course good old-fashioned smackyface leading to their deaths. His justification? "You got to understand, although it seems harsh, the Iraqis they only understand force. If you try to talk to them one on one as a normal person, they won't respect you, they won't do what you want, prisoner or just normal person on the street. So you've got to be forceful with them in some ways."[425] If you don't beat them, they won't do what you want: the key to understanding our culture's relationship ethos in one phrase.

Slotkin could have predicted his justification. By now we should be able to as well.

But that's not really why I bring it up now. I bring it up now because I don't want to fall into the same trap Slotkin describes. In some ways this is similar to my concern over claims to virtue: a daily round of self-examination. I don't want to say, "Just this once I need to deviate from my peaceful ways to enter into defensive warfare" unless I'm sure that a) my ways really are peaceful; b) the warfare really is defensive, and c) this deviation really is a need. At the same time I don't want to be narcissistic and short-sighted enough to presume that my own sense of self-righteousness—*After all*, says the pacifist™, *I choose the moral high ground*—is more important than the survival of salmon, murrelets, migratory songbirds, my nonhuman neighbors whose land this was long before I was born. Nor do I want to choose my own self-righteousness over the survival, ultimately, of human beings. Because if we continue on this same path, it is not only murrelets who will be exterminated. Human beings will not survive.

❨ ❨ ❨

Sometimes I think we think too much. Sometimes I think we don't think very clearly. Usually I think it's both at the same time. Our thinking, which so often isn't thinking, makes us crazy, ties us in knots. This is not accidental. It is common to abusive situations. As Lundy Bancroft, former codirector of Emerge, the nation's first therapeutic program for abusive men, writes in his book *Why Does He Do That? Inside the Minds of Angry and Controlling Men*, "In one important way, an abusive man works like a magician. His tricks largely rely on getting you to look off in the wrong direction, distracting your attention so that you won't notice where the real action is. . . . He leads you into a convoluted maze, making your relationship with him a labyrinth of twists and turns. He wants you to puzzle over him, to try to figure him out, as though he were a wonderful but broken machine for which you need only to find and fix the malfunctioning parts to bring it roaring to its full potential. His desire, though he may not admit it even to himself, is that you wrack your brain in this way so that you won't notice the patterns and logic of his behavior, the consciousness behind the craziness."[426]

As I tried to make clear in *Language* and *Culture*, nearly everything in civilization leads us away from being able to think clearly and from being able to feel. If we were able to do either, we would not allow those in power to kill the world, to kill our nonhuman neighbors, to kill humans we love, to kill us. And once we have been inculcated into this thinking that is not thinking, this feeling that is not feeling, the culture does not need to do much to continue to confuse us. We will continue to confuse ourselves with all of our not-thinking and not-feeling. We will do this gladly, because if we did not confuse ourselves, if we allowed ourselves to think in a way that really was thinking and to feel in a way that really was feeling, we would suddenly understand that we need to stop the horrors that surround us, and we would suddenly understand that we *can* stop the horrors that surround us, and we would suddenly understand what we need to do in order to stop the horrors—the problems are not cognitively challenging—and we would start to do it.

I do not think the nonhuman mothers I mentioned earlier entered into philosophical debates on the purity of their motives.[427] They just knew in their bodies what they needed to do. As we know in ours.

The Chinese poet Sengtsan wrote, "The more talking and thinking, the farther from the truth."[428] I sometimes think he was talking about us.

Several thousand years of inculcation and ideology all aimed at driving us equally out of our minds and our bodies, away from any realistic sense of self-defense, have gotten us to identify not with our bodies and our landbases, but with our abusers, with governments, with civilization. This misidentification is a marker of our insanity, and it is one of the things that drives us further insane, that leads to further confusion, that leads to further inaction.

Break that identification, and one's course of action becomes so much clearer.

SHOULD WE FIGHT BACK?

Kind-hearted people might, of course, think there were some ingenious way to disarm or defeat an enemy without too much bloodshed, and might imagine that this is the true goal of the art of war. Pleasant as it sounds, it is a fallacy that must be exposed.

Carl von Clausewitz

A BIG ARGUMENT BROKE OUT RECENTLY ON THE DERRICK JENSEN discussion group, between those who believe that civilization must be brought down *now* by any means necessary—and they mean any means necessary—and those who "will not budge," to use their phrase, from the belief that no human blood should ever be shed, and especially, to once again use a phrase of theirs, no "innocent" blood. Members of this latter camp state—again and again—that if only we feel sufficient compassion for those who are killing the planet, then they will, by basking in the reflected glow of our own shining and munificent love, come to see the error of their ways and stop all this silly destruction. The pacifists say that no one should ever under any circumstances, for example, kidnap Charles Hurwitz, nor especially his children, even if that could somehow force him to stop deforesting. The others counter by asking about all the nonhuman innocents murdered so Hurwitz can make a buck. They ask as well about the humans whose water supplies are trashed by Hurwitz's activities. Where, they ask, is accountability? How do we stop him?

I'll tell you the part of the discussion I've found most interesting: I've been imagining the thousands of somewhat similar conversations—some even more heated than this one—held around thousands of campfires and in thousands of longhouses by members of hundreds or thousands of indigenous tribes as they desperately strove (and strive) to figure out strategies and tactics that would (and will) save their lives and their ways of life. I see them standing around fires in forests in Europe, preparing as a people to face down Greek phalanxes or later the legions of Rome or still later priests and missionaries (and still later merchants and traders: what would now be called businesspeople and resource specialists) carrying the same message: submit or die. I see them in the forests and plains of China choosing whether to fight against an encroaching civilization—is there any other kind?—or to be dispossessed, then given that same choice of assimilation (submission) or death. Or maybe they'll move away, then move again, and again, each time being pushed away by civilization's insatiable lust for land, for conquest, for control, for expansion, each time being pushed onto the land of other of the indigenous. Or maybe their choice will be to simply disappear, evaporate like mist in the heat of this other culture.

I see them standing outside the forts of the Dutch or Portuguese in Africa, wondering whether they should try to talk these strange people from across the sea into stealing no more of their land—as they have tried time and again to talk to them, all to no end—or if they should attempt to stop them by force.

I see and hear these conversations in Aotearoa,[429] Mosir,[430] Hbun Squmi,[431] Chukiyawu,[432] Yondotin,[433] iTswani,[434] and in thousands of other places whose real names are not now remembered. I see and hear people having these conversations in great communal gatherings in maraes and longhouses, and I see them having these conversations singly, with friends, brothers, grandmothers. I see men (and women) sharpening their arrowheads and honing the edges of their tomahawks. I see them preparing for war, and I see the determination in their eyes and in the set of their jaw. I see also sorrow, for what's been lost, and joy and exuberance, excitement and clarity at the prospect of finally fighting back. They are of all races, from all places, getting ready to fight to defend their lives and the land they love. I see others wrapping their weapons in skins, putting them away, vowing to bring them out again only to hunt, but to fight the civilized no more forever.

I can hear those who argue against fighting back. I hear the Choctaw Pushmataha, for example. The night is warm. Fall has not yet fully arrived in this land. The fire is low. It is late. Pushmataha says, "The question before us now is not what wrongs they have inflicted upon our race, but what measures are best for us to adopt in regard to them; and though our race may have been unjustly treated and shamefully wronged by them, yet I shall not for that reason alone advise you to destroy them, unless it was just and expedient for you so to do; nor, would I advise you to forgive them, though worthy of your commiseration, unless I believe it would be to the interest of our common good. We should consult more in regard to our future welfare than our present. What people, my friends and countrymen, were so wise and inconsiderate as to engage in a war of their own accord, when their own strength, and even with the strength of others, was judged unequal to the task?"[435] We should not fight, he says, because we cannot win.[436]

Now I hear another also argue against fighting back. It is the Santee Sioux Taóyatedúta. His people are starving to death because his tribe has been forced onto a reservation—forced into dependency—and the food they were promised in exchange for giving up their land has (of course) not arrived. Most of the Santee are ready to go to war. Taóyatedúta warns against this, for reasons as pragmatic as Pushmataha's, though in language more direct: "See!—the white men are like the locusts when they fly so thick that the whole sky is a snow-storm. You may kill one—two—ten; yes, as many as the leaves

in the forest yonder, and their brothers will not miss them. Kill one—two—
ten, and ten times ten will come to kill you. Count your fingers all day long and
white men with guns in their hands will come faster than you can count. . . .
Yes; they fight among themselves, but if you strike at them they will all turn
on you and devour you and your women and little children just as the locusts
in their time fall on the trees and devour all the leaves in one day. . . . You will
die like the rabbits when the hungry wolves hunt them in the Hard Moon
[January]." After saying all of this, Taóyatedúta looks at the faces of those
around him. He again begins to speak. He thinks those who clamor for war
are fools, but if his people are foolish enough to go to war against these overwhelming odds, he says, "Taóyatedúta is not a coward: he will die with you."[437]

I see and hear others who do not counsel caution or cooperation with those
who are killing them, but who wish to strike back, and strike back hard. Standing at the same low fire as Pushmataha, the great Shawnee Tecumseh states, "If
there is one here tonight who believes that his rights will not sooner or later be
taken from him by the avaricious American pale-faces, his ignorance ought to
excite pity, for he knows little of the character of our common foe. And if there
be one among you mad enough to undervalue the growing power of the white
race among us, let him tremble in considering the fearful woes he will bring
down upon our entire race, if by his criminal indifference he assists the designs
of our common enemy against our common country. Then listen to the voice
of duty, of honor, of nature and of your endangered country. Let us form one
body, one heart, and defend to the last warrior our country, our homes, our liberty, and the graves of our fathers."[438]

In my heart and mind I follow Tecumseh village to village, as he speaks a
voice of desperation and truth that stirs something deep inside me that makes
me want to stand and join him in fighting what he and I both see as a war that
is necessary for the survival of people and landbases against an incomprehen-
sively implacable enemy. I hear Tecumseh say to the Osages, "The blood of many
of our fathers and brothers has run like water on the ground, to satisfy the
avarice of the white men. We, ourselves, are threatened with a great evil; nothing will pacify them but the destruction of all the red men.

"*Brothers*—When the white men first set foot on our grounds, they were
hungry; they had no place on which to spread their blankets, or to kindle their
fires. They were feeble; they could do nothing for themselves. Our fathers com-
miserated their distress, and shared freely with them whatever the Great Spirit
had given his red children. They gave them food when hungry, medicine when
sick, spread skins for them to sleep on, and gave them grounds, that they might

hunt and raise corn. Brothers, the white people are like poisonous serpents: when chilled they are feeble and harmless; but invigorate them with warmth, and they sting their benefactors to death.

"The white people came among us feeble; and now we have made them strong, they wish to kill us, or drive us back, as they would wolves and panthers.

"*Brothers*—The white men are not friends to the Indians: at first, they only asked for land sufficient for a wigwam; now, nothing will satisfy them but the whole of our hunting grounds, from the rising to the setting sun.

"*Brothers*—The white men want more than our hunting grounds; they wish to kill our warriors; they would even kill our old men, women, and little ones. . . .

"*Brothers*—My people wish for peace; the red men all wish for peace: but where the white people are, there is no peace for them, except it be on the bosom of our mother.

"*Brothers*—The white men despise and cheat the Indians; they abuse and insult them; they do not think the red men sufficiently good to live.

"The red men have borne many and great injuries; they ought to suffer them no longer. My people will not; they are determined on vengeance; they have taken up the tomahawk; they will make it fat with blood; they will drink the blood of the white people.

"*Brothers*—My people are brave and numerous; but the white people are too strong for them alone. I wish you to take up the tomahawk with them. If we all unite, we will cause the rivers to stain the great waters with their blood.

"*Brothers*—If you do not unite with us, they will first destroy us, and then you will fall an easy prey to them. They have destroyed many nations of red men because they were not united, because they were not friends to each other. . . .

"*Brothers*—Who are the white people that we should fear them? They cannot run fast, and are good marks to shoot at; they are only men; our fathers have killed many of them."[439]

Tecumseh is tireless. He knows what he has to do to leverage the power of his own people, and he sets out to do it. He sets out to recruit those who will fight back. He says, that same night before that same fire speaking to those same Choctaws and Chickasaws, "Have we not courage enough remaining to defend our country and maintain our ancient independence? Will we calmly suffer the white intruders and tyrants to enslave us? Shall it be said of our race that we knew not how to extricate ourselves from the three most to be dreaded calamities— folly, inactivity and cowardice? But what need is there to speak of the past? It speaks for itself and asks, 'Where today is the Pequot? Where are the Narragansetts, the Mohawks, Pocanokets, and many of the other once powerful tribes of our race?' They have vanished before the avarice and oppression of the white men, as snow before a summer sun. In the vain hope of alone defending their ancient possessions, they have fallen in the wars with the white men. Look abroad over their once beautiful country, and what see you now? Naught but the ravages of the pale-face destroyers meet your eyes. So it will be with you Choctaws and Chickasaws! Soon your mighty forest trees, under the shade of whose wide spreading branches you have played in infancy, sported in boyhood, and now rest your wearied limb after the fatigue of the chase, will be cut down to fence in the land which the white intruders dare to call their own. Soon their broad roads will pass over the graves of your fathers, and the place of their rest will be blotted out forever. . . . Think not, brave Choctaws and Chickasaws, that you can remain passive and indifferent to the common danger, and thus escape the common fate. Your people too will soon be as falling leaves and scattering clouds before their blighting breath. You too will be driven away from your native land and ancient domains as leaves are driven before the wintry storms. Sleep not longer, O Choctaws and Chickasaws, in false security and delusive hopes. Our broad domains are fast escaping from our grasp. Every year our white intruders become more greedy, exacting, oppressive and overbearing. Every year contentions spring up between them and our people and when blood is shed we have to make atonement whether right or wrong, at the cost of the lives of our greatest chiefs, and the yielding up of large tracts of our lands. Before the pale-faces came among us, we enjoyed the happiness of unbounded freedom, and were acquainted with neither riches, wants, nor oppression. How is it now? Wants and oppression are our lot; for are we not controlled in everything, and dare we move without asking, by your leave? Are we not being stripped day by day of the little that remains of our ancient liberty? Do they not even now kick and strike us as they do their black-faces? How long will it be before they tie us to a post and whip us, and make us work for them in their corn fields as they do them? Shall we wait for that moment or shall we die fighting before submitting to such ignominy? Have we not for years had before our eyes a sample of their designs, and are they not sufficient harbingers of their future determinations? Will we not soon be driven from our respective countries, and the graves of our ancestors? Will not the bones of our dead be plowed up and their graves be turned into fields? Shall we calmly wait until they become so numerous that we will no longer be able to resist oppression? Will we wait to be destroyed in our turn, without making an effort worthy of our race? Shall we give up our homes, our country, bequeathed to us by the Great Spirit, the graves of our dead, and everything that is dear and sacred to us, without a struggle? I know you will cry with me. Never! Never! Then let us by unity of action destroy them all, which we now can do, or drive them back whence they came. War or extermination is now our only choice. Which do you choose?"[440]

I hear Tecumseh speaking to the Creeks. I cannot tell if his voice is more full of rage, sorrow, excitement, determination, or reason. He says, in clear thoughts echoed by wild humans everywhere, "Let the white race perish! They seize your land, they corrupt your women, they trample on your dead! Back! whence they came, upon a trail of blood, they must be driven! Back! back—ay, into the great water whose accursed waves brought them to our shores. Burn their dwellings! Destroy their stock! Slay their wives and children! The red-man owns the country, and the pale-face must never enjoy it! War now! War forever! War upon the living! War upon the dead! Dig their very corpses from the graves! Our country must give no rest to a white man's bones."[441]

No matter where I go, no matter whom I listen to, from continent to continent, people to people, the reasons given for fighting back are always the same. I hear the words the Sauk Makataimeshiekiakiak (Black Hawk) said of himself in the third person to the whites who captured him, "He has done nothing for which an Indian should be ashamed. He has fought for his countrymen, the squaws and papooses, against white men, who came, year after year, to cheat them and take away their lands. You know the cause of our making war. It is known to all white men. They ought to be ashamed of it. The white men despise the Indians, and drive them from their homes. But the Indians are not deceitful. The white men speak bad of the Indian, and look at him spitefully. But the Indian does not tell lies; Indians do not steal. An Indian who is as bad as the white men could not live in our nation; he would be put to death, and eat up by the wolves. The white men are bad schoolmasters; they carry false looks, and deal in false actions; they smile in the face of the poor Indian to cheat him; they shake them by the hand to gain their confidence, to make them drunk, to deceive them, to ruin our wives. We told them to let us alone, and keep away from us; but they followed on, and beset our paths, and they coiled among us, like the snake. They poisoned us by their touch. We were not safe. We lived in danger. We were becoming like them, hypocrites and liars, adulterers, lazy drones, all talkers, and no workers.... Things were growing worse. There were no deer in the forest. The opossum and beaver were fled; the springs were drying up, and our squaws and papooses without victuals to keep them from starving; we called a great council, and built a large fire. The spirit of our fathers arose and spoke to us to avenge our wrongs or die.... We set up the war-whoop, and dug up the tomahawk; our knives were ready, and the heart of

Black Hawk swelled high in his bosom when he led his warriors to battle. He is satisfied. He will go to the world of the spirits contented. He has done his duty. His father will meet him there, and commend him. . . . [Black Hawk] cares for his nation and the Indians. They will suffer. He laments their fate. The white men do not scalp the head; but they do worse—they poison the heart; it is not pure with them.—His countrymen will not be scalped, but they will, in a few years, become like the white men, so that you can't trust them, and there must be, as in the white settlements, nearly as many officers as men to take care of them and keep them in order."[442]

The indigenous of Europe, Africa, Oceania, the Americas tell me of meeting the civilized, welcoming them, feeding them, saving their lives, then learning too late that welcoming, helping, trusting, saving the civilized is a fatal error, and so people after people determine to fight them.[443] Listen to the words of the Mandan Mato Tope (The Four Bears), dying of introduced small-pox, "Ever since I can remember, I have loved the Whites. I have lived With them ever since I was a Boy, and to the best of my Knowledge, I have never wronged a White Man, on the Contrary, I have always Protected them from the insults of Others, Which they cannot deny. The 4 Bears never saw a White Man hungry, but what he gave him to eat, Drink, and a Buffaloe skin to sleep on, in time of Need. I was always ready to die for them, Which they cannot deny. I have done everything that a red Skin could do for them, and how have they repaid it! With ingratitude! I have Never Called a White Man a Dog, but to day, I do Pronounce them to be a set of Black hearted Dogs, they have deceived Me, them that I always considered as Brothers have turned Out to be My Worst enemies. I have been in Many Battles, and often Wounded, but the Wounds of My enemies I exalt in. But to day I am Wounded, and by Whom, by those same White Dogs that I have always Considered, and treated as Brothers. I do not fear *Death* my friends. You Know it, but to *die* with my face rotten, that even the Wolves will shrink with horror at seeing Me, and say to themselves, that is The 4 Bears, the friend of the Whites—

"Listen well what I have to say, as it will be the last time you will hear Me. think of your Wives, Children, Brothers, Sisters, Friends, and in fact all that you hold dear, are all Dead, or Dying, with their faces all rotten, caused by those dogs the whites, think of all that My friends, and rise all together and Not leave one of them alive. The 4 Bears will act his part—"[444]

Voice after voice tells us the same story. In 1540, the Timucua Acuera stated, "Others of your accursed race have, in years past, disturbed our peaceful shores. They have taught me what you are. What is your employment? To wander about like vagabonds from land to land, to rob the poor, to betray the confiding, to murder in cold blood the defenseless. No! With such a people I want no peace,—no friendship. War, never-ending war, exterminating war, is all the boon I ask.... Keep on, robbers and traitors: in Acuera and Apalachee we will treat you as you deserve. Every captive will we quarter and hang up to the highest tree along the road."[445]

In the 1640s the Narraganset Miantinomo said: "You know our fathers had plenty of deer and skins, and our plains were full of deer and turkeys, and our coves and rivers were full of fish. But, brothers, since these English have seized upon our country, they cut down the grass with scythes, and the trees with axes. Their cows and horses eat up the grass, and their hogs spoil our beds of clams; and finally we shall starve to death! Therefore, stand not in your own light, I beseech you, but resolve with us to act like men. All the sachems both to the east and west have joined with us, and we are all resolved to fall upon them, at a day appointed.... And, when you see the three fires that will be made at the end of 40 days hence, in a clear night, then act as we act, and the next day fall on and kill men, women and children, but no cows; they must be killed as we need them for provisions, till the deer come again."[446]

Yet another voice. It is the Hunkpapa Sioux Tatanka Yotanka (Sitting Bull): "This land belongs to us, for the Great Spirit gave it to us when he put us here. We were free to come and go, and to live in our own way. But white men, who belong to another land, have come upon us, and are forcing us to live according to their ideas. That is an injustice; we have never dreamed of making white men live as we live.

"White men like to dig in the ground for their food. My people prefer to hunt the buffalo as their fathers did. White men like to stay in one place. My people want to move their tepees here and there to the different hunting grounds. The life of white men is slavery. They are prisoners in towns or farms. The life my people want is a life of freedom. I have seen nothing that a white man has, houses or railways or clothing or food, that is as good as the right to move in the open country, and live in our own fashion. Why has our blood been shed by your soldiers? ... The white men had many things that we wanted, but we could see that they did not have the one thing we liked best,—freedom. I would rather live in a tepee and go without meat when game is scarce than give up my privileges as a free Indian, even though I could have all that white men have. We marched across the lines of our reservation, and the soldiers followed us. They attacked our village, and we killed them all. What would you do if your home was attacked? You would stand up like a brave man and defend it. That is our story. I have spoken."[447]

❨ ❨ ❨

Tecumseh's elder brother Chiksika put the problem plainly: "When a white man kills an Indian in a fair fight it is called honorable, but when an Indian kills a white man in a fair fight it is called murder.[448] When a white army battles Indians and wins it is called a great victory, but if they lose it is called a massacre and bigger armies are raised. If the Indian flees before the advance of such armies, when he tries to return he finds that white men are living where he lived. If he tries to fight off such armies, he is killed and the land is taken anyway. When an Indian is killed, it is a great loss which leaves a gap in our people and a sorrow in our heart; when a white is killed three or four others step up to take his place and there is no end to it. The white man seeks to conquer nature, to bend it to his will and to use it wastefully until it is all gone and then he simply moves on, leaving the waste behind him and looking for new places to take. The whole white race is a monster who is always hungry and what he eats is land."[449]

❨ ❨ ❨

As he lay dying from wounds suffered fighting the whites, Tecumseh's father Pucksinwah made his son Chiksika promise that neither he nor Tecumseh would ever make peace with the whites. His last words were, "They only wish to devour us."[450]

What would happen if we were to fully internalize his last words? What would happen if we were to abide by this same promise he extracted from his sons?

❨ ❨ ❨

Notice that I said the arguments in the Derrick Jensen discussion group were *somewhat* similar to those I imagine have been held by countless of the indigenous. There are several significant differences.

The first of course is that the conversations among the indigenous took (and take) place within functioning communities of the uncivilized, that is, people who are free, that is, people who are not slaves. There is a world of difference between free men and women—free creatures of any sort—deciding whether to fight to defend their freedom, whether to fight to not be forced into slavery; and slaves deciding whether to fight to gain a freedom they've never known at all. The latter are less likely to fight, because their default, their experience, the state by which all others will be judged, is that of submission. They breathe it in from childhood, and drink it in their mother's milk, consume it at the table, and learn it from their fathers. Gaining freedom in this case requires a long and arduous series of conscious and willful acts, many of which will be opposed not only by their owners but perhaps more effectively by all of their training as slaves, by the myriad ways they've internalized the needs and desires (and psychoses) of their owners, and more effectively still by all of the ways they've come to accept the status quo, the default, the existence of the system of slavery as anything other than what it is: a system of slavery.

Far less likely to fight back even than slaves are those so deeply and thoroughly enslaved that they no longer perceive their own slavery. This is what we today would call normal. As Frank Garvey wrote, "In this country people are rarely imprisoned for their ideas because they're already imprisoned by their ideas. The wage-slaves of today aren't ripe for revolt because they don't know that they're slaves and no more free than the slaves of yore, despite the fact that they think so. . . . You can't get rid of slave culture until the slaves know that they are slaves, and are proud of the historical responsibility it gives them to be the agent of social change."[451]

It's not too much to say that most of us have essentially no understanding of what it would be like to live free. A few years ago I interviewed Vine Deloria, American Indian author of such books as *God Is Red, Custer Died For Your Sins*, and *Red Earth, White Lies*. He commented that we all—and most especially American Indians—are now living at a very hazardous time, because most of the current Indian elders "probably reached adulthood in the 1930s. This means their grandfathers were the guys who fought Custer and Miles, and who in the '30s were sitting on their reservations getting ready to die. Those people had been brought up in freedom. They had not had reservation experiences in their early years. We're now losing the last people who ever spoke to the last people who were free."[452]

Black Hawk's fears have come true: "They poisoned us by their touch. We were not safe. We lived in danger. We were becoming like them, hypocrites and liars, adulterers, lazy drones, all talkers, and no workers."

If many Indians have become civilized, how much tighter, then, are civilization's chains on those of us who are further removed from freedom? I know parts of my genealogy back several hundred years, and though I count a U.S. Secretary of State (William Seward) and Danish royalty among my relatives, there is not a free man or woman as far back as I can see. Far from freedom flowing through my veins and permeating every cell and informing every step and breath I take, if I wish to be free I must endeavor to squeeze out every drop of slave's blood as I find it, straining and pushing hard against everything the culture taught me: how to submit, how to not make waves, how to fear authority, how to fear perceiving my submission as submission, how to fear my feelings, how to fear perceiving the killing of those I love as the killing of those I love (or perhaps I should say the killing of those I would love had I not been taught to fear love, too), how to fear stopping by any means necessary those who are killing those I love, how to fear and loathe freedom, how to cherish and rely on insane moral structures stamped into me since birth. It's a lot of work to try to cleanse oneself of several thousand years of inculcation, even when this inculcation is into a society so obviously self- and other-destructive as this one, which is one reason so many people fail to make this effort.

Another way to say all of this is that a difference between the conversation on the discussion group and those around the campfires is that most of the participants around the campfires probably weren't insane. Sadly, the same cannot be said for the rest of us. (In related news, the front page of yesterday's *San Francisco Chronicle* carried the first installment of a *thirty-nine* part series. The subject of this in-depth coverage? Global warming? The biodiversity crisis? The murder of the oceans? Sorry, no. The series is on wine. But in the interests of full disclosure I must mention that the paper *did* cover something environmental that day: a buried article stated that since albacore tuna have less mercury in them, conscientious consumers may wish to choose to purchase them over other species. No mention was made of why any tuna have mercury in them at all.)

The good news is that, beyond and beneath that several thousand years of inculcation into this culture of slavery, our bodies carry deep inside them memories of the freedom that is the birthright of all of us, whether we are animal, plant, rock, river, or anything else.

Another difference between the conversations in the discussion group and those held by the indigenous is that the former were held in "cyberspace," which means in no place at all, but were instead entirely abstracted from place, from our bodies, from each other.

Further, most of us today have never experienced a healthy natural community. We have all been born into a world of wounds, a world being murdered, and we simply don't know what it would be like to be beneficial and welcome partners in the ongoing creation that is the daily life of a forest, river, mountain, desert, and so on. Recall the person who wrote to me stressing the need for us to remember, who said, "I've realized that outside of radical activist circles and certain indigenous peoples, the majority has completely forgotten about the passenger pigeon, completely forgotten about salmon so abundant you could

fish with baskets. I've met many people who think if we could just stop destroying the planet right now, that we'll be left with a beautiful world. It makes me
wonder if the same type of people would say the same thing in the future even
if they had to put on a protective suit in order to go outside and see the one tree
left standing in their town. Would they also have forgotten? Would it still be a
part of mainstream consciousness that there used to be whole forests teeming
with life?" When Tecumseh warned that "Soon your mighty forest trees, under
the shade of whose wide spreading branches you have played in infancy, sported
in boyhood, and now rest your wearied limb after the fatigue of the chase, will
be cut down to fence in the land which the white intruders dare to call their
own," he could presume that most of the people he addressed had not only seen
"mighty forest trees" but had formed intimate personal relationships with them
and with the other parts of the landbases they were being called upon to defend.
They knew the cycles of the insects and the cycles of the birds. They knew the
places where elk bedded down and the paths where panthers pass across. They
learned from and loved the big woodpeckers and the tiny voles. These were their
relations. Now, rare indeed are the people who have seen "mighty forest trees,"
much less participated in long-term relationships with them. This is not only
because Tecumseh's warning has come true and the forests have been cut down,
but because for the most part we humans have been metabolized fully enough
into this narcissistic culture that by now we spend far more time with machines
and other human creations than with wild beings of any sort. Years ago John A.
Livingston, author of *The Fallacy of Wildlife Conservation* and *One Cosmic
Instant: Man's Fleeting Supremacy*, told me, "Nowadays most of us live in cities.
That means most of us live in an insulated cell, completely cut off from any
kind of sensory information or sensory experience that is not of our own manufacture. Everything we see, hear, taste, smell, touch, is a human artifact. All the
sensory information we receive is *fabricated*, and most of it is mediated by
machines. I think the only thing that makes it bearable is the fact that our sensory capacities are so terribly diminished—just as they are in all domesticates—
that we no longer know what we're missing. The wild animal is receiving
information for all of the senses, from an uncountable number of sources, every
moment of its life. We get it from one only—ourselves. It's like doing solitary
in an echo chamber. People doing solitary do strange things. And the common
experience of victims of sensory deprivation is hallucination. I believe that our
received cultural wisdom, our anthropocentric beliefs and ideologies, can easily be seen as institutionalized hallucinations."[453]

(In related news, the stock market rose sharply today in heavy trading.)

Put another way, having long laid waste our own sanity, and having long forgotten what it feels like to be free, most of us too have no idea what it's like to live in the real world. Seeing four salmon spawn causes me to burst into tears. I have never seen a river full of fish. I have never seen a sky darkened for days by a single flock of birds. (I have, however, seen skies perpetually darkened by smog.) As with freedom, so too the extraordinary beauty and fecundity of the world itself: It's hard to love something you've never known. It's hard to convince yourself to fight for something you may not believe has ever existed.

Another difference between conversations now about stopping the culture versus those happening before is that civilization's stranglehold over life has grown stronger. It's always easier to stop invaders before they establish a beachhead, and it would have been a good thing had someone been able to warn the Indians not to trust and help the civilized. Maybe the Atlantic Ocean would have held them at bay for a lot longer, and without the resources from the Americas civilization might not have been able to keep expanding, and so might have collapsed. In any case, many of the pleas by Indians trying to get other Indians to join them in the fight stressed the need to strike soon, before the civilized became even more numerous and the world and its people so much weaker.

Well, we all know by now that the civilized have pretty much insinuated themselves into all the nooks and crannies. We've already discussed the number of soldiers and cops at the disposal of the rulers. And we can't forget the technologies such as video cameras, DNA banks, predator drones, RFID chips, all of which increase the control by those in power. In some ways we'll need a far bigger lever to stop civilization than we would have needed a couple of hundred years ago.

That's the bad news.

The good news is that we might not. Civilization, with its relentless drive for standardization and absolute need to destroy diversity, has made itself extremely vulnerable to certain forms of attack. Any diverse system will by definition have far fewer bottlenecks, and those it does have will be far less crucial: diversity creates alternatives and leads to adaptability. If for some reason the salmon failed to return one season, the Tolowa could have eaten the abundant elk and even more abundant crabs and even more abundant lamprey. Standardized systems, while superficially more efficient, by their very nature are more susceptible to bottlenecks, and the bottlenecks they do have are far more constraining. By now, if oil supplies get cut off, the people who live in this occupied Tolowa territory will starve to death: the salmon, elk, crabs, and lamprey are gone, along with the knowledge of how to feed ourselves.[454] Further, a globally interdependent economy will, once again by definition, be subject to far more and greater bottlenecks. Remember all the fools it takes to cut down just one big tree. Break a link in this chain of fools (chain of supply), and the chainsaws will fall silent.

For all its fancy surveillance software and bunker buster bombs, for all the propaganda pumped continuously into our homes and into our hearts, for all the massive prison complexes waiting for when the propaganda systems fail, the whole system is, as we'll explore in *Volume II*, far more vulnerable than it was at the time of Tecumseh, or than it has been at any time since its wretched beginnings. In its haste to control and destroy the world, civilization has handed us some very long levers, and pointed us toward some very well-placed and solid fulcrums. In case you are wondering, that's a very good thing indeed.

☾ ☾ ☾

I need to mention one more striking difference between arguments among the civilized and among the indigenous about whether to fight back. It's an absolutely crucial difference: Only rarely do the indigenous argue on moral grounds against fighting back. Sometimes they'll make moral arguments against fighting back in this or that case, because they feel the particular injuries they're discussing do not merit a violent response, and some tribes are extremely pacifistic among themselves (and even sometimes among other tribes), but almost never do the indigenous attempt to argue that one should on moral grounds never fight back against someone—let's be precise, kill someone—who is stealing your land and killing your people.

So far I've only found one clear example of an indigenous person counseling that one should never under any circumstances fight back. It's an article written by a Cheyenne Chief named Lawrence Hart.[455] Hart describes what he calls the Cheyenne Peace Tradition, the essence of which is, according to Hart, the following teaching: "If you see your mother, wife, or children being molested or harmed by anyone, you do not go and seek revenge. Go, sit and smoke and do nothing, for you are now a Cheyenne chief." To make sure we get his point, he repeats this word-for-word (and bolded) seven times in less than four pages. He also describes the actions of three Cheyenne he suggests we should all strive to emulate. The first of these was Lean Bear, who went to Washington, D.C., to meet with Abraham Lincoln. For this he was given a "peace medal," and documents that "would show that he was a friendly, that he had made a treaty with the United States. A peace treaty, if you will." Soon after he got home, he was out riding with some other Cheyenne and came upon a column of soldiers. He approached them. The soldiers shot him. He died clutching the documents showing he was a friendly. The second person we are supposed to emulate was White Antelope, who also had gone to D.C., and who also had received a peace medal. Lawrence Hart doesn't mention whether White Antelope was holding this medal on the morning of November 29, 1864, as Colonel Chivington's troops began the Sand Creek Massacre. White Antelope shouted in English at the white troops, "Stop! Stop!" This shouting worked no better at stopping the slaughter of Indians by whites than peace treaties had. When he finally realized the troops were attacking in earnest, he did not fight back, but folded his arms and sang his death song, "Nothing lives long/Except the earth/And the mountains."[456] The third person Hart wants us to emulate was also present at Sand Creek. Black Kettle somehow survived, and somehow continued to want to make peace with the whites. But he met the same end as the others, murdered along with his wife by Custer and the boys at the Washita massacre.

I have to be honest and tell you Hart's examples didn't compel me to want to become a moral pacifist,[457] and I have to be even more honest and tell you that I found the notion of standing by while someone molests or harms one's children or other loved ones to be profoundly immoral and irresponsible—despicable even. Many traditional Indians would have agreed. The response by Shawnees to members of their own tribe who refused to fight the whites[458] was to sneer at their weakness and fright,[459] to evince disgust and anger.[460] Of one of those who wanted peace with the whites it was written that he "was generally considered to be an inconsequential chief with nothing of any great consequence between his ears, [who] was very inclined to attend the proposed peace treaty talks and wished to grasp the American offerings of peace irrespective of at what cost."[461]

I didn't, however, find Hart's pacifism surprising, for two related reasons. The first was that the article was written in 1981,[462] long after Black Hawk's fears were realized, long after many Indians had taken on the mantle of their oppressors. The second reason has to do with how the mantle in this case manifests. Even more important to my understanding of Hart's statements is the fact that he's a Christian: a Mennonite pastor. The most direct (and so far only) argument I've seen for absolute moral pacifism by an Indian was written by a Christian. *Of course.* *Of course* a Christian would counsel pacifism and accommodation in the face of oppression. *Of course* a Christian would explicitly suggest that nothing be done to stop violence that flows down a hierarchy, even when this violence is done to one's family. *Of course* a Christian would counsel that withdrawal and contemplation (sitting and smoking and doing nothing) are appropriate *and moral* responses to molestations and harms that could be stopped. That's the point. A purpose of Christianity is and always has been to rationalize submission to those in power. Those in power conquer under the sign of the cross, while the rest of us count on getting our rewards in heaven. Or maybe we'll get some rewards here: If only we're meek enough, we're told, with a barely perceptible smile and the hint of a wink, we may someday get to inherit (the wreckage of) this world.

Now, I could understand Hart's story and teachings if he presented them as simply one part of a community's spiritual life, a part that is necessary to the health of the community, but no more nor less necessary than appropriate counsel for war, appropriate counsel for hunting, for child rearing, for where to place your communal latrines. The Shawnee, for example, had five clans, each of which served functions for the Shawnee as a whole. Two clans dealt with political matters both within and without the tribe, one dealt with matters of health and medicine, one with spiritual matters, and one provided the majority of warriors and war chiefs.[463] They all worked together. Further, I can see how it's appropriate for people to think clearly under as many circumstances as possible (but where does feeling enter Hart's description?), and I can see how it may be appropriate for some people in the community to attempt to think clearly and contemplatively under all circumstances, even the most personally trying. And I can see how others in the community may serve other roles, as appropriate. But it is simplistic, absurd, unrealistic, unnatural, and just plain incorrect to suggest, as Hart seems to,[464] that absolute pacifism is a better, more effective, more moral, or more adaptive way to structure a community, or that it is an appropriate response to the deathliness of civilization. It is also simply untrue to ascribe universal moral pacifism to the Cheyenne. Certainly the famous Cheyenne fighter Roman Nose (Woo-Kay-Nay or Arched Nose) would have been surprised to learn that the Cheyenne were or are moral pacifists. So would the Cheyenne who fought alongside Red Cloud, and those who fought alongside Sitting Bull and Crazy Horse. Far from being in any way pacifistic, the Cheyenne had at least seven full-fledged military societies: the Kit-Fox Men (Woksihitaneo); Red Shields (Mahohivas); Crazy Dogs (Hotamimasaw); Crooked Lance Society (Himoiyoqis), known by the ethnohistorian George Grinnell as the Elks; Bowstring Men (Himatanohis); Wolf Warriors (Konianutqio); and the famous Dog Soldiers (Hotamitaneo).[465]

In fact, instead of disproving my point about traditional indigenous peoples not advocating absolute moral pacifism, I think instead Lawrence Hart's article—given his Christianity—supports it.

Having said all this, I think we all know the real reason behind the paucity of speeches in support of moral pacifism by the indigenous, which is that absolute moral pacifism is a product of civilization. It is, as we'll soon explore in *Volume II*, a response by the exploited to their trauma. It is an unnatural state. It is a state that is nurtured by exploiter and victim alike, to perpetuate their exploitative and destructive relationship.

STAR WARS

Twaddle, rubbish, and gossip is what people want, not action. . . . The secret of life is to chatter freely about all one wishes to do and how one is always being prevented—and then do nothing.

Soren Kierkegaard[466]

I WENT TO SEE STAR WARS WHEN I WAS IN HIGH SCHOOL, WHICH SEEMS about the right time to see it. I liked it a lot. I wasn't one of those people who saw it a hundred times or anything. I wasn't *that* much of a nerd. Besides, I was too busy playing *Dungeons and Dragons*. I saw it again recently. It's not so good as I remember. In fact it's pretty bad. The characters are flat, the dialog hokey, the acting nondescript. But I still loved the ending, where Luke remembers to "use the force" to blow up the Death Star. For those of you who may have forgotten, the Death Star (according to the official *Star Wars* website) "was the code name of an unspeakably powerful and horrific weapon developed by the Empire. The immense space station carried a weapon capable of destroying entire planets. The Death Star was to be an instrument of terror, meant to cow treasonous worlds with the threat of annihilation. While the massive station is evidence of the evil that was the Galactic Empire, it was also proof of the New Order's greatest weakness—the belief that technology and terror were superior to the will of oppressed beings fighting for freedom." That's all pretty interesting stuff, and of course applicable to the discussion at hand: civilization as Death Star.

The website also says, "The Death Star was a battle station the size of a small moon. It had a formidable array of turbolasers and tractor beam projectors, giving it the firepower of greater than half the Imperial Starfleet. Within its cavernous interior were legions of Imperial troops and fightercraft, as well as all manner of detention blocks and interrogation cells. The Death Star was spherical, and dark gray in color. Located on the Death Star's northern hemisphere was a concave disk housing the station's main laser weapon. . . . In a brutal display of the Death Star's power, Grand Moff Tarkin targeted its prime weapon at the peaceful world of Alderaan. [Rebel princess] Leia Organa, an Imperial captive at the time, was forced to watch as the searing laser blast split apart her beloved world, turning the planet and its populace into orbital ash and debris." I'm not sure if you feel a stab of recognition at being a captive of the empire, forced to watch your beloved world and its (human and nonhuman) populace turned into orbital ash and debris. I do.

The website continues, "Using . . . stolen technical data, [rebel] Alliance tacticians were able to pinpoint a crucial flaw in the Death Star's design. A small ray-shielded thermal exhaust port led directly from the surface of the station into the heart of its colossal reactor. If the port could be breached by proton torpedoes, then the resulting chain reaction would destroy the station."[467] We all know what happened next: By using the force, and with the help of Han Solo and Chewbacca, as well as the spirit of Obi-Wan Kenobi, Luke Skywalker was able to drop a proton torpedo right down the tiny port, and blow up the Death Star.

One small proton torpedo destroyed the Death Star. This would be a prime example of leveraging your power by using a properly placed fulcrum. In our case, to switch metaphors, where do we place the charges? Where is the correct thermal exhaust port? How do we start a chain reaction that will cause the "Death Star" before us to self-destruct?

《 《 《

You know, don't you, that this wasn't the movie's original ending. I have in my hands an extremely rare early draft of the *Star Wars* film script, never before published.[468] It may surprise you to learn that the early drafts were written by environmentalists.[469] In this version, the rebels do not of course blow up the Death Star, but instead prefer to use other tactics to slow the intergalactic march of Empire. For example, they set up programs for people on planets about to be destroyed to produce luxury items like hemp hacky sacks and gourmet coffee for sale to inhabitants of the Death Star. Audience members will also discover that there are plans afoot to encourage loads of troopers and other citizens of the Empire to take ecotours of doomed planets. The purpose will be to show to one and all that these planets are economically important to the Empire and so should not be destroyed. In a surprise move that will rivet viewers to the edges of their seats, other groups of rebels file lawsuits against the Empire, attempting to show that the Environmental Impact Statement Darth Vader was required to file failed to adequately support its decision that blowing up this planet would cause "no significant impact." Viewers will thrill to learn of plans to boycott items produced by corporations that have Darth Vader on the board of directors, and will leap to their feet in theaters worldwide when they see bags full of letters written directly to Mr. Vader himself asking that he please not blow up anymore planets. (Scribbled in the margin is a note from one of the screenwriters: "For accuracy's sake, when we show examples of these letters, it is *imperative* that all letters to Mr. Vader be respectful and courteous, and that they stress the need to find cooperative solutions to the differences between the rebels and the Empire. Under no circumstances should the letters be such that they would alienate or anger Mr. Vader. If the letters upset Mr. Vader, the rebels' letter campaign to the Grand Moff Tarkin would certainly fail as well.") Other plans include sending petitions and filing lawsuits.

Now, you and I both know that all of this should be sufficient not only to bring the Empire to its knees but to make a damn fine and exciting movie. The thing is: there's more. Thousands of renegade rebels, unhappy with what they perceive as toadying on the part of the mainstream rebels, decide, in a scene guaranteed to bring tears to the eyes of even the most cold-hearted theatergoers, to stand on the planets to be destroyed, link arms (or, in some cases, tentacles), and sing "Give Peace a Chance." They send DVDs of this to both Darth Vader and his boss the Grand Moff Tarkin, to whom they also send wave after wave of lovingkindness™. Some few rebels sneak aboard the Death Star and lock themselves down to various pieces of equipment. (Early in this draft of the film, the screenwriters included a long scene showing the extensive training in nonviolent communication that is a prerequisite to joining the rebels. Most writers had originally, by the way, called it a rebel army, but several objected to the violence inherent in that word. Next came "rebel force," but nearly as many objected to that word as well. In any case, the nuanced scene of nonviolence training was dropped in later drafts and the infamous [and horribly violent] Cantina scene was, incomprehensibly to some, put in its place.[470]) Stirring debates are held onscreen among these rebels as to whether they should voluntarily surrender on approach of the troopers, or whether they should remain locked down to the end. In a brilliant and brave touch of authenticity, the rebels are never able to come to consensus.

The writers themselves entered into a debate as to whether the troopers should decapitate the locked-down rebels on or off screen, with one writer pleading that instead rebels must be explicitly shown being taken alive to interrogation cells: "Showing," he wrote in the margin, "or even implying that the troopers would ever commit these acts of violence, even in response to such obvious challenges to their authority as rebels *invading* their space and doing *violence* to their machinery by interfering with that machinery's lawful use would send absolutely the wrong message to theatergoers, and would give the wrong impression of Mr. Vader's ultimately peaceful intentions."

Once inside the Death Star, a splinter group breaks off from those about to lock themselves down. They rush down long hallways, somehow avoiding the myriad troopers. They burn a couple of transport ships, and use chemicals to etch "Galaxy Liberation Front" on the walls of the Death Star. This group miraculously escapes back to the planet about to be destroyed, where they're held by the peaceful protesters so they can be immediately and rightly turned over to troopers. That same writer comments in the margin, "Not only is it vital, once again, that the right message be sent to audience members by showing these rebels being put in a position to take responsibility for their actions, but it would also be terribly unrealistic to expect these peaceful rebels to put up with these actions that would simply give Darth Vader the excuse he needs to blow up the planet. The disrespectful hooligans *must* be turned over to the Empire promptly and without question."

Near the end of the movie another debate is held among the rebels. (One problem I had with this environmentalist screenplay was that there was a bit too much debate and not quite enough action.[471]) As the Death Star looms directly overhead, a few of the rebels advocate picking up weapons to fight back. These rebels are generally shouted down by pacifist rebels, who argue that attacking those who run the Death Star is "just another example of the Empire's harmful philosophy coming in by the back door." They state that the rebels who want to fight back are simply being co-opted by the need to control things. If we want to change Darth Vader, they say, we must all first *become* the change. To change Darth Vader's heart, we must first change our own. We must above all else have compassion for Darth Vader, and remember that he, too, was once a child. One writer put in the margins: "Excellent! This will be sure to moisten the cheeks of sensitive people everywhere!" He did not mention whether or not these tears would be of frustration. Finally Leia, Luke, Han, Chewbacca, and a couple of robots show up and tell these others they've found a way to blow up the whole Death Star. The rest of the rebels—even those who'd previously been in favor of surgical strikes aimed at "removing" Darth Vader—are horrified. They point out that blowing up the Death Star will do nothing to change the hearts and minds of those who create Death Stars, and so will accomplish nothing. Han Solo replies, "It will stop this Death Star from destroying this planet." The pacifist rebels are unmoved. They remind the unruly four that the Death Star has a crew of 265,675, plus 52,276 gunners, 607,360 troops, 25,984 stormtroopers, 42,782 ship support staff, and 167,216 pilots and support crew.[472] Each of these people on the Death Star has a family. Do you want to make their children orphans? The pacifists themselves begin to cry. (That same screenwriter comments: "If that doesn't yank the tears out of audience members' tiny ducts, I don't know what will!") They say, voices firm behind the sobs, "You cannot blow up the Death Star. What about the custodial engineers? What about the cooks? What about the people who work the shopping malls? What about those who joined the empire's armed services just so they could go to college? You—Leia, Han, Luke, and Chewbacca—are heartless and cruel."

In the exciting final scene of the environmentalist version, a scuffle breaks out between Leia, Luke, Han, and Chewbacca on one side, and the pacifists on the other. At last the pacifists chase those four from the room and from the film. They're never seen again, which isn't really important since in this version they're minor characters anyway. The Death Star looms closer and closer. Audience members chew their fingernails as they wait to see whether the letters and petitions and lawsuits will work their magic. Viewers see lasers inside the Death Star warming up to destroy the planet. The lasers glow a hellish red. The camera switches to cover the endangered planet. Suddenly a cheer will rise up from the audience as they see a small bright speck emerge from the planet's surface and speed into space. "Yes!" they will roar, as they learn that all of the intrepid environmentalist protesters were able to get off the planet moments before it got blown up!

Coda: The final shot of the movie, revealing what a complete triumph this was for the rebels, will be a still showing an article on the lower-left of page forty-three of the *New Empire Times* devoting a full three sentences to the destruction of the planet. Yes! The protesters got some press![473]

❆ ❆ ❆

During the Q & A of a talk I gave last week, someone asked, "How many environmentalists does it take to change a lightbulb?"

"I'll bite," I said, "How many?"

"None," he replied. "They just sit in the dark and whine about fossil fuel emissions."

I didn't get it. Evidently, neither did anyone else in the audience. Nobody laughed. I, as well as the rest of the audience, ended up more or less scratching our heads.

Later that night, an answer came to me: Ten. One to write the lightbulb a letter requesting that it change. Four to circulate online petitions. One to file a lawsuit demanding it change. One to send the lightbulb lovingkindness™, knowing that this is the only way real change occurs. One to accept the lightbulb precisely the way it is, clear in the knowledge that to not accept another is to do great harm to oneself. One to write a book about how and why the lightbulb needs to change. And finally, one to smash the fucking lightbulb, because we all know it's never going to change.

Acknowledgments

AS ALWAYS, MY FIRST AND MOST IMPORTANT DEBT OF GRATITUDE is to the land where I live, which sustains and supports me. Equal to this is my debt to the muse, who gives me these words, and without whom I cannot imagine my life. Thanks also to the source of my dreams.

My thanks to the redwoods, red and Port Orford cedars, alders, and cascara; the Del Norte, Olympic, slender, and Pacific giant salamanders; the Pacific tree and northern red-legged frogs; the rough-skinned newts; the spotted and barred owls, the phoebes, pileated woodpeckers, hummingbirds, herons, mergansers, and so many others; the coho and chinook salmon, the steelhead; the banana slugs, the flying ants, and the solitary bees. Thank you to the reeds, rushes, sedges, grasses, and ferns, the huckleberries, thimbleberries, salal, and salmonberries. The chanterelles, turkey tails, amanitas, and so many others. My thanks to the gray foxes, the black bears, Douglas squirrels, the moles, the shrews, the bats, the woodrats, the mice, the porcupines, and the shy aplodontia. My thanks to all the others who graciously allow me to share their home, and who teach me how to be human.

There are others who have helped with this book. They include, among others: Melanie Adcock, Roianne Ahn, Anthony Arnove, Tammis Bennett, Gabrielle Benton, Werner Brandt, Karen Breslin, Julie Burke, Leha Carpenter, George Draffan, Bill and Mary Gresham, Felicia Gustin, Alex Guillotte, Nita Halstead, Tad Hargraves, Phoebe Hwang, Mary Jensen, Lierre Keith, Casey Maddox, Marna Marteeny, Mayana, Aric McBay, Dale Morris, Theresa Noll, John Osborn, Sam Patton, Peter Piltingsrud, Karen Rath, Remedy, Tiiu Ruben, Terry Shistar and Karl Birns, Dan Simon, Julianne SkaiArbor, Shahma Smithson, Jeff and Milaka Strand, Becky Tarbotton, Luke Warner, Bob Welsh, Belinda, Rob, Brian, Dean, my military friends, John D., Narcissus, Amaru, Yeti, Persephone, Shiva, Emmett.

All these acknowledgments are in a sense premature. It is customary after finishing a book for authors to acknowledge in print all those who helped bring the book to fruition. But this book isn't yet finished. If it is to be more than mere words, this book will only be complete when this culture of death no longer imperils life on earth. At that point the acknowledgments and gratitude will flow like rivers rushing through canyons once blocked by now-crushed dams.

Notes

1. Bonhoeffer, 298.

Apocalypse

2. This is of course premise four of this book. We can say the same thing for police or the military killing regular people versus those people fighting back.
3. Eckert, 176.

Five Stories

4. McIntosh, 46.
5. Combs, 2. I'm sorry about the masculine-pronoun use here.
6. *San Francisco Chronicle*, September 13, 2001, 1.
7. "Media March to War."
8. Ibid.
9. *Z Magazine*, 62. Even here, however, Cohen was being disingenuous; because of corporate welfare programs "we" generally provide the investment as well.
10. Edwards, *Burning All Illusions*, 141.
11. There is a fifth version I did not include, which is that the bombings on 9/11 were at the very least committed with the foreknowledge (if not connivance) of those in power, and have served as a pretext to ratchet up repression and state and personal power, à la the Reichstag Fire. The last half of this equation—that the bombings have served as pretext—is undeniable, while the first half is quite possible.
12. Jefferson, 345.
13. George Draffan, Endgame Research Services: A Project of the Public Information Network, http://www.endgame.org (accessed July 10, 2004).
14. Of course it is manifestly true that the corporate managers, stock brokers, financial analysts, FBI and CIA employees, and so on (I am explicitly excluding the janitors, food service workers, temp workers, undocumented employees, firemen, and so on) who worked in the World Trade Center and the Pentagon did extraordinary harm —far more than mere frontline soldiers—to most of the humans and nonhumans in the world. As I wrote in *The Culture of Make Believe*, "It is possible to kill a million people without personally shedding a drop of blood. It is possible to destroy a culture without being aware of its existence. It is possible to commit genocide or ecocide from the comfort of one's drawing room. Presumably, the people who profit from the manufacture of ozone-depleting substances are fine and upstanding men and women. Presumably, the people who profit from the manufacture of weapons of mass destruction are well respected within their communities. Most of the horrors we are forced to live with have been caused by respectable—even great—men who themselves most often have clean hands. Warren Anderson, responsible for so many deaths in Bhopal, killed not a single Indian. The owners of Carbide who ordered the expansion of the Hawk's Nest Tunnel killed not a single black man. Thomas Jefferson killed no Indians (the same cannot be said for Andrew Jackson). If you are a god you can kill from afar, and if you kill from afar you can maintain in your own mind the objectivity to believe that those you are killing are objects, or better, you can think of them not at all." And I also wrote, "The oftentimes physical and psychic distance between financier and the activities which are financed in no way lessens their mutually reinforcing relationship. This must be understood if one is to fully apprehend the inhumanity of our culture. Most people do not

cut down forests, pollute rivers, force indigenous peoples off their land and commit genocide, or exploit workers out of a conscious sense of hatefulness (conscious perhaps being the operant word); they do it for money. Money fuels economic activities, and at the same time is the reward for participation in a culturally valued enterprise, causally linking financier to activity; without venture capital there can be no capitalist venture, and without monetary reward no venture capital will be provided. Another way to say this is that slavery would not have been viable without loans from bankers like Junius Morgan, and while Junius never once wielded the whip he undeniably, and from a distance, benefited from the lashings. It's very simple: our culture allows, even encourages (demands would probably be the best word) someone to profit—to gain power, material possessions, or prestige—at the expense of another's misery" (Jensen, *Culture*, 408, 410).

15. See, for example, Lewis Mumford, Farley Mowat, R. D. Laing, and Derrick Jensen.

Civilization

16. Diamond, 1.
17. *Webster's New Twentieth-Century Dictionary of the English Language*, 2nd ed., s.v. "civilization."
18. *Oxford English Dictionary*, compact ed., s.v. "civilization."
19. Stannard, 4.
20. Ibid.
21. Mies, 98.
22. Mumford, *Technics*, 186.
23. Diamond, 1.
24. Mumford, *Technics*, 186. There is awkwardness in the original, even though Mumford is normally an exquisite stylist.
25. Diamond, 4.
26. Turner, 182.
27. Faust, 293.

Clean Water

28. Personal communication, December 11, 1998.
29. George W. Bush and others stated in response to the World Trade Center attack that it was our patriotic duty to go out and shop: "Take your family," Bush also said, "down to Disneyworld."
30. Most women I know consider those numbers to be low, with actual numbers approaching unity. Many women tell me they know of no women who have not been sexually assaulted.
31. Caputi, *Age of Sex Crime*, 91.
32. Ibid., 160.
33. I'm speaking theoretically on this one: I love doing research, but my love does know bounds.
34. Mullan and Marvin, 157.
35. The work of Charlene Spretnak was important to my understanding here.

Catastrophe

36. Paz, 212.
37. Mowat, Stannard, Drinnon, and Turner, for example.
38. Laing, 58.
39. For example, that the damn New York Yankees go to the World Series every year; oh, scratch that: the Yankees in the Series *is* about as inevitable as you can get.
40. Previous paragraphs cobbled together from Scheffer, et al.; "Gradual Change"; and "Accumulated Change."
41. Cited in Vidal, 19.
42. Mumford, *Pentagon*, plate 24.

Violence

43. Peter, 115.
44. Mies, 99.
45. Grassroots ESA Coalition, http://nwi.org/GrassrootsESA.html (accessed January 16, 2002).
46. "Fast Facts about Wildlife Conservation Funding Needs," http://www.nwf.org/naturefunding/wildlifeconservationneeds.html (accessed January 16, 2002).
47. "States Get $16 Million."
48. Center for Defense Information, http://www.cdi.org/ (accessed January 16, 2002). It's very hard to find old budgets on their website, but the numbers are just as startling, if not more so, in more recent budgets.
49. Stark and Stark.
50. The CIA's *World Factbook*, s.v. "Afghanistan," http://www.odci.gov/cia/publications/factbook/geos/af.html (accessed November 19, 2001).
51. Ibid.
52. "MK84," FAS Military Analysis Network, http://www.fas.org/man/dod-101/sys/dumb/mk84.htm (accessed November 19, 2001).
53. Walker and Stambler, 23.
54. Ibid., 24.
55. Matus.
56. Edward Herman, 24.
57. Matus; Walker and Stambler; and "BLU-82B," FAS Military Analysis Network, http://www.fas.org/man/dod-101/sys/dumb/blu-82.htm (accessed November 19, 2001).
58. Anderson.
59. Tomlinson.
60. Cockburn, "Left," 1.
61. "CNN Says Focus."
62. "Fox: Civilian Casualties."
63. Ibid.
64. "Victims."
65. Oxborrow.
66. Watson.
67. "Information on Depleted Uranium," Sheffield-Iraq Campaign, http://www.syuergynet.co.uk/sheffield-iraq/articles/du.htm (accessed January 23, 2002).
68. I first accidentally typed *hell*, but this couldn't have been a Freudian slip, could it?
69. "Biased Process" and "Coming Your Way."
70. Cobbled together from Andreas Schuld, "Dangers Associated with Fluoride," EcoMall: A Place to Save the Earth, http://www.ecomall.com/greenshopping/fluoride2.htm (accessed January 21, 2002); Citizens for Safe Drinking Water, http://www.nofluoride.com/ (accessed January 21, 2002); and "Facing Up to Fluoride."
71. "Fluoride Conspiracy," Northstarzone, http://www.geocities.com/northstarzone/FLUORIDE.html (accessed January 21, 2002), and many others.
72. "What Is Depleted Uranium?" http://www.web-light.nl/VISIE/depleted_uranium1.html (accessed January 23, 2002).
73. Ibid.
74. "Information on Depleted Uranium."
75. "Cancers and Deformities," one part of the extraordinary Fire This Time site, http://www.wakefieldcam.freeserve.co.uk/cancersanddeformitites.htm (accessed January 26, 2002).
76. "Information on Depleted Uranium."
77. Kershaw.
78. "Extreme Deformities," Fire This Time, http://www.wakefieldcam.freeserve.co.uk/extremedeformitites.htm (accessed January 26, 2002). See also "Cancer and Deformities" on the same site.
79. Davidson.

80. *San Francisco Chronicle*, February 16, 2002, 1–3.
81. Mowat, 27.

Irredeemable

82. Root, 7.
83. Judith Herman, 33.
84. Ibid., 34.
85. Ibid., 35.
86. Ibid., chap. 2.
87. Ibid., 121.
88. Bright and Ryle.
89. And who have bought into the notion that "resources" are in fact resources at all.
90. Cottin.
91. Chomsky, 33.
92. Ibid.
93. Ibid., 48.
94. Flounders, 5.
95. See, for example, Garamone.

Counterviolence

96. Jeff Sluka, "National Liberation Movements in Global Context," Tamil Nation, http://www.tamil nation.org/selfdetermination/fourthworld/jeffsluka.htm (accessed October 10, 2004).
97. I'm not saying, of course, that all spirituality is abstract, but merely that for some people, and indeed for some entire traditions, spirituality can certainly be a way to *transcend*, i.e., avoid, embodied responses.
98. ACME Collective.
99. Ibid.
100. Ibid.
101. Ibid.
102. Ibid.
103. "Socially Responsible Shopping Guide," Global Exchange, http://www.globalexchange.org/ economy/corporations/sweatshops/ftguide.html (accessed March 16, 2002).
104. Ibid.
105. Ibid.
106. "Fair Trade: Economic Justice in the Marketplace," Global Exchange, http://www.globalexchange.org/ campaigns/fairtrade/stores/fairtrade.html (accessed March 16, 2002).
107. "Global Exchange Reality Tours," Global Exchange, http://www.globalexchange.org/tours/, and follow links from there for the other information (accessed March 16, 2002).
108. "Sweating for Nothing," Global Exchange, http://www.globalexchange.org/economy/corporations/ (accessed March 16, 2002).
109. "Report to the Seattle City Council," 7, n. 5.
110. "Frequently Asked Questions about Anarchists at the 'Battle for Seattle' and N30," Infoshop, http://www.infoshop.org/octo/a_faq.html (accessed March 16, 2002).
111. "Anarchists and Corporate Media." The Global Exchange activist denies saying this to the *New York Times* reporter.
112. Loïc Wacquant, "Ghetto, Banlieue, Favela: Tools for Rethinking Urban Marginality," http://sociology.berkeley.edu/faculty/wacquant/condpref.pdf (accessed March 16, 2002).
113. "Frequently Asked Questions."
114. Ibid.
115. "NMFS Refuses to Protect Habitat for World's Most Imperiled Whale: Despite Six Years of Continuous Sightings in SE Bering Sea, NMFS Claims It Can't Determine Critical Habitat for Right Whale,"

Center for Biological Diversity, http://www.biologicaldiversity.org/swcbd/press/right2-20-02.html (accessed March 20, 2002).

116. Dimitre, 10.

117. Locke, 48.

Listening to the Land

118. Personal communication, October 30, 2001.

119. Planck, 33–34.

120. An argument I've heard too often having to do with this is that because I was abused as a child, I am not, in fact, angry at the culture but rather at my father. According to this argument, all of my impassioned defense of the land where I live is a displacement of the defense of myself I wish I could have made as a child. The people who say this nearly always have a look on their face that suggests they're saying something incredibly profound, the possibility of which has never occurred to me. Of course I've long since sorted this one out. I despise my father because of his own despicable actions, not because of the actions of the industrial economy. I despise the industrial economy not because of my father but because of the despicable actions of the industrial economy and because of its effects on those I love. They're entirely separable.

I always respond to this argument: "I could have had the best childhood in the world, and 90 percent of the large fish would still be gone from the oceans. Salmon would still be in trouble. Dioxin would still be in every mother's breast milk."

The argument is a transparent attempt to avoid looking at the real issues.

121. Roycroft, 8.

122. Note that I said *raise* not *give birth to*. There are already too many industrial humans, and there are plenty of unwanted industrial humans already here who need plenty of love.

Carrying Capacity

123. Catton, vii.

124. Ibid., 39.

125. Ibid., 41.

126. Mumford, *City*, 38.

127. Ibid., 36.

128. Ibid.

129. Catton, 43.

130. Ibid., 42.

131. Ibid., 43.

132. Ibid., 52.

133. Faust, 81.

134. Neal Hall, personal communication, November 1, 2002.

135. "Population Increases and Democracy," http://www.eeeee.net/sd03048.htm (accessed September 23, 2002).

136. One could argue, of course, that because the constant importation of resources is necessary to the perpetuation of cities (and their political/historical successors: empires and nation/states), and because those who have fully internalized the values of these ever-expanding cities have come to identify their own survival with the perpetuation of the city/empire—instead of identifying with their survival as human animals within a community that includes both humans and a landbase—*all* wars of empire are in some twisted sense wars of self-defense.

137. From the perspective of the activists, this may not be a bad thing: not only do they get job security, but they get to pretend they're doing something meaningful while not threatening their own identity—or privilege—as civilized.

138. "Witch Hunting and Population Policy," http://www.geocities.com/iconoclastes.geo/witches.html (accessed September 23, 2002), referencing Krag and Devereux.

139. Genesis 1:28.
140. Pearce, 5.

The Needs of the Natural World

141. Cited in Catton, 93.
142. The night I gave that talk I was suffering a flare-up of Crohn's disease. After I got home I collapsed and was in bed for months. I took Western medicines. Now, as I put the final editing touches on this book, I am in the midst of another flare-up, this one far worse. I have been sick for five months. At first I tried ignoring the sickness, hoping it would go away on its own. That didn't work. Then I tried herbs from an extraordinary Chinese herbalist, but they didn't work either. I tried Western meds. They aren't helping. Now I am set up to take a high-tech drug in two weeks (twelve-and-a-half days, actually, or, to be more precise, 289 hours, not that I'm counting). It's supposed to work wonders. My point in mentioning this is that these are not abstract questions. I am fully aware that without these high-technology drugs I would quite possibly die within the next month or two. This disease—a disease of civilization—would probably kill me. I am also aware that the fact that these drugs will probably save my life is not a good enough reason to not take down civilization. Years ago I interviewed someone who had been an anti-civilization philosopher until open heart surgery saved his life. That changed his perspective. Saving my own skin is not a good enough reason to kill the planet. I am also aware that I am taking this high-tech medicine, with all of its attendant costs to the planet, to vivisected animals, and so on. If not taking it would by itself stop the horrors, I would of course not take it. But my not taking this particular medicine will not stop the horrors. I will take it.
143. Bacher, 1.
144. *San Francisco Chronicle*, September 26, 2002.
145. St. Clair and Cockburn. See also "Get the Facts and Clear the Air," National Campaign Against Dirty Power, http://cta.policy.net/dirtypower/ (accessed September 3, 2004).

Predator and Prey

146. *Anderson Valley Advertiser*, November 19, 2003, 2.
147. Jensen, *Culture*, 60.
148. Ibid., 87–89.
149. Faust, 184.
150. In other words, I didn't know the answer to this question, but because I am male I was required by law to still give an answer.
151. I can, however, say that when I've shared the stage with traditional indigenous peoples, I've noticed that they have nodded in agreement when I've mentioned this predator/prey relationship.

Choices

152. *Anderson Valley Advertiser*, August 18, 2004, 8. I'm sorry again for the male specificity.
153. The National Science Foundation to the Center for Biological Diversity, October 16, 2002, http://www.biologicaldiversity.org/swcbd/species/beaked/NSFResponse.pdf (accessed October 26, 2002). Information about whales and decibels cobbled together from the following sites: http://www.biologicaldiversity.org/swcbd/press/beaked10-15-2002.html (accessed October 26, 2002); http://actionnetwork.org/campaign/whales/explanation (accessed October 26, 2002); http://www.faultline.org/news/2002/10/beaked.html (accessed October 27, 2002); and from lots of sources on sound. The authors of the sources on whales would almost undoubtedly be (at least publicly) horrified by my next paragraph, and I need to make clear that my response is just that: my response.
154. Actually the river flows are a little more complex: while what I said about the Russian River being tapped out because of vineyards is true, that effect is actually invisible because of the diversion of the Eel River into the Russian River. In fact, the Russian River used to close off seasonally, but since

the Eel diversion, it hasn't. There's actually more water flowing down the Russian River today than historically. There's even a project under way to decrease the amount of water to make it resemble natural flows more closely. So far, so good. But it gets sticky really fast: part of the reason for the diversion from the Eel is that the town of Santa Rosa dumps its treated sewage water into the Russian River, and wants to follow the mantra of "the solution to pollution is dilution"; the other part of the reason is recreational. Drinking water and water for vineyards is all taken from groundwater—that underground part of the river we don't see. We see the effects of this mainly in secondary streams, which dry up, and in the drying up of people's wells. But it may also be that were the flow from the Eel shut off entirely, the Russian River would probably go to levels lower than its natural flow: you can't take water from a region without affecting water flow. The bottom line: people have messed up what nature was doing with these rivers—the Eel is depleted by diversion, and the Russian River floods worse every year, partly because of the diversion and partly because it has been boxed in and has had so much sediment dumped into it. I am grateful to Leha Carpenter for this analysis.

155. My thanks to Sean Tanner for this paragraph.
156. "U.S. Military Spending," 9.
157. *Oxford English Dictionary*, compact ed., s.v. "addict."
158. Engels, 668.

Abuse

159. Murray.
160. Fisk.
161. "Signs to Look for in a Battering Personality," Projects for Victims of Family Violence, Inc., http://www.angelfire.com/ca6/soupandsalad/content13.htm (accessed November 17, 2002).
162. Exodus 20:5.
163. Deuteronomy 6:14–15.
164. And for those of you naïve enough to think Capitalism is guided by some mythic invisible hand of some mythical Free Market™, I give you the words of someone who should know, Dwayne Andreas, CEO of Archer Daniels Midland, an agricorporation that has done as much as almost any other to destroy the lives of family farmers the world over: "There's not one grain of anything in the world that is sold in the free market. The only place you see a free market is in the speeches of politicians" (Barsamian).
165. Baran, xvii.
166. New World Vistas, 89.
167. Safire.
168. I've dealt with this at length elsewhere. See, for example, Jensen, *Culture*, 174–85.
169. My thanks to Deda Bea for teaching me about toxic mimics.
170. Glaspell, 188.
171. Jensen and Draffan, *Strangely Like War*.
172. My thanks to Jeff and Milaka Strand for this analysis.

A Culture of Occupation

173. *Estrogen Effect.*
174. My thanks to Nita Halstead for this analysis.
175. Oregon State Senate Bill 742, 72nd Legislative Assembly.
176. Severn, 8.

Why Civilization is Killing the World, Part I

177. *Anderson Valley Advertiser*, April 2, 2003, 9.
178. George Draffan and I used the following analysis in our book *Welcome to the Machine.*
179. Quotes from the *Human Resource Exploitation Training Manual—1983.*

180. Ibid.

181. Ibid.

182. "Report on the School of the Americas," Federation of American Scientists, http://www.fas.org/irp/congress/1997_rpt/soarpt.htm (accessed May 12, 2003).

183. Ibid.

184. "Weapons of American Terrorism: Torture," http://free.freespeech.org/americanstateterrorism/weapons/US-Torture.html (accessed May 12, 2003).

185. Ibid.

186. Ibid.

187. Ibid.

188. Cockburn, untitled, 9.

189. Priest and Gellman.

190. Rutten.

191. Kirby. His statement makes me wonder whether Dr. Bruno Broughton has enough of a brain to feel pain. I'm sure that there are honeybees that would be willing to sting his lips and see if this leads to "anomalous behavior."

192. Rand. For what it's worth, she absolutely emphasizes the word *white* in the last sentence.

193. He says this like it's a bad thing.

194. Ah, now there's the point!

195. I know this was a rhetorical question, but the answer is *quite often.*

196. Note that following premise four of this book, the only blood that matters, the only lives that matter, the only broken hearts that matter are those of the exploiters.

197. U.S. Congress, *Congressional Record* (56th Congress, 1st sess., 1900), vol. 23, 704, 711–12. Note that this speech was met with applause in the galleries. The entire speech is worth reading for its jingoism as well as its seamless applicability to today's invasions and the continued march of empire.

198. Ledeen, "Creative Destruction."

199. Ibid.

200. Mersereau.

201. Ledeen, "Machiavelli."

202. Beeman.

203. Ibid.

204. Ledeen, "Faster, Please."

205. Ledeen, "Scowcroft Strikes Out."

206. Ledeen, "Iranian Comedy Hour."

207. Ledeen, "Temperature Rises."

208. Ledeen, "Heart of Darkness."

209. Ledeen, "Willful Blindness."

210. Ledeen, "Lincoln Speech."

211. Leggett, 173–74.

212. St. Clair, "Santorum."

213. Scherer.

214. Regular.

215. *Anderson Valley Advertiser*, July 2, 2003, 6. The speaker is Karen Hughes.

Why Civilization is Killing the World, Part II

216. I peeled this one off numerous websites.

217. Heinen.

218. I'm thinking of songs and stories that teach us how to live in place. There's no reason to let members of the cult of the machine claim ownership of the word and concept *technology*.

219. Genesis 1:28.

220. Tiger salamanders, for crying out loud! As recently as during my childhood in the sixties, tiger salamanders were still as common as, well, as common as the toads that are now just as endangered.

221. Jensen, *Culture*, 382.
222. I was going to call them masturbational, but at least masturbation feels good.
223. Weiss.
224. Llanos.
225. Ibid.
226. Weiss.
227. That was perhaps the central point of my book *The Culture of Make Believe.*
228. I've talked about this in my other books, too, but I think it's a point that bears repeating.
229. "About PNAC," Project for the New American Century, http://www.newamericancentury.org/aboutpnac.htm (accessed June 1, 2003).
230. "Rebuilding America's Defenses."
231. Ibid.
232. "Get the Facts and Clear the Air," National Campaign Against Dirty Power, http://cta.policy.net/dirtypower/ (accessed September 3, 2004).
233. Mokhiber, 16–17.
234. Ibid., 3–4.
235. Reckard.
236. Baker.
237. Which means only those civilians killed by those who oppose the policies of the U.S and its allies.
238. Paul Richardson, "Hojojutsu—The Art of Tying," Sukisha Ko Ryu: Bringing Together All the Elements of the Ninjutsu & Samuraijutsu Takamatsu-den Traditions, http://homepages.paradise.net.nz/sukisha/hojojutsu.html (accessed June 4, 2003).
239. Jensen, *Culture*, 107.
240. Crévecoeur, 214.
241. Johansen.
242. Franklin, 481–82.
243. Axtell, 303.
244. *American Cynic.*
245. Axtell, 327.
246. *American Cynic.*
247. Ibid.
248. Stannard, 105.
249. Morgan, 74.
250. Ibid.
251. Ibid.
252. *Alcatraz.*

Bringing Down Civilization, Part I

253. Reich, 3–4.
254. Malakoff.
255. Dvorak.
256. Schor, 19.
257. Wilkinson.
258. Most of my account of the Bolt Weevils comes from Losure.
259. Kinda makes you proud to be an American, don't it?
260. Welcome to the club.
261. Sadovi.
262. Wikle. See also Towerkill, http://www.towerkill.com, for a very good exploration of how towers kill birds.

A History of Violence

263. Miller.
264. "A Study of Assassination." This can be found at innumerable websites (well, a Google search shows 138). One version, complete with drawings, is at http://www.gwu.edu/~nsarchiv/NSAEBB/NSAEBB4/ciaguat2.html (accessed July 7, 2003).
265. Diamond, 1.
266. Mowat, 92–94.
267. Ibid., 49.
268. Ibid., 61.
269. Ibid., 63–64.
270. Cokinos, 102–4
271. "B.C. Court."
272. "B.C.'s Spotted."
273. Mowat, 174.
274. Alaska Fisheries Science Center, "A Ghastly View of Fish Squeezed through the Net by the Tons of Fish Trapped within the Main Body of the Net," NOAA Photo Library, http://www.photolib.noaa.gov/fish/fish0167.htm (accessed July 10, 2003).
275. "Why Is Everybody." This article shows me *yet again* why this culture is killing the planet, and why I hate it. The magazine is supposedly pro-environmental, but the authors and editors make this whole disturbing discussion into a big joke. I don't know what their fucking problem is. I was as disgusted by this article as by the "people" the author was writing about.
276. Seekers of the Red Mist, http://seekersoftheredmist.com (accessed July 10, 2003). This particular comment was posted April 25, 2003, at 8:31 a.m.
277. White, 3. This issue also contains a very favorable review of an extraordinary book called *The Culture of Make Believe* by an extraordinarily cool guy who happens to have the same name as this author.
278. I would add, to women and to the natural world. And women, too, are alienated, of course.

Hatred

279. Fromm, 114–15.
280. *Merriam-Webster's Collegiate Dictionary*, electronic ed., vers. 1.1, s.v. "samsara."
281. Ming Zhen Shakya, "What Is Zen Buddhism, Part II—Samsara and Nirvana," http://www.hsuyun.org/Dharma/zbohy/Literature/essays/mzs/whatzen2.html (accessed July 14, 2003).
282. Richard Hooker, "India Glossary," s.v. "samsara," World Civilizations: An Internet Classroom and Anthology, Washington State University, http://www.wsu.edu:8080/~dee/GLOSSARY/SAMSARA.HTM (accessed July 14, 2003).
283. Jay Morgan, "Monks Always Get the Coolest Lines," Ordinary-Life, http://www.ordinary-life.net/blog/archives/002058.php (accessed July 29, 2003). I've made minor modifications to his version to match other versions I've read.
284. Robbins, 86.
285. I'm indebted to Ward Churchill for this phrase.
286. What's the difference between God granting his chosen people dominion over every living thing and the U.S. military seeking full-spectrum dominance?
287. Laing, 106–7.
288. Ibid., 107.

Love Does Not Imply Pacifism

289. Guevara, 225.
290. Doing her part to contribute to overshooting carrying capacity.

291. Perhaps by talking about it she has gone to the "dark side."
292. Laing, 36–37.
293. Elliott, 12. The italics are in the original.
294. Goleman, 177.

It's Time to Get Out

295. *Anderson Valley Advertiser*, April 28, 2004, 12. I'm sorry about the masculine pronoun.
296. I am indebted to Becky Tarbotton for the previous several paragraphs.
297. Gruen, 62.
298. Brown, 273, 449.
299. Orwell, 210.
300. Weber, 156.
301. Mallat.
302. "Living in Reality," 21. That same Indian gave his own answer as to what they could do: "We Aymara carry rebellion in our blood. Bolivia is totally corrupt, not just the mayor. All of them should be finished the same way, if not burnt then drowned or strangled or pulled apart by four tractors. . . . It's the only way they are going to learn."
303. Melançon.

Courage

304. I love the book enough that I published it myself, along with the extraordinary designer Tiiu Ruben.
305. As opposed to (this article would say if its author had a shred of integrity) the terrorism that characterizes those in power, and in fact is how they gain and maintain that power.
306. Wilson.
307. Fox, "Largest Arctic Ice Shelf."
308. Perlman.
309. My sister's got ovaries (the female equivalent of balls, I suppose).
310. I am indebted to my dreamgiver for these dreams and their interpretations.
311. Just last night I had dinner with a couple of mainstream environmentalists and a bunch of other people. My mom was there, too. The mainstreamers spent much of the dinner putting forward precisely these arguments. They said, and this is a direct quote, "Things will be all right if only we take back the Senate." I don't think they meant by storm, and by *we* I don't think they meant normal human beings but Democrats, the left wing of the corporate party. They then said, "And it would be great if we could take over the White House, too." My mom then said, "It doesn't matter whether the Democrats or Republicans are in the White House. The government would still be run by the big corporations." Wrinkled noses all around. Something stunk. What *was* that awful smell? Then lots of very fast sentences spilling from the mouths of the mainstream environmentalists, anything to make the moment disappear. Earlier in the evening they'd taken a different approach to someone Saying Something That Shouldn't Be Said. I was giving a talk the next day, and someone asked what I would talk about. I said, "How to take down civilization." The same awkward silence. The same wrinkling of noses. But this time the next thing that was said was, "Could you please hand me the hummus? It's awfully good. And this soup is simply delightful." Down the old memory hole.

Hope

312. Wheatley, 19.
313. There now, wasn't that easy?
314. Actually, *every* time I get on a plane, I hope it doesn't crash.
315. Well, a social life would be nice, but let's leave that aside.
316. I've lost patience with those who use *any* excuse for inaction.
317. Kind of like a belief in a Christian God or a Christian heaven.

318. Goldsmith.
319. I mean by this both that the culture is killing the planet and killing us, and more specifically that to follow the morality generated by this culture contributes to the killing of the planet and to the killing of ourselves.
320. "Sardar Kartar Singh Saraba," Gateway to Sikhism, http://allaboutsikhs.com/martyrs/sarabha.htm (accessed December 29, 2003). Cites Jagdev Singh Santokh, *Sikh Martyrs* (Birmingham, England: Sikh Missionary Resource Centre, 1995).

The Civilized Will Smile As They Tear You Limb From Limb

321. Cited in LeGuin, 45.
322. These poll results are of course jokes.
323. This person is a combination of several people I have known.
324. Doesn't that feel good just to admit that? That realization was extraordinarily liberating for me! Now I can just get on with the work.
325. And don't give me any shit about how the wants of most Americans aren't in opposition to the needs of their landbase. That's just crazy. Sure, we can talk about their deep-down desire for connection, but you and I both know that's not what I'm talking about here.
326. Jensen, *Culture*, 105–6.
327. Ibid., 106–7.
328. And even that was a farce: using a rigid definition of slavery, there are more slaves in the world today than came across on the Middle Passage, and of course that number swells to even more unimaginable proportions when we include sweatshops, wage slavery, and dispossession. For a compelling examination of modern slavery, see Bales.
329. Jensen, *Culture*, 110–12.
330. Jefferson, 345.

Their Insanity Was Permanent

331. Forbes, 31–32, 135.
332. It's hard to snare a fly when you're six feet under.
333. I've never understood why more people don't do protests. They don't really accomplish anything, but they're pretty darn fun.
334. Of course we can say the same thing about the Cleveland Indians and many other sports teams.
335. As complete and permanent as that of the vivisectors themselves.
336. Rape racks are most commonly used to impregnate factory sows.
337. It didn't help to write them letters begging them to do the right thing. Nor did it help to hold signs and placards. Nor did it help to burn candles, say prayers, and send them lovingkindness™. It didn't even work to send them faxes.
338. Planck.
339. My thanks to Gabrielle Benton.
340. Even more important than GNP. Even more important than the Dow. Even more important than jobs. Even more important than that really sexy man or woman who will be attracted to you if you buy the right toothpaste.

Romantic Nihilism

341. Nizza Thobi, "Chanah Senesh," http://www.nizza-thobi.com/Senesh_engl/html (accessed December 3, 2004).
342. Did that former agent even read any of the book?
343. For an entertaining article about this phrase, see Shulman. Here's what Emma Goldman actually wrote:

"At the dances I was one of the most untiring and gayest. One evening a cousin of Sasha [Alexander Berkman], a young boy, took me aside. With a grave face, as if he were about to announce the death of a dear comrade, he whispered to me that it did not behoove an agitator to dance. Certainly not with such reckless abandon, anyway. It was undignified for one who was on the way to become a force in the anarchist movement. My frivolity would only hurt the Cause.

"I grew furious at the impudent interference of the boy. I told him to mind his own business, I was tired of having the Cause constantly thrown into my face. I did not believe that a Cause which stood for a beautiful ideal, for anarchism, for release and freedom from conventions and prejudice, should demand the denial of life and joy. I insisted that our Cause could not expect me to become a nun and that the movement should not be turned into a cloister. If it meant that, I did not want it. 'I want freedom, the right to self-expression, everybody's right to beautiful, radiant things.' Anarchism meant that to me, and I would live it in spite of the whole world—prisons, persecution, everything. Yes, even in spite of the condemnation of my own comrades I would live my beautiful ideal" (Goldman, 56).

Neck Deep in Denial

344. *The Sun*, March 2004, 48.
345. For a stirring and extraordinary description of this by two exceedingly amazing guys, see Jensen and Draffan, *Machine*.
346. Including "Fair Trade: Economic Justice in the Marketplace," Global Exchange, http://www.globalexchange.org/campaigns/fairtrade/stores/fairtrade.html (accessed March 16, 2002).
347. Some sources say fifty.
348. My description of these four groups is from Jeannette Armstrong, by way of a personal communication with Zenobia Barlow, July 15, 2001.
349. Mumford, *Technics*, 186.
350. Apologies to Joseph Heller.
351. (in all the bad movies)

Making It Happen

352. *Anderson Valley Advertiser*, October 1, 2003, 5.
353. *U.S. et al. v. Goering et al.*
354. Tokyo War Crimes Trial Decision.
355. We can know precisely what many of these tortures were, not only because many members of the Nazi "security" agencies were later recruited by other governments (notably the U.S.) to continue plying their trade for "freedom and democracy,"™ and not only because the Nazis sometimes recorded their atrocities in as compulsive detail as sometimes do the Americans, but also because at least one of the members of the resistance—Major Fabian von Schlabrendorff—survived through a miraculous string of events (just one part of which, to give you a taste, was that the Allies happened to bomb the court and kill the judge just as Schlabrendorff was about to be sentenced to death (the judge was found clutching Schlabrendorff's file in his cold hands). In his monumental (and essential) *The History of the German Resistance, 1933–1945*, Peter Hoffmann describes it in horrifying detail:

"In the first stage Schlabrendorff's hands were tied behind his back and his fingers encased in a contraption in which spikes penetrated into the fingertips; with the turning of a screw they penetrated deeper. When this produced no answer the prisoner was strapped down on a sort of bedstead and his legs were encased in tubes covered on the inside with sharp metal spikes; the tubes were slowly drawn tighter so that the spikes gradually penetrated deeper into the flesh. During this process his head was pushed into a sort of metal hood and covered with a blanket to muffle his screams. Meanwhile he was belaboured with bamboo canes and leather switches. In the third stage, using the same bedstead, his body was stretched either violently and in jerks or gradually. If he lost consciousness he was revived with douches of cold water. These tortures had extracted no confession from

Schlabrendorff and so another method was tried. He was trussed up, bent forwards so that he could move neither backwards nor forwards, and then beaten from behind with a heavy club; with each blow he fell forward on his face with his full weight. All these tortures were applied to Schlabrendorff on the same day but the only result was that he lost consciousness. The next day he had a heart attack and could not move for several days. As soon as he had recovered, however, the tortures were repeated. Finally Schlabrendorff decided to say something. . . . [He implicated only a dead man the Nazis already knew about. As Hoffmann says on a previous page, "In fact Fabian von Schlabrendorff was so severely tortured that he was ultimately forced to abandon his initial policy of total silence and make statements implicating himself and those already dead. Only in this way, he thought, could he avoid tortures during which he might lose control of himself and his tongue."] This seemed to temporarily satisfy the *Gestapo*; it was enough to prove complicity" (Hoffmann, 522, 521).

Those in power will stop at nothing: torture is routinely used by those in power today, as is known by those who already experience it. Would that we could emulate the courage of those who already resist, knowing the potential consequences.

356. Although pagans take a lot of heat because of Hitler's dabblings in pseudo-paganism, the truth is that Hitler, and his atrocities, owe infinitely more to the mainstream Christians who supported him than to any pagans.

357. Previous paragraph from Hoffmann, and Mulholland. The Tresckow quote is from Mulholland, 8, although you can find it in many places. There are of course many heroes and heroines who resisted. And for right now for obvious reasons I've confined my list to those who tried to bring down Germany from within. There were also countless partisans and others in the occupied countries, and there were famous (and unknown) uprisings among those at concentration camps and among those about to be sent to concentration camps.

358. Montgomery, 62.

359. A modified version of the first two paragraphs appears in my book *A Language Older Than Words*.

360. Churchill, *Struggle for the Land*, 73.

361. The following several paragraphs were also used in a different form elsewhere.

362. I have to note that there were many who saw from the get-go that Hitler had to be killed, and there were many who tried to do something about this.

363. Hoffmann, 251.

364. Ibid., 253. Mason, 62, gives Hitler's next two lines: "There is no time to lose! War must come in my lifetime."

365. Of course, had Hitler been killed, the German economy would have trundled on as well.

Fulcrums

366. *The Sun*, October 2003, 48.

367. Fulcra, for those Latin aficionados keeping score at home.

368. One common way is through amassing money, but this power seeking takes many forms.

369. Or even talk about fighting back.

370. Hoffmann, 258.

371. Formerly the Bürgerbräu restaurant.

372. Mason, 80. I want to emphasize, *it was as easy as that*.

373. Some accounts make this a 180 mm shell.

374. Mason, 81–82.

375. Mason's account states he set it for 9:40, and the timer malfunctioned, going off twenty minutes early. In any case, Hitler survived.

376. Account put together from http://www.joric.com/Conspiracy/Center.html (a great site on the conspiracies to kill Hitler); Mason; and Hoffmann, 257–58.

377. Hastings, 227.

378. Ibid.

379. Keegan, 430.

380. *Effects of Strategic Bombing*, 13.
381. As were aviation fuel refineries: no aviation fuel, no airborne defenses (Dowling, 198).
382. Keegan, 430.
383. I've received a large number of letters, by the way, with cogent and radical analysis of the culture from people in the military.
384. And as I did in *A Language Older Than Words*.
385. I need to say something else here that doesn't really fit in the book but is crucial to the discussion, and this is a good place to raise it. I'm often asked if I'm afraid of getting arrested or killed by feds because of my writing. I always answer, "Absolutely. But I'm far more afraid of what this culture is doing to the planet and to all of us. It's as Robert E. Lee said, when asked why he so often attacked even when outnumbered, 'We must decide between the risk of action versus the positive loss of inaction.'"

I'll tell you my fantasy, which is that as some fed reads this book, perhaps with an increasing sense of outrage, that instead of ordering me arrested or killed, he disproves me. I would like nothing more than to be shown conclusively that my premises are wrong and that we do not have as difficult a path ahead of us as I know we do.

Show me how a way of life based on the use of nonrenewable resources can be sustainable. Show me how a way of life based on perceiving those living beings around us (and often ourselves) as resources can be sustainable. Show me how civilization can and does benefit landbases. Show me how civilization isn't based on systematic and widespread violence. (As Ursula K. LeGuin writes, "All civilization does is hide the blood and cover up hate with pretty words" [*The Sun*, March 2004, 48].) Convince me. I don't think you can do it.

I mean, by the way, really convince me. I don't mean throw at me your angry and absurd roadblocks to understanding, tossed at me simply because you are too afraid of the implications not only to allow yourself to examine them but to allow anyone else to examine them either (see R. D. Laing's Jack and Jill above). I get enough of that already. For example, after a recent talk someone emailed me with this question: "If you don't like civilization and all it brings, why don't you and your liberal [*sic*] friends just move someplace else?" I mentioned this at a talk I did a couple of nights later, and a woman in the audience exclaimed, "By Christ, tell me where I can go! The fucking culture is everywhere. I can't get away from it. The poisons are in my cells, and they're in the cells of everyone everywhere. Civilization is killing the planet!"

That email is one example of what I'm talking about. Here's another. Immediately after another show I did, an older man with a gray ponytail and loose-knit sweater rushed the stage. He demanded, "Do you have a bank account?"

"Why do you ask?"

"Because if you do, I can discount everything you say."

I stared at him, eyes wide, dumbstruck.

"All through your talk, I kept wondering whether you're a hypocrite. If you participate in the system, you're a hypocrite, and then nothing you say matters at all."

I pointed to his sweater. "Where do you think this was made? And your pants? Your shoes? My shoes? My backpack? Just because we're immersed in this culture that systematically eliminates alternatives doesn't mean—"

He cut me off, looked smug. "Ah, ha! So you feel defensive. You do have a bank account then."

I just shook my head and walked away.

Back to the feds and other cops reading this book. If you don't like what I say, disprove me. I don't think it can be done. And if you can't disprove me, don't simply act out your denial and kill or arrest me. Join me. Do the real work. Protect your landbase. I'm sure we could use your skills.

I want to be clear, by the way, that this is not a general invitation to debate my life or work in private. I do enough of that in public and have no interest in doing it in private. And frankly, more or less all of the "attempts to disprove me" I've seen have been nothing more than these roadblocks I mentioned or, even more often than that, plain old bursts of anger (and especially passive aggressiveness) because people are afraid, and so they lash out (never once, of course, admitting they're lashing out). This paragraph is instead a very specific invitation for servants of power not to fall back on force to defend that power, but to try real discourse. And for them to seriously examine the premises of this book. If they honestly find errors in my premises or thinking, I'd be willing to reexamine everything I'm saying, on the condition that if they cannot find errors, they not only seriously examine their own role in the ongoing apocalypse that is industrial civilization, but help stop the apocalypse—help bring down civilization.

Violence

386. Nopper.
387. I am indebted to Alex Guillotte for this definition.
388. My thanks to Redwood Leaverish for this definition.
389. Williams.
390. Conot, 384–85, citing *Trial of the Major War Criminals, Volume 5*, 118.
391. Cook.
392. Of course it's not unusual for corporate/capitalist journalism, and that, I guess, is the point.
393. Douglas firs, by the way, do not viably reproduce until they are eighty years old. Soon there will be none or extremely few of reproductive age on the entire continent.
394. Jensen and Draffan, *Strangely Like War*, 49.
395. This might be a good place to mention Stossel's self-proclaimed reasons for switching from consumer to corporate protection. "I just got sick of it. I also now make so much money I just lost interest in saving a buck on a can of peas." When confronted with this statement, Stossel denied making it. But it's caught on tape. Russell Mokhiber and Robert Weissman, "Stossel Tries to Scam His Public," Essential Information, http://lists.essential.org/pipermail/corp-focus/2004/000177.html (accessed April 8, 2004).
396. Jensen, *Language*, 2. The version here is slightly different than in the book because I never liked the way that paragraph was edited. Also, just in case people are interested, until I was about three-quarters of the way through *Language*, that paragraph was actually the first one.

Spending Our Way to Sustainability

397. He forgot military.
398. Sale.
399. We can pretty much say the same thing about sex, eating, feeling, or many other things. Just plug the word in for violence and the paragraph works as well.
400. It depends on who "we" are. I don't think members of the French resistance would have included the German occupiers or the French collaborators in a similar statement. Similarly, I'm not in this with Charles Hurwitz or John Stossel. Yes, they're killing the planet they live on, too, but I'm trying to stop them. I'm not on their side.
401. Well, the real point is fear. It's far less scary to not purchase an airline ticket than to blow up a dam. And we still get to say, "Ha! I delivered a blow against the machine!"
402. Just last month I bought a bunch of heirloom apple trees from a very small grower. The trees will eventually pay part of my rent to the bears and deer and birds and insects whose home this was long before I moved in.
403. And why?
404. If I may change this cliché so it finally makes sense.
405. Barsamian.
406. J. Bradford DeLong, "The Corporations as a Command Economy," http://www.j-bradford-delong.net/Econ_Articles/Command_Corporations.html (accessed March 17, 2004).
407. Don't laugh. It's been done.
408. Too bad, darn it.
409. I want them to not be created.

Empathy and Its Other

410. Silko, 94–95.
411. For a brilliant analysis of this, see Livingston's *Fallacy of Wildlife Conservation*.
412. Can you imagine a vivisector or deforester with empathies intact?
413. Hell, Hitler was nice to his dog, although that may not mean as much as it could, since more than one of his girlfriends committed suicide, an overwhelming indicator that he was emotionally and possibly physically abusive.
414. Jensen, *Listening*, 144.
415. Griffin.
416. Maybe even good and great.
417. The word they use is "our," but in this case "our" really just means theirs.
418. More on the Missoula Flood later.
419. We so often shy away even from using "violent" language, at the same time that those in power are killing us all.
420. Which I suppose could be a weapon if people would smack someone upside the head with them.
421. Moodie, part 1, 205.
422. Drinnon, 314.
423. John Moore, 7:187.
424. *San Francisco Chronicle*, September 13, 2001, 1.
425. "New Iraq Abuse."
426. Bancroft, 21.
427. And if you're one of those strange people who unaccountably thinks nonhumans can't think, then I would suggest that this "thinking" that civilized humans do at this point is worse than useless. If it causes us to hesitate to protect those we love, it is pathetic, and if it causes us to fail to protect our landbase, it is evolutionarily maladaptive.
428. It's from his *Hsin Hsin Ming: Inscribed on the Believing Mind*. See Blyth, 68.

Should We Fight Back?

429. Maori: New Zealand.
430. Ainu: Hokkaido.
431. Atayal: Taipei.
432. Aymara: La Paz.
433. Wyandott: Detroit.
434. Xhosa: Pretoria.
435. Blaisdell, 54.
436. Pushmataha said this in response to Tecumseh's declaration of solidarity with other Indians and war against the whites, and Pushmataha was probably jealous of the influence that Tecumseh wielded. It's also important to note that Pushmataha said that his people the Choctaw were at peace with the whites, and so had nothing to fear. He was, as later events unfortunately showed, wrong. That Pushmataha was no moral pacifist (and further, that he played right into the hands of the whites) is shown by the fact that he threatened to kill anyone who sided with Tecumseh or who otherwise fought against the whites. See Eckert, 548.
437. Gordon, 343–44.
438. Blaisdell, 52.
439. Hunter, 30–31.
440. Blaisdell, 50–52.
441. Brice, 193–94.
442. Blaisdell, 84–85.
443. Nonhumans of course follow the same pattern.
444. Abel, 124–25.
445. Francis S. Drake, 34.

446. Blaisdell, 6.
447. Creelman, 299–302.
448. This is of course premise four of this book. We can say the same thing for police or the military killing regular people versus those people fighting back.
449. Eckert, 176.
450. Ibid., 86.
451. *Anderson Valley Advertiser*, March 24, 2004, 11.
452. Jensen, "Where the Buffalo Go."
453. Jensen, *Listening*, 61.
454. Isn't it wonderful to live in such a "high stage of social and cultural development"?
455. Liddell Hart, 4–7.
456. Evidently White Antelope had never seen an open-pit mine.
457. Note, by the way, that I am in no way condemning the actions of Lean Bear, White Antelope, and Black Kettle, but merely saying that their actions do not make me want to fight no more forever.
458. Not on moral grounds, of course, but because they feared they could not win.
459. Eckert, 76.
460. Ibid., 107.
461. Ibid., 279.
462. Yes, 1981, not 1881. 1981. The best example I can find of a dogmatic pacifist indigenous person claiming to speak for that indigenous tradition is from the late twentieth century.
463. Eckert, 683, n. 30.
464. So do other Christian pacifist writers. See, for example, Juhnke and Schrag.
465. Richard S. Grimes, "Cheyenne Dog Soldiers," Manataka American Indian Council, http://www.manataka.org/page164.html (accessed February 23, 2005). Note that some ethnohistorians consider the Bowstring Men and the Wolf Warriors to be the same group.

Star Wars

466. *The Sun*, October 2003, 48.
467. *Star Wars*, http://www.starwars.com/databank/location/deathstar/ (accessed April 23, 2004).
468. Of course I'm making this up.
469. The draft doesn't exist.
470. They also titled the movie *Star Protest* instead of *Star Wars*.
471. That was to be an example of art imitating life.
472. *Star Wars*, http://www.starwars.com/databank/location/deathstar/?id=eu (accessed April 24, 2004).
473. It's a joke! There's no script!

Bibliography

Abel, Annie Heloise. *Chardon's Journal at Fort Clark, 1834–1839*. Lincoln: University of Nebraska Press, 1997.

"About FEMA." FEMA. http://www.fema.gov/about/ (accessed July 21, 2004).

"Accumulated Change Courts Ecosystem Catastrophe." *Science Daily*, October 12, 2001. http://www.sciencedaily.com/releases/2001/10/011011065827.htm (accessed November 29, 2001).

ACME Collective. "N30 Black Bloc Communique." Infoshop, December 4, 1999. http://www.infoshop.org/octo/wto_blackbloc.html (accessed March 16, 2002).

Alcatraz: The Whole Shocking Story. Directed by Paul Krasny. 1980.

American Cynic 2, no. 32 (August 11, 1997). http://www.americancynic.com/08111997.html (accessed June 7, 2003).

"Anarchists and Corporate Media at the Battle of Seattle." *Global Action: May Our Resistance Be as Transnational as Capital*, December 4, 1999. http://flag.blackened.net/global/1299anarchistsmedia.htm (accessed March 16, 2002).

Anderson Valley Advertiser. http://www.theava.com.

Anderson, Zack. "Dark Winter." *Anderson Valley Advertiser*, November 7, 2001.

"Antisocial Personality Disorder." Mental Health Matters. http://www.mental-health-matters.com/disorders/dis_details.php?disID=8 (accessed August 6, 2004).

Atcheson, John. "Ticking Time Bomb." *Baltimore Sun*, December 15, 2004. http://www.commondreams.org/views04/1215-24.htm (accessed February 9, 2005).

Axtell, James. *The Invasion Within: The Contest of Cultures in Colonial North America*. Oxford: Oxford University Press, 1985.

Bacher, Dan. "Bush Administration Water Cuts Result in Massive Fish Kill on Klamath." *Anderson Valley Advertiser*, October 2, 2002, 1.

"Index of Comments for *A Boy and His Dog*." Badmovies. http://www.badmovies.org/comments/?film=185 (accessed September 17, 2004).

Baker, David R. "Living a Fantasy (League)." *San Francisco Chronicle*, September 21, 2004, F1.

Bales, Kevin. *Disposable People: New Slavery in the Global Economy*. Berkeley: University of California Press, 1999.

Bancroft, Lundy. *Why Does He Do That? Inside the Minds of Angry and Controlling Men*. New York: Berkley Books, 2002.

Baran, Paul. *The Political Economy of Growth*. New York: Monthly Review, 1957.

Barringer, Felicity. "U.S. Rules Out Dam Removal to Aid Salmon." *New York Times*, December 1, 2004. http://www.nytimes.com/2004/12/01/politics/01fish.html?ex=1102921137&ei=1&en=1ba893433747ec91 (accessed December 1, 2004).

Barsamian, David. "Expanding the Floor of the Cage, Part II: An Interview with Noam Chomsky." *Z Magazine*, April 1997.

Bauman, Zygmunt. *Modernity and the Holocaust.* Ithaca, NY: Cornell University Press, 1989.

"B.C. Court OKs Logging in Endangered Owl Habitat." *CBC News*, July 9, 2003. http://www.cbc.ca/storyview/CBC/2003/07/08/owl_spotted030708 (accessed July 10, 2003).

"B.C.'s Spotted Owl Faces Extinction Scientists Warn." *CBC News*, October 7, 2002. http://www.cbc.ca/storyview/CBC/2002/10/07/spotted_owls021007 (accessed July 10, 2003).

Beeman, William O. "Colin Powell Should Make an Honorable Exit." *La Prensa San Diego*, March 14, 2003. http://www.laprensa-sandiego.org/archieve/march14-03/comment2.htm (accessed June 20, 2003).

Bettelheim, Bruno. Introduction to *Auschwitz: A Doctor's Eyewitness Account*, by Miklos Nyiszli. New York: Frederick Fell, 1960.

"Biased Process Promotes Forced Exposure to Nuclear Waste; Radioactive Materials Could Be Released into Consumer Goods, Building Supplies." *Public Citizen*, March 26, 2001. http://www.citizen.org/pressroom/release.cfm?ID=600 (accessed January 21, 2002).

Blaisdell, Bob, ed. *Great Speeches by Native Americans*. Mineola, NY: Dover, 2000.

Blakeslee, Sandra. "Minds of Their Own: Birds Gain Respect." *New York Times*, February 1, 2005.

Blyth, Reginald Horace. *Zen and Zen Classics*. Tokyo: The Hokuseido Press, 1960.

"BLU-82B." FAS Military Analysis Network. http://www.fas.org/man/dod-101/sys/dumb/blu-82.htm (accessed November 19, 2001).

Bonhoeffer, Dietrich. *Dietrich Bonhoeffer: Letters and Papers from Prison: The Enlarged Edition*. Edited by Bethge Eberhard. New York: The MacMillan Company, 1953.

BP, Frank, Ellen, and Griffin. http://www.bp.com/genericarticle.do?categoryId=2010104&contentId=2001196 (accessed June 21, 2004).

BP, Steph. http://www.bp.com/genericarticle.do?categoryId=2010104&contentId=2001092 (accessed June 21, 2004).

Brandon, William. *New Worlds for Old: Reports from the New World and Their Effect on the Development of Social Thought in Europe, 1500–1800*. Athens: Ohio University Press, 1986.

Brice, Wallace A. *History of Fort Wayne: From the Earliest Known Accounts of This Point, to the Present Period*. Fort Wayne, IN: D. W. Jones and Son, 1868.

Bright, Martin, and Sarah Ryle. "United Kingdom Stops Funding Batterers Program." *Guardian*, May 27, 2000.

Bromley, Chris, and Michael Kelberer. *The Alumni Channel: A Newsletter for Alumni and Friends of St. Anthony Falls Laboratory.* February 2004. http://www.safl.umn.edu/newsletter/alumni_channel_2004-12.html (accessed July 13, 2004).

Brown, Dee. *Bury My Heart at Wounded Knee: An Indian History of the American West.* New York: Holt, Rinehart, and Winston, 1970.

Bruno, Kenny. "BP: Beyond Petroleum or Beyond Preposterous?" *CorpWatch: Holding Corporations Accountable*, December 14, 2000. http://www.corpwatch.org/article.php?id=219 (accessed June 22, 2004).

Burroughs, William S., and David Odlier. *The Job: Interviews with William S. Burroughs.* New York: Penguin, 1989.

Burton, Bob. "Packaging the Beast: A Public Relations Lesson in Type Casting." *PRWatch* 6, no. 1 (1999): 12. http://www.prwatch.org/prwissues/1999Q1/beast.html (accessed June 21, 2004).

Cancers and Deformities. One part of the extraordinary "The Fire This Time" site, http://www.wakefieldcam.freeserve.co.uk/cancersanddeformities.htm (accessed January 26, 2002).

Caputi, Jane. *The Age of Sex Crime.* London: The Woman's Press, 1987.

————. *Gossips, Gorgons, & Crones: The Fates of the Earth.* Santa Fe, NM: Bear & Company, 1993.

Catton Jr., William R. *Overshoot: The Ecological Basis of Revolutionary Change.* Chicago: University of Illinois Press, 1982.

Center for Defense Information. http://www.cdi.org/ (accessed January 16, 2002). It's very hard to find old budgets on their Web site, but the numbers will be just as startling, if not more so, in more recent budgets.

"Child Sexual Abuse: Information from the National Clearinghouse on Family Violence." The National Clearinghouse on Family Violence (Ottawa, Canada), January 1990, revised February 1997. Available in pdf format at http://www.phac-aspc.gc.ca/ncfv-cnivf/family violence/nfntsabus_e.html (accessed March 13, 2006).

Chomsky, Noam. *Year 501: The Conquest Continues.* Boston: South End Press, 1993.

Churchill, Ward. "Appreciate History in Order to Dismantle the Present Empire." *Alternative Press Review: Your Guide Beyond the Mainstream*, August, 17, 2004. http://www.altpr.org/modules.php?op=modload&name=News&file=article&sid=272&mode=thread&order=0&thold=0 (accessed August 23, 2004).

————. "The New Face of Liberation: Indigenous Rebellion, State Repression, and the Reality of the Fourth World." In *Acts of Rebellion: The Ward Churchill Reader.* New York: Routledge, 2003.

————. *Pacifism as Pathology: Reflections on the Role of Armed Struggle in North America.* Winnipeg, Canada: Arbiter Ring, 1998.

————. *Struggle for the Land: Indigenous Resistance to Genocide, Ecocide, and Expropriation in Contemporary North America.* Monroe, ME: Common Courage, 1993.

Clausewitz, Carl von. *On War.* Translated by Michael Howard and Peter Paret. New Brunswick, NJ: Princeton University Press, 1976.

"CNN Says Focus on Civilian Casualties Would Be 'Perverse.'" *Fairness and Accuracy in Reporting,* November 1, 2001. http://www.fair.org/index.php?page=1670 (accessed March 11, 2006).

Cockburn, Alexander. *Anderson Valley Advertiser,* April 2, 2003, 9.

———. "The Left and the 'Just War.'" *Anderson Valley Advertiser,* October 31, 2001, 1.

———. "London and Miami: Cops in Two Cities." *Anderson Valley Advertiser,* November 26, 2003, 5.

Cokinos, Christopher. *Hope Is the Thing with Feathers: A Personal Chronicle of Vanished Birds.* New York: Jeremy P. Tarcher, 2000.

Combs, Robert. *Vision of the Voyage: Hart Crane and the Psychology of Romanticism.* Memphis: Memphis State University Press, 1978.

"Coming Your Way: Radioactive Garbage." *Rachel's Hazardous Waste News,* no. 183, May 30, 1990. http://www.ejnet.org/rachel/rhwn183.htm (accessed January 21, 2002).

Conot, Robert E. *Justice at Nuremberg.* New York: Carroll & Graf, 1983.

Cook, Kenneth. "Give Us a Fake: The Case Against John Stossel." *TomPaine.com,* August 15, 2000. http://www.tompaine.com/feature.cfm/ID/3481 (accessed March 13, 2004).

Cottin, Heather. "Scripting the Big Lie: Pro-War Propaganda Proliferates." *Workers World Newspaper,* November 29, 2001. http://groups.yahoo.com/group/MainLineNews/message/20262.

Crane, Jeff. "The Elwha Dam: Economic Gain Wins Out Over Saving Salmon Runs." *Columbia Magazine* 17, no. 3 (Fall 2003). http://www.washingtonhistory.org/wshs/columbia/articles/0303-a2.htm (accessed July 8, 2004). The Washington State Historical Society publishes this journal.

Creelman, James. *On the Great Highway: The Wanderings and Adventures of a Special Correspondent.* Boston: Lothrop Publishing Co., 1901.

Crévecoeur, Hector St. John de. *Letters from an American Farmer and Sketches of Eighteenth-Century America.* Edited with an introduction by Albert E. Stone. New York: Penguin, 1981.

Dam Removal: Science and Decision Making. Washington, DC: The H. John Heinz III Center for Science, Economics, and the Environment, 2002.

Dams and Development: A New Framework for Decision-Making. The Report of the World Commission on Dams. London: Earthscan, November 2000.

Davidson, Keay. "Optimistic Researcher Draws Pessimistic Reviews: Critics Attack View That Life Is Improving." *San Francisco Chronicle,* March 4, 2002, A4.

"Deepsea Fishing Nets Devastating the World's Sea Beds, Greenpeace Says." *CBC News.* http://www.cbc.ca/cp/world/040618/w061818.html (accessed June 20, 2004).

DeLong, J. Bradford. "The Corporations as a Command Economy." http://www.j-bradford -delong.net/Econ_Articles/Command_Corporations.html (accessed March 17, 2004).

Densmore, Frances. *Teton Sioux Music.* Bureau of American Ethnology, bulletin 61. Washington, DC: Smithsonian Institution, 1918.

DeRooy, Sylvia. "Before the Wilderness." *Wild Humboldt* 1 (Spring/Summer 2002): 12.

Devereux, George. *A Study of Abortion in Primitive Society.* New York, 1976.

Diamond, Stanley. *In Search of the Primitive: A Critique of Civilization.* Somerset, NJ: Transaction Publishers, 1993.

Dimitre, Tom. "Salamander Extinction?" *Econews: Newsletter of the Northcoast Environmental Center,* March 2002, 10.

Disinfopedia, s.v. "BP." http://www.disinfopedia.org/wiki.phtml?title=BP (accessed June 22, 2004).

Douglass, Frederick. *The Frederick Douglass Papers.* Edited by John Blassingame. Series 1 (Speeches, Debates, and Interviews), vol. 3 (1855–63). New Haven, CT: Yale University Press, 1985.

Dowling, Nick. "Can the Allies Strategic Bombing Campaigns of the Second World War Be Judged a Success or Failure?" *Historic Battles: History Revisited Online.* http://www.historic-battles.com/Articles/can_the_allies_strategic_bombing.htm (accessed March 5, 2004).

Draffan, George. Endgame Research Services: A Project of the Public Information Network. http://www.endgame.org (accessed July 10, 2004).

Drake, Francis S. *The Indian Tribes of the United States: Their History, Antiquities, Customs, Religion, Arts, Language, Traditions, Oral Legends, and Myths.* Vol. 2. Philadelphia: J. B. Lippincott and Co., 1884.

Drake, Samuel G. *Biography and History of the Indians of North America, from Its First Discovery.* 11th ed. Boston: Benjamin B. Mussey & Co., 1841.

Drinnon, Richard. *Facing West: The Metaphysics of Indian-Hating & Empire-Building.* Norman: University of Oklahoma Press, 1997.

Dvorak, Petula. "Cell Phones' Flaws Imperil 911 Response." *Washington Post,* March 31, 2003, B1. http://www.washingtonpost.com/ac2/wp-dyn?pagename=article&node=&contentId= A54802-2003Mar30&notFound=true (accessed June 14, 2003).

Eckert, Allan W. *A Sorrow in Our Heart: The Life of Tecumseh.* New York: Bantam Books, 1992.

Edwards, David. *Burning All Illusions.* Boston: South End Press, 1996.

———. *The Compassionate Revolution: Radical Politics and Buddhism.* Devon, U.K.: Green Books, 1998.

The Effects of Strategic Bombing on the German War Economy. The United States Strategic Bombing Survey, Overall Economic Effects Division, October 31, 1945.

Elliott, Rachel J. "Acts of Faith: Philip Berrigan on the Necessity of Nonviolent Resistance." *The Sun,* no. 331, July 2003, 4–13.

Engels, Frederick. *Herr Eugen Dühring's Revolution in Science.* Moscow: Cooperative Publishing Society of Foreign Workers in the U.S.S.R., 1934.

"Learn about EPRI." EPRI. http://www.epri.com/about/default.asp (accessed July 22, 2004).

Estes, Ralph. *Tyranny of the Bottom Line: Why Corporations Make Good People Do Bad Things.* San Francisco: Berrett-Koehler, 1996.

The Estrogen Effect: Assault on the Male. Written and produced by Deborah Cadbury for the British Broadcasting Corporation, 1993. Televised by the Discovery Channel, 1994.

Extreme Deformities. One part of the extraordinary "The Fire This Time" site, http://www.wakefieldcam.freeserve.co.uk/extremedeformities.htm (accessed January 26, 2002).

"Facing Up to Fluoride: It's in Our Water and in Our Toothpaste. Should We Worry?" *The New Forest Net.* http://www.thenewforestnet.co.uk/alternative/newforest-alt/jan2fluoride.htm (accessed January 21, 2002).

"Fair Trade: Economic Justice in the Marketplace." Global Exchange. http://www.globalexchange.org/stores/fairtrade.html (accessed March 16, 2002).

Farrell, Maureen. "A Brief (but Creepy) History of America's Creeping Fascism." *Buzzflash*, December 5, 2002. http://www.buzzflash.com/contributors/2002/12/05_Fascism.html (accessed July 21, 2004).

"Fast Facts about Wildlife Conservation Funding Needs." http://www.nwf.org/naturefunding/wildlifeconservationneeds.html (accessed January 16, 2002).

Faust, Drew Gilpin. *The Ideology of Slavery.* Baton Rouge: Louisiana State University Press, 1981.

Fischer, Louis. *The Life of Mahatma Gandhi.* New York: Harper, 1983.

Fisk, Robert. "Iraq Through the American Looking Glass: Insurgents Are Civilians. Tanks That Crush Civilians Are Traffic Accidents. And Civilians Should Endure Heavy Doses of Fear and Violence." *Independent*, December 26, 2003. http://fairuse.1accesshost.com/news1/fisk4.html (accessed October 15, 2004).

"Flack Attack." *PR Watch* 6, no. 1 (1999): 1. http://www.prwatch.org/prwissues/1999Q1/ (accessed July 22, 2004).

Flounders, Sara. Introduction to *NATO in the Balkans: Voices of Opposition*, by Ramsey Clark, Sean Gervasi, Sara Flounders, Nadja Tesich, Thomas Deichmann, et al. New York: International Action Center, 1998.

"Fluoride Conspiracy." The Northstar Foundation. http://www.geocities.com/northstarzone/FLUORIDE.html (accessed January 21, 2002).

Forbes, Jack D. *Columbus and Other Cannibals: The Wétiko Disease of Exploitation, Imperialism and Terrorism.* Brooklyn: Autonomedia, 1992.

"Fox: Civilian Casualties Not News." *Fairness and Accuracy in Reporting*, November 8, 2001. http://www.fair.org/index.php?page=1668 (accessed March 11, 2006).

Fox, Maggie. "Largest Arctic Ice Shelf Breaks Up." *ABC News* (Australia), September 23, 2003. http://www.abc.net.au/science/news/stories/s952044.htm (accessed October 28, 2003).

Fox News Sunday, June 17, 2001.

Franklin, Benjamin. *The Papers of Benjamin Franklin.* Vol. 4, *July 1, 1750–June 30, 1753.* Edited by Leonard W. Labaree, Whitfield J. Bell, Helen C. Boatfield, and Helene H. Fineman. New Haven, CT: Yale University Press, 1961.

"The Free Trade Area of the Americas and the Threat to Water." International Forum on Globalization. http://www.ifg.org/reports/ftaawater.html (accessed September 27, 2004).

"Frequently Asked Questions about Anarchists at the 'Battle for Seattle' and N30." *Infoshop.org.* http://www.infoshop.org/octo/a_faq.html (accessed March 16, 2002).

Fromm, Erich. *The Sane Society.* New York: Fawcett, 1967.

Gandhi, Mohandas K. *Gandhi on Non-Violence.* Edited by Thomas Merton. New York: New Directions, 1964.

Gantenbein, Douglas. "Swimming Upstream." *National Parks Conservation Association Magazine,* Summer 2004. http://www.npca.org/magazine/2004/summer/salmon3.asp (accessed July 10, 2004).

Garamone, Jim. "*Joint Vision 2020* Emphasizes Full-Spectrum Dominance." *American Forces Information Service News Articles,* June 2, 2000. http://www.defenselink.mil/news/Jun2000/n06022000_20006025.html (accessed March 8, 2002).

Gaura, Maria Alicia. "Curbing Off-Road Recreation: Asbestos, Rare Plants Threaten Free-wheeling Bikers in the Clear Creek Management Area." *San Francisco Chronicle,* June 13, 2004, B1.

Genesis 1:28. The Bible, silly. What did you think?

"Get the Facts and Clear the Air." Clear the Air, National Campaign Against Dirty Power. http://cta.policy.net/dirtypower/ (accessed September 3, 2004).

"A Ghastly View of Fish Squeezed through the Net by the Tons of Fish Trapped within the Main Body of the Net." NOAA Photo Library. http://www.photolib.noaa.gov/fish/fish0167.htm (accessed July 10, 2003).

Gibbon, Edward. *The Decline and Fall of the Roman Empire: Complete and Unabridged in Three Volumes.* Vol. 3, *The History of the Empire from A.D. 1135 to the Fall of Constantine in 1453.* New York: The Modern Library, n.d.

Gide, André. *André Gide: Journals.* Vol. 4, *1939–1949.* Translated by Justin O'Brien. Champaign: University of Illinois Press, 2000.

Glaspell, Kate Eldridge. "Incidents in the Life of a Pioneer." *North Dakota Historical Quarterly,* 1941, 187–88.

Global Exchange Reality Tours http://www.globalexchange.org/tours/, and follow links from there for the other information (accessed March 16, 2002).

Goldman, Emma. *Living My Life.* New York: New American Library, 1977.

Goldsmith, Zac. "Chemical-Induced Puberty." *Ecologist,* January 2004, 4.

Goleman, Daniel. *Healing Emotions.* Boston: Shambhala, 1997.

Gordon, H. L. *The Feast of the Virgins and Other Poems.* Chicago: Laird and Lee, 1891.

"Gradual Change Can Push Ecosystems into Collapse." *Environmental News Network*, October 12, 2001. http://www.enn.com/news/enn-stories/2001/10/10122001/s_45241.asp (accessed November 29, 2001).

Grassroots ESA. http://nwi.org/GrassrootsESA.html (accessed January 16, 2002).

Griffin, Susan. *A Chorus of Stones: The Private Life of War*. New York: Doubleday, 1992.

Grimes, Richard S. "Cheyenne Dog Soldiers." Manataka American Indian Council. http://www.manataka.org/page164.html (accessed February 23, 2005).

Gruen, Arno. *The Insanity of Normality: Realism as Sickness: Toward Understanding Human Destructiveness*. Translated by Hildegarde and Hunter Hannum. New York: Grove Weidenfeld, 1992.

Guevara, Ernesto Che. *Che Guevara Reader: Writings on Politics & Revolution*. 2nd ed. Edited by David Deutschmann. Melbourne: Ocean Press, 2003.

Handler, Marisa. "Indigenous Tribe Takes on Big Oil: Ecuadoran Village Refuses Money, Blocks Attempts at Drilling on Ancestral Land." *San Francisco Chronicle*, August 13, 2004. http://www.sfgate.com/cgi-bin/article.cgi?file=/chronicle/archive/2004/08/13/MNGHB86B4V1.DTL (accessed August 19, 2004).

Harden, Blaine. "Bush Would Give Dam Owners Special Access: Proposed Interior Dept. Rule Could Mean Millions for Industry." *San Francisco Chronicle*, October 28, 2004, A1, A4. http://www.sfgate.com/cgi-bin/article.cgi?file=/chronicle/archive/2004/10/28/MNGIE9HQ6U1.DTL (accessed October 31, 2004).

Hart, Lawrence. "Cheyenne Peace Traditions." *Mennonite Life*, June 1981, 4–7.

Hastings, Max. *Bomber Command*. New York: Touchstone, 1979.

Havoc Mass. "Electric Funeral: An In-Depth Examination of the Megamachine's Circuitry." *Green Anarchy*, no. 15, Winter 2004.

Hawley, Chris. "World's Land Turning to Desert at Alarming Speed, United Nations Warns." *SFGate.com*, June 15, 2004. http://sfgate.com/cgi-bin/article.cgi?file=/news/archive/2004/06/15/international1355EDT0606.DTL (accessed June 20, 2004).

Heinen, Tom. "Prophecy Believers Brace for Armageddon: Many Think Apocalyptic Battle between Jesus and the Anti-Christ Could Loom in Not-Too-Distant Future." *Milwaukee Journal-Sentinel Online*, December 31, 1999 (appeared in print January 1, 2000). http://www.jsonline.com/news/metro/dec99/apoc01123199a.asp (accessed May 18, 2003).

Henry, Robert Selph. *"First with the Most" Forrest*. Jackson, TN: McCowat-Mercer Press, 1969.

Herman, Edward S. "Nuggets from a Nuthouse." *Z Magazine*, November 2001, 24.

Herman, Judith Lewis. *Trauma and Recovery: The Aftermath of Violence—from Domestic Abuse to Political Terror*. New York: Basic Books, 1992.

"Herring and Salmon." Raincoast Research Society. http://www.raincoastresearch.org/herring-salmon.htm (accessed July 16, 2004).

Hicks, Sander. "Fearing FEMA." *Guerilla News Network*. http://www.guerrillanews.com/war_on_terrorism/doc1611.html (accessed July 21, 2004).

Hoffmann, Peter. *The History of the German Resistance, 1933–1945*. Translated by Richard Barry. Cambridge, MA: The MIT Press, 1977.

Hooker, Richard. "Samsara." *World Civilizations*. http://www.wsu.edu:8080/~dee/GLOSSARY/SAMSARA.HTM (accessed July 14, 2003).

Hopper, Jim. "Child Abuse: Statistics, Research, and Resources." Last revised February 25, 2006. http://www.jimhopper.com/abstats/ (accessed August 19, 2004).

Hoskins, Ray. *Rational Madness: The Paradox of Addiction*. Blue Ridge Summit: Tab Books, 1989.

Human Resource Exploitation Training Manual—1983. CIA, 1983. http://www.gwu.edu/~nsarchiv/NSAEBB/NSAEBB27/02-02.htm (accessed March 11, 2006).

Hume, David. "On the First Principles of Government." You can find this in any library or all over the internet.

Hunn, Eugene S. "In Defense of the Ecological Indian." Paper presented at the Ninth International Conference on Hunting and Gathering Societies, Edinburgh, Scotland, September 9, 2002. http//:www.abdn.ac.uk.chaggs9/1hunn.litm (accessed May 30, 2004).

Hunter, John D. *Memoirs of a Captivity among the Indians of North America, from Childhood to the Age of Nineteen*. Edited by Richard Drinnon. New York: Schocken Books, 1973.

Huntington, Samuel. *The Clash of Civilizations and the Remaking of World Order*. New York: Simon and Schuster, 1997.

Hurdle, Jon. "Lights-Out Policies in Cities Save Migrating Birds." *Yahoo! News*, June 10, 2004. http://story.news.yahoo.com/newtmpl=story&cid=572&e=8&u=/nm/life_birds_dc (accessed July 6, 2004).

"In His Own Words: What Bush Told the Convention." *San Francisco Chronicle*, September 9, 2004, A14.

"Information on Depleted Uranium: What Is Depleted Uranium?" Sheffield-Iraq Campaign, 6 Bedford Road, Sheffield S35 0FB, 0114-286-2336. http://www.synergynet.co.uk/sheffield-iraq/articles/du.htm (accessed January 23, 2002).

Jefferson, Thomas. *The Writings of Thomas Jefferson* Edited by Andrew A. Lipscomb and Albert Ellery Bergh. Vol. 11. Washington, DC: Thomas Jefferson Memorial Association, 1903.

Jensen, Derrick. *The Culture of Make Believe*. White River Junction, VT: Chelsea Green, 2002.

———. "Free Press for Sale: How Corporations Have Bought the First Amendment: An Interview with Robert McChesney." *The Sun*, September 2000.

———. *A Language Older Than Words*. White River Junction, VT: Chelsea Green, 2000.

———. *Listening to the Land*. White River Junction, VT: Chelsea Green, 2004.

———. *Walking on Water: Reading, Writing, and Revolution*. White River Junction, VT: Chelsea Green, 2004.

———. "Where the Buffalo Go: How Science Ignores the Living World: An Interview with Vine Deloria." *The Sun*, July 2000.

Jensen, Derrick, and George Draffan. *Strangely Like War: The Global Assault on Forests.* White River Junction, VT: Chelsea Green, 2003.

————. *Welcome to the Machine: Science, Surveillance, and the Culture of Control.* White River Junction, VT: Chelsea Green, 2004.

Johansen, Bruce E. *Forgotten Founders: Benjamin Franklin, the Iroquois and the Rationale for the American Revolution.* Ipswich, MA: Gambit Incorporated, 1982. Also available in pdf format at http//:www.ratical.org/many_worlds/6Nations/FF.pdf (accessed June 7, 2003).

"John Trudell: Last National Chairman of AIM." Redhawks Lodge. http://siouxme.com/lodge/trudell.html (accessed September 12, 2004).

Joint Vision 2020. Approval Authority: General Henry H. Shelton, Chairman of the Joint Chiefs of Staff; Office of Primary Responsibility: Director for Strategic Plans and Policy, Strategy Division. Washington, DC: U.S. Government Printing Office, June 2000.

http://www.joric.com/Conspiracy/Center.htm. A great site on the conspiracies to kill Hitler. To find the information on Elser, go to this site, then follow the link to The Lone Assassin, and continue to follow links to the end of the story.

Juhnke, James C., and Valerie Schrag. "The Original Peacemakers." *Fellowship*, May/June 1998, 9–10.

Keegan, John. *The Second World War.* 1st Amer. ed. New York: Viking Penguin, 1990.

Kennedy, Harold. "Marines Sharpen Their Skills in Hand-to-Hand Combat." *National Defense Magazine*, November 2003. http://www.nationaldefensemagazine.org/article.cfm?Id=1263 (accessed September 5, 2004).

Kennedy, Nancy. "Outrage-ous." *Shield* (the international magazine of the BP Amoco Group, U.S. ed.), Summer 1992. http://www.psandman.com/articles/shield.htm (accessed June 21, 2004).

Kershaw, Andy. "A Chamber of Horrors So Close to the 'Garden of Eden: In Foreign Parts in Basra, Southern Iraq.'" *Independent*, December 1, 2001. http://news.independent.co.uk/world/middle_east/story.jsp?story=107715 (accessed January 27, 2002).

Killian, Lewis M. *The Impossible Revolution? Black Power and the American Dream.* New York: Random House, 1968.

Kirby, Alex. "Fish Do Feel Pain, Scientists Say." *BBC News*, April 30, 2003. http://news.bbc.co.uk/2/hi/science/nature/2983045.stm (accessed May 12, 2003).

Koopman, John. "Interpreter's Death Rattles Troops: Iraqi Woman Became Close Friend of U.S. Soldiers." *San Francisco Chronicle*, August 1, 2004, A1. http://www.sfgate.com/cgi-bin/article.cgi?file=/chronicle/archive/2004/08/01/MNGJ57UGB826.DTL (accessed August 16, 2004).

Kopytoff, Verne. "Google Goes Forth into Great Beyond—Who Knows Where?" *San Francisco Chronicle*, May 2, 2004, A1.

Krag, K. *Plants Used as Contraceptives by the North American Indians.* Cambridge, MA: Harvard University Press, 1976.

Laing, R. D. *The Politics of Experience.* New York: Ballantine Books, 1967.

Lame Deer, John (Fire), and Richard Erdoes. *Lame Deer: Seeker of Visions.* New York: Simon and Schuster, 1972.

Larsen, Janet. "Dead Zones Increasing in World's Coastal Waters." Earth Policy Institute, June 16, 2004. http://www.earth-policy.org/Updates/Update41.htm (accessed June 20, 2004).

Lean, Geoffrey. "Why Antarctica Will Soon Be the *Only* Place to Live—Literally." *Independent,* May 2, 2004. http://news.independent.co.uk/world/environment/story.jsp?story=517321 (accessed May 6, 2004).

Ledeen, Michael. "Creative Destruction: How to Wage a Revolutionary War." *National Review Online,* September 20, 2001. http://www.nationalreview.com/contributors/ledeen092001.shtml (accessed May 17, 2003).

———. "Faster, Please." *National Review Online,* February 7, 2005. http://www.nationalreview.com/ledeen/ledeen200502070850.asp (accessed March 11, 2006).

———. "The Heart of Darkness: The Mullahs Make Terror Possible." *National Review Online,* December 12, 2002. At the Benador Associates website, http://www.benador associates.com/article/161 (accessed May 18, 2003).

———. "The Iranian Comedy Hour: In the U.S., the Silence Continues." *National Review Online,* October 23, 2002. At the Benador Associates website, http://www.benador associates.com/article/112 (accessed May 18, 2003).

———. "The Lincoln Speech." *National Review Online,* May 2, 2003. http://www.national review.com/ledeen/ledeen050203.asp (accessed May 18, 2003).

———. "Machiavelli on Our War: Some Advice for Our Leaders." *National Review Online,* September 25, 2001. http://www.nationalreview.com/contributors/ledeen092501.shtml (accessed May 18, 2003).

———. "Scowcroft Strikes Out: A Familiar Cry." *National Review Online,* August 18, 2002. At the Benador Associates website, http://www.benadorassociates.com/article/71 (accessed May 18, 2003).

———. "The Temperature Rises: We Should Liberate Iran First—Now." *National Review Online,* November 12, 2002. At the Benador Associates website, http://www.benador associates.com/article/130 (accessed May 18, 2003).

———. "The Willful Blindness of Those Who Will Not See." *National Review Online,* February 18, 2003. http://www.nationalreview.com/ledeen/ledeen021803.asp (accessed May 18, 2003).

Leggett, Jeremy. *The Carbon War.* New York: Routledge, 2001.

LeGuin, Ursula. "Women/Wildness." In *Healing the Wounds.* Edited by Judith Plant. Philadelphia: New Society, 1989.

Liddell Hart, B. H., ed. *The Rommel Papers.* Translated by Paul Findlay. New York: Harcourt, Brace, and Company, 1953.

"Living in Reality: Indigenous and Campesino Resistance." *Green Anarchy,* no. 19, Spring 2005.

Livingston, John A. *The Fallacy of Wildlife Conservation.* Toronto: McClelland & Stewart, 1981.

Llanos, Miguel. "Study: Big Ocean Fish Nearly Gone." *MSNBC News,* May 14, 2003. http://www.msnbc.com/news/913074.asp?ocl=cR#BODY (accessed May 31, 2003).

Locke, John. *The Second Treatise on Government.* Edited with an introduction by J. W. Gough. New York: The MacMillan Company, 1956.

Lorde, Audre. "The Master's Tools Will Never Dismantle the Master's House." In *Sister/Outsider.* Trumansburg: The Crossing Press, 1984.

Losure, Mary. "Powerline Blues." *Minnesota Public Radio,* December 9, 2002. http://news.mpr .org/features/200212/08_losurem_powerline/ (accessed July 2, 2003).

Lyons, Dana. *Turn of the Wrench* (CD). Bellingham, WA: Reigning Records.

Malakoff, David. "Faulty Towers." *Audubon,* October 2001. http://magazine.audubon.org/ features0109/faulty_towers.html (accessed June 12, 2003).

Mallat, Chibli. "New Ways Out of the Arbitration Deadlock." *Daily Star,* December 19, 1996. http://www.soas.ac.uk/Centres/IslamicLaw/DS19-12-96EuroArabChib.html (accessed October 8, 2004).

Mann, Charles C. "1491." *Atlantic Monthly,* March 2002, 41–53. http://www.theatlantic.com/ issues/2002/03/mann.htm (accessed May 39, 2004).

Marcos, Subcomandante. *Our Word Is Our Weapon: Selected Writings of Subcomandante Insurgente Marcos.* New York: Seven Stories, 2001.

Martin, Brian. "Sabotage." Chap. 8 in *Nonviolence Versus Capitalism.* London: War Resisters' International, 2001. http://www.uow.edu.au/arts/sts/bmartin/pubs/01nvc/nvc08.html (accessed August 27, 2004).

Martin, Glen. "Battle of Battle Creek: Which Way to Save Salmon?" *San Francisco Chronicle,* March 15, 2004, A1, A11.

Martin, Harry V., with research assistance from David Caul. *FEMA: The Secret Government.* 1995. http://www.globalresearch.ca/articles/MAR402B.html (accessed March 12, 2006). This is all over the internet.

Marufu, L. T., B. F. Taubman, B. Bloomer, C. A. Piety, B. G. Doddridge, J. W. Stehr, and R. R. Dickerson. "The 2003 North American Electrical Blackout: An Accidental Experiment in Atmospheric Chemistry." *Geophysical Research Letters,* vol. 31, L13106, doi:10.1029/2004GL 019771, 2004. http://www.agu.org/pubs/crossref/2004/2004GL019771.shtml (accessed September 22, 2004).

Mason Jr., Herbert Molloy. *To Kill the Devil: The Attempts on the Life of Adolf Hitler.* New York: W. W. Norton & Company, 1978.

Matus, Victorino. "Big Bombs Are Best." *The Weekly Standard,* November 9, 2001. http://www.weeklystandard.com/Content/Public/Articles/000/000/000/514obizp.asp (accessed November 19, 2001).

McCarthy, Michael. "Disaster at Sea: Global Warming Hits UK Birds." *Independent*, 30 July 2004. http://news.independent.co.uk/uk/environment/story.jsp?story=546138 (accessed August 2, 2004).

————. "Greenhouse Gas 'Threatens Marine Life.'" *Independent*, February 4, 2005. http://news.independent.co.uk/world/environment/story.jsp?story=607579 (accessed February 9, 2005).

McConnell, Howard. "Remove the Dams on the Klamath River." *Eureka Times-Standard*, July 25, 2004. http://www.times-standard.com/Stories/0,1413,127~2906~2294032,00.html (accessed July 25, 2004).

McIntosh, Alistair. *Soil and Soul.* London: Aurum Press, 2002.

"Media March to War." *Fairness and Accuracy in Reporting*, September 17, 2001. http://www.fair.org/index.php?page=1853 (accessed March 11, 2006).

Melançon, Benjamin Maurice, with Vladimir Costés. "Landless Movement Regional Leader Jailed." *Narcosphere*, August 12, 2004. http://narcosphere.narconews.com/story/2004/8/13/0229/42902 (accessed October 8, 2004).

Merriam-Webster's Collegiate Dictionary, electronic ed., vers. 1.1, 1994–1995.

Mersereau, Adam. "Why Is Our Military Not Being Rebuilt? The Case for a Total War." *National Review Online*, May 24, 2002. http://www.nationalreview.com/comment/comment-mersereau052402.asp (accessed May 18, 2003).

Mies, Maria. *Patriarchy and Accumulation on a World Scale.* London: Zed Books, 1999.

Miller, Arthur. "Why I Wrote *The Crucible*." *The New Yorker*, October 21, 1996, 158–64. http://www.newyorker.com/archive/content/?020422fr_archive02 (accessed December 1, 2004).

Ming Zhen Shakya. "What Is Zen Buddhism, Part II—Samsara and Nirvana." http://www.hsuyun.org/Dharma/zbohy/Literature/essays/mzs/whatzen2.html (accessed July 14, 2003).

Mitford, Jessica. *The American Prison Business.* New York: Penguin Books, 1977.

"MK84." FAS Military Analysis Network. http://www.fas.org/man/dod-101/sys/dumb/mk84.htm (accessed November 19, 2001).

Mokhiber, Russell. *Corporate Crime and Violence.* San Francisco: Sierra Club Books, 1988.

Mokhiber, Russell, and Robert Weissman. "Stossel Tries to Scam His Public." Essential Information, April 7, 2004. http://lists.essential.org/pipermail/corp-focus/2004/000177.html (accessed April 8, 2004).

Montgomery, David R. *King of Fish: The Thousand-Year Run of Salmon.* Boulder, CO: Westview, 2003.

Moodie, Donald. *The Record, or a Series of Official Papers Relative to the Condition and Treatment of the Native Tribes of South Africa.* Amsterdam: A. A. Balkema, 1960.

Moore, John Bassett. *A Digest of International Law.* Vol. 7. Washington, DC: Government Printing Office, 1906.

Moore, Kathleen Dean, and Jonathan W. Moore. "The Gift of Salmon." *Discover*, May 2003, 45–49.

Morgan, Edmund S. *American Slavery, American Freedom: The Ordeal of Colonial Virginia.* New York: W. W. Norton & Company, 1975.

Morgan, Jay. "Monks Always Get the Coolest Lines." Ordinary-Life.net. http://www.ordinary -life.net/blog/archives/002058.php (accessed July 29, 2003).

Mowat, Farley. *Sea of Slaughter.* Toronto: Seal, 1989.

Mulholland, Virginia. "The Plot to Assassinate Hitler." *Strategy & Tactics*, November/December 1976, 4–15.

Mullan, Bob, and Garry Marvin. *Zoo Culture: The Book About Watching People Watch Animals.* 2nd ed. Chicago: University of Illinois Press, 1999.

Mumford, Lewis. *The City in History: Its Origins, Its Transformations, and Its Prospects.* New York: Harcourt, Brace & World, 1961.

———. *The Myth of the Machine: Technics and Human Development.* New York: Harcourt Brace Jovanovich, 1966.

———. *The Myth of the Machine: The Pentagon of Power.* New York: Harcourt Brace Jovanovich, 1970.

Murray, Andrew. "Hostages of the Empire." *Guardian Unlimited*, Special Report: Iraq, July 1, 2003. http://www.guardian.co.uk/Iraq/Story/0,2763,988418,00.html (accessed October 10, 2004).

Letter from the National Science Foundation to the Center for Biological Diversity, October 16, 2002. http://www.biologicaldiversity.org/swcbd/species/beaked/NSFResponse.pdf (accessed October 26, 2002).

Neeteson, Kees. "The Dutch Low-Threshold [sic] Drugs Approach." http://people.zeelandnet. nl/scribeson/DutchApproach.html (accessed September 18, 2004).

New Columbia Encyclopedia. 4th ed. New York: Columbia University Press, 1975.

"New Iraq Abuse Pictures Surface." *Aljazeera.net*, May 6, 2004. http://english.aljazeera .net/NR/exeres/901052D2-7E43-49C3-A3F9-B0C3690CF59F.htm (accessed May 6, 2004).

"New World Vistas." Air Force Scientific Advisory Board, 1996, ancillary vol. 15.

"NMFS Refuses to Protect Habitat for World's Most Imperiled Whale: Despite Six Years of Continuous Sightings in SE Bering Sea, NMFS Claims It Can't Determine Critical Habitat for Right Whale." Center for Biological Diversity, February 20, 2002. http://www.biological diversity.org/swcbd/press/right2-20-02.html (accessed March 20, 2002).

NoFluoride. 2002. http://www.nofluoride.com/ (accessed January 21, 2002).

Nopper, Tamara Kil Ja Kim. "Yuri Kochiyama: On War, Imperialism, Osama bin Laden, and Black-Asian Politics." *AWOL Magazine*, 2003. http://awol.objector.org/yuri.html (accessed October 13, 2004).

Notes from Nowhere, eds. *We Are Everywhere: The Irresistible Rise of Global Anticapitalism.* New York: Verso, 2003.

Online Etymology Dictionary. http://www.etymonline.com/index.html (accessed August 14, 2004).

Oregon State Senate Bill 742, 72nd Legislative Assembly.

Orwell, George. *1984.* New York: New American Library, 1961.

Oxborrow, Judy. *The Oregonian,* January 20, 2002, F1.

Oxford English Dictionary, compact ed. Oxford: Oxford University Press, 1985.

Patton Boggs, "Profile," http://www.pattonboggs.com/AboutUs/index.html (accessed July 22, 2004).

Paulson, Michael. "Deal Clears Way to Buy Elwha Dams: Dicks, Gorton and Babbitt Agree on Planning for Their Demolition." *Seattle PI,* October 20, 1999. http://seattlepi.nwsource. com/local/elwh2o.shtml (accessed July 10, 2004).

Paz, Octavio. *The Labyrinth of Solitude.* New York: Grove Press, 1985.

Pearce, Joseph Chilton. *Magical Child.* New York: Plume, 1992.

Perlman, David. "Decline in Oceans' Phytoplankton Alarms Scientists: Experts Pondering Whether Reduction of Marine Plant Life Is Linked to Warming of the Seas." *San Francisco Chronicle,* October 6, 2003, A-6. http://sfgate.com/cgi-bin/article.cgi?f=/c/a/2003/10/06/ MN31432.DTL&type=science (accessed October 28, 2003).

Peter, Laurence J. *Peter's Quotations: Ideas for Our Time.* New York: William Morrow and Company, 1977.

Pitt, William Rivers. "Kenny-Boy and George." *Truthout,* July 7, 2004. http://www.truthout.org/ docs_04/070804A.shtml (accessed July 20, 2004).

Planck, Max. *Scientific Autobiography and Other Papers.* Translated by Frank Gaynor. New York: Philosophical Library, 1949.

"Population Increases and Democracy." http://www.eecce.net/sd03048.htm (accessed September 23, 2002).

Priest, Dana, and Barton Gellman. "U.S. Uses Torture on Captive Terrorists: CIA Doesn't Spare the Rod in Interrogations." *San Francisco Chronicle,* December 26, 2002, A1, A21.

Project for the New American Century. "About PNAC." http://www.newamericancentury.org/ aboutpnac.htm (accessed June 1, 2003).

PR Watch, Center for Media and Democracy. http://www.prwatch.org/cmd/ (accessed July 22, 2004).

Ramsland, Katherine. "Dr. Robert Hare: Expert on the Psychopath." Chap. 5, "The Psychopath Defined." *Court TV's Crime Library: Criminal Minds and Methods.* http://www.crimelibrary .com/criminal_mind/psychology/robert_hare/5.html?sect=19 (accessed August 6, 2004).

Rand, Ayn. Talk given at the United States Military Academy at West Point, NY, March 6, 1974. Nita Crabb, who has been of invaluable assistance in tracking down sources, was able to obtain a CD of this talk with the help of some extraordinary librarians at West Point. The sound quality is poor, but where you can hear what she says, it really is quite appalling.

"Rebuilding America's Defenses: Strategy, Forces and Resources for a New Century." A report of the Project for the New American Century, September 2000. http://www.newamerican century.org/RebuildingAmericasDefenses.pdf.

Reckard, E. Scott. "FBI Shift Crimps White-Collar Crime Probes: With More Agents Moved to Anti-Terrorism Duty, Corporate Fraud Cases are Routinely Put on Hold, Prosecutors Say." *Los Angeles Times*, August 30, 2004.

Regular, Arnon. "'Road Map Is a Life Saver for Us,' PM Abbas Tells Hamas." *Ha'aretz*, June 27, 2003. Also at *Unknown News* (and many other sites, though of course not in US capitalist newspapers), http://www.unknownnews.net/insanity7.html#quote (accessed June 30, 2003).

Reich, Wilhelm. *The Murder of Christ: The Emotional Plague of Mankind*. New York: Farrar, Strauss and Giroux, 1953.

Reitlinger, Gerald. *The Final Solution: The Attempt to Exterminate the Jews of Europe, 1939–1945*. 2nd ed. New York: Thomas Yoseloff, 1961.

Remedy. "Mattole Activists Assaulted, Arrested after Serving Subpoena for Pepper Spray Trial." *Treesit Blog*, August 27, 2004. http://www.contrast.org/treesit/ (accessed August 27, 2004).

"Report on the School of the Americas." Federation of American Scientists, March 6, 1997. http://www.fas.org/irp/congress/1997_rpt/soarpt.htm (accessed May 12, 2003).

"Report to the Seattle City Council WTO Accountability Committee by the Citizens' Panel on WTO Operations." Citizens' Panel on WTO Operations, September 7, 2000. http://www.cityofseattle.net/wtocommittee/panel3final.pdf (accessed March 17, 2002).

"Reviving the World's Rivers: Dam Removal." Part 4, Technical Challenges. International Rivers Network. http://www.irn.org/revival/decom/brochure/rrpt5.html (accessed July 11, 2004).

Revkin, Andrew C. "Bad News (and Good) on Arctic Warming." *New York Times*, October 30, 2004. The article is also at http://www.iht.com/bin/print_ipub.php?file=/articles/2004/10/29/news/arctic.html (accessed October 30, 2004).

Richardson, Paul. "Hojojutsu—The Art of Tying." Sukisha Ko Ryu: Bringing Together All the Elements of the Ninjutsu & Samuraijutsu Takamatsu-den Traditions. http://homepages.paradise.net.nz/sukisha/hojojutsu.html (accessed June 4, 2003).

"Rivers Reborn: Removing Dams and Restoring Rivers in California." Friends of the River. http://www.friendsoftheriver.org/Publications/RiversReborn/main3.html (accessed July 11, 2004).

Robbins, Tom. *Still Life with Woodpecker*. New York: Bantam, 1980.

Rogers, Lois. "Science Turns Monkeys into Drones—Humans Are Next, Genetic Experts Say." *Ottawa Citizen*, October 17, 2004. http://www.canada.com/ottawa/ottawacitizen/news/story.html?id=14314591-ee96-440f-8c83-11a9822d3d42 (accessed October 22, 2004).

Root, Deborah. *Cannibal Culture: Art, Appropriation, & the Commodification of Difference*. Boulder, CO: Westview Press, 1996.

Roycroft, Douglas. "Getting Well in Albion." *Anderson Valley Advertiser*, March 6, 2002, 8.

Russell, Diana E. H. *The Secret Trauma: Incest in the Lives of Girls and Women.* New York: Basic Books, 1986.

————. *Sexual Exploitation: Rape, Child Sexual Abuse, and Sexual Harassment.* Beverly Hills, 1984.

Rutten, Tim. "Cheney's History Needs a Revise." *Los Angeles Times,* November 26, 2005. http://fairuse.1accesshost.com/news3/latimes163.html (accessed November 28, 2005).

"Sabotage Blamed for Power Outage: Bolts Removed from 80-Foot Wisconsin Tower." *CNN.com.* http://www.cnn.com/2004/US/10/11/wisconsin.blackout.ap/ (accessed October 15, 2004).

Sadovi, Carlos. "Cell Phone Technology Killing Songbirds, Too." *Chicago Sun-Times,* November 30, 1999. Also at http://www.rense.com/politics5/songbirds.htm (accessed July 5, 2003).

Safire, William. "You Are a Suspect." *New York Times,* November 14, 2002.

Sale, Kirkpatrick. "An Illusion of Progress." *Ecologist,* June 2003. http://www.theecologist.org/archive_article.html?article=430&category=45 (accessed October 13, 2004).

San Francisco Chronicle. http://www.sfgate.com .

"Sardar Kartar Singh Saraba." Gateway to Sikhism. http://allaboutsikhs.com/martyrs/sarabha.htm (accessed December 29, 2003). Citing Jagdev Singh Santokh., *Sikh Martyrs.* Birmingham, England: Sikh Missionary Resource Centre, 1995.

Savinar, Matt. "Life After the Oil Crash: Deal with Reality, or Reality Will Deal with You." http://www.lifeaftertheoilcrash.net/PageOne.html (accessed September 21, 2004).

Scheffer, Marten, Steve Carpenter, Jonathan A. Foley, Carl Folke, and Brian Walker. "Catastrophic Shifts in Ecosystems." *Nature,* October 11, 2001, 591–96.

Scherer, Glenn. "Religious Wrong: A Higher Power Informs the Republican Assault on the Environment." *E Magazine,* May 5, 2003, 35–39. Also available online under the title "Why Ecocide Is 'Good News' for the GOP." http://www.mindfully.org/Reform/2003/Ecocide-Is-Good-News5may03.htm (accessed May 21, 2003).

Schmitt, Diana. "Weapons in the War for Human Kindness: Why David Budbill Sits on a Mountaintop and Writes Poems." *The Sun,* March 2004.

Schor, Juliet B. *The Overworked American: The Unexpected Decline of Leisure.* New York: Basic Books, 1991.

Schuld, Andreas. "Dangers Associated with Fluoride." EcoMall: A Place to Save the Earth. http://www.ecomall.com/greenshopping/fluoride2.htm (accessed January 21, 2002).

Seekers of the Red Mist. http://www.seekersoftheredmist.com/ (accessed July 10, 2003). This particular comment was posted April 25, 2003, at 8:31 am.

Severn, David. "Vine Watch." *Anderson Valley Advertiser,* April 2, 2003, 8.

Shirer, William. *The Rise and Fall of the Third Reich: A History of Nazi Germany.* Greenwich: Fawcett Crest, 1970.

Shulman, Alix Kates. "Dances with Feminists." The Emma Goldman Papers, University of California, Berkely. http://sunsite.berkeley.edu/Goldman/Features/dances_shulman.html (accessed October 8, 2004). First published in the *Women's Review of Books* 9, no. 3 (December 1991).

"Signs to Look for in a Battering Personality." Projects for Victims of Family Violence, Inc. http://www.angelfire.com/ca6/soupandsalad/content13.htm (accessed November 17, 2002).

Silko, Leslie Marmon. "Tribal Councils: Puppets of the U.S. Government." In *Yellow Woman and a Beauty of the Spirit*. New York: Touchstone Books, 1997.

Sluka, Jeff. "National Liberation Movements in Global Context." Tamilnation.org. http://www.tamilnation.org/selfdetermination/fourthworld/jeffsluka.htm (accessed October 10, 2004).

Socially Responsible Shopping Guide. Global Exchange. http://www.globalexchange.org/economy/corporations/sweatshops/ftguide.html (accessed March 16, 2002).

Spretnak, Charlene. *States of Grace*. San Francisco: HarperSanFrancisco, 1991.

Stannard, David. *American Holocaust: Columbus and the Conquest of the New World*. Oxford: Oxford University Press, 1992.

Stark, Lisa, and Michelle Stark. "$100 Million Wasted: While Some Soldiers Paid Their Own Way, Thousands of Pentagon Airline Tickets Went Unused." *ABC News*. http://abcnews.go.com/sections/WNT/YourMoney/wasted_airline_tickets_040608-1.html (accessed June 9, 2004).

Star Wars. http://www.starwars.com/databank/location/deathstar/ (accessed April 23, 2004).

Star Wars 2. http://www.starwars.com/databank/location/deathstar/?id=eu (accessed April 24, 2004).

"States Get $16 Million for Endangered Species." *Environmental News Network*, September 28, 2001. http://www.enn.com/news/enn-stories/2001/09/09282001/s_45096.asp (accessed January 16, 2002).

St. Clair, Jeffrey. "Santorum: That's Latin for Asshole." *Anderson Valley Advertiser*, April 30, 2003, 8.

St. Clair, Jeffrey, and Alexander Cockburn. "Born Under a Bad Sky." *Anderson Valley Advertiser*, September 18, 2002, 5.

Steele, Jonathan. "Bombers' Justification: Russians Are Killing Our Children, So We Are Here to Kill Yours: Chechen Website Quotes Bible to Claim Carnage as Act of Legitimate Revenge." *Guardian Unlimited*, September 6, 2004. http://www.guardian.co.uk/russia/article/0,2763,1298075,00.html (accessed September 6, 2004).

"A Study of Assassination." This can be found at innumerable Web sites (well, a Google search shows 138). One version complete with drawings is at http://www.gwu.edu/~nsarchiv/NSAEBB/NSAEBB4/ciaguat2.html (accessed July 7, 2003).

"Study Says Five Percent of Greenhouse Gas Came from Exxon." *Planet Ark*, January 30, 2004. http://www.planetark.com/dailynewsstory.cfm/newsid/23638/story.htm (accessed June 22, 2004).

Sunderland, Larry T. B. "California Indian Pre-Historic Demographics." Four Directions Institute. http://www.fourdir.com/california_indian_prehistoric_demographics.htm. See also the map "Native American Cultures Populations Per Square Mile at Time of European Contact." http://www.fourdir.com/aboriginal_population_per_sqmi.htm (accessed June 4, 2004).

Sweating for Nothing. Global Exchange, *Global Economy*. http://www.globalexchange.org/economy/corporations/ (accessed March 16, 2002).

Taber, Robert. *The War of the Flea*. 1st paperbound ed. New York: The Citadel Press, 1970.

Tebbel, John, and Keith Jennison. *The American Indian Wars*. Edison, NJ: Castle Books, 2003.

"Third National Incidence Study of Child Abuse and Neglect." Centers for Disease Control.

Thobi, Nizza. "Chanah Senesh." http://www.nizza-thobi.com/Senesh_engl.html (accessed December 3, 2004).

Thomas, Emory M. *The Confederate State of Richmond: A Biography of the Capital*. Austin: University of Texas Press, 1971.

Thomson, Bruce. "The Oil Crash and You." Great Change. http://greatchange.org/ov-thomson ,convince_sheet.html (accessed September 28, 2004).

Thompson, Don. "Klamath Salmon Plight Worsens: State: Fish Kill May Be Double Previous Estimates." *The Daily Triplicate* (Crescent City, CA), July 31, 2004, A1, A10.

Tillich, Paul. *Systematic Theology*. Vol. 3, *Life and the Spirit; History and the Kingdom of God*. Chicago: University of Chicago Press, 1963.

Tokyo War Crimes Trial Decision. The International Military Tribunal for the Far East, May 3, 1946, to November 12, 1948.

Tomlinson, Chris. "Evidence of U.S. Bombs Killing Villagers." *San Francisco Chronicle*, December 6, 2001, A10.

"Too Hot for Uncle John's Bathroom Reader." Triviahalloffame. http://www.triviahall offame.com/gandhi.htm (accessed August 8, 2004).

Towerkill. http://www.towerkill.com/ (accessed July 5, 2003).

Trial of the Major War Criminals before the International Military Tribunal, Nuremberg, 14 November 1945–1 October 1946. Nuremberg, 1947–1949.

Trudell, John. *Green Anarchy*, Fall 2003, 15.

Turner, Frederick. *Beyond Geography: The Western Spirit Against the Wilderness*. New Brunswick, NJ: Rutgers, 1992.

Udall, Stewart, Charles Conconi, and David Osterhout. *The Energy Balloon*. New York: McGraw-Hill, 1974.

U.S. Congress. *Congressional Record*. 56th Cong., 1st sess., 1900. Vol. 33, 704, 711–12.

U.S. et al. v. Goering et al. [The Nuremberg Trial]. Extra Lexis 1, 120, 6 F.R.D. 69 (1947) (accessed December 16, 2005).

"U.S. Military Spending to Exceed Rest of World Combined!" *Nexus*, September/October 2005, 9.

"The Victims: The Fight Against Terrorism." *The Oregonian*, January 16, 2002, A2.

Vidal, Gore. *The Decline and Fall of the American Empire*. Berkeley: Odonian Press, 1995.

Wacquant, Loïc. "Ghetto, Banlieue, Favela: Tools for Rethinking Urban Marginality." http://sociology.berkeley.edu/faculty/wacquant/condpref.pdf (accessed March 16, 2002).

Walker, Paul F., and Eric Stambler. ". . . And the Dirty Little Weapons: Cluster Bombs, Fuel-Air Explosives, and 'Daisy Cutters,' not Laser-Guided Weapons, Dominated the Gulf War." *Bulletin of the Atomic Scientists* 47, no. 4 (May 1991): 21–24. Also available at http://www.bullatomsci.org/issues/1991/may91/may91walker.html (accessed November 19, 2001).

Watson, Paul. "Report from the Galapagos." *Earth First! Journal*, Samhain/Yule 2003, 38.

"Weapons of American Terrorism: Torture." http://free.freespeech.org/americanstateterrorism/weapons/US-Torture.html (accessed May 12, 2003).

Weber, Max. *Max Weber: The Theory of Social and Economic Organization*. Translated by A. M. Henderson and Talcott Parsons. Edited with an introduction by Talcott Parsons. Oxford: Oxford University Press, 1947.

Webster's New Twentieth Century Dictionary of the English Language, 2nd ed. New York: Simon and Schuster, 1979.

Weiss, Rick. "Major Species Annihilated by Fishing, Says Study." *San Francisco Chronicle*, May 15, 2003, A13.

Weizenbaum, Joseph. *Computer Power and Human Reason: From Judgment to Calculation*. San Francisco: W. H. Freeman, 1976.

Whales. http://www.biologicaldiversity.org/swcbd/press/beaked10-15-2002.html, http://action network.org/campaign/whales/explanation, sites visited October 26, 2002. Also, http://www.faultline.org/news/2002/10/beaked.html, site visited October 27, 2002.

"What Is Depleted Uranium." http://www.web-light.nl/VISIE/depleted_uranium1.html (accessed January 23, 2002).

"What's the Dam Problem?" *The Why Files: The Science Behind the News*, January 16, 2003. http://whyfiles.org//169dam_remove/index.html (accessed July 11, 2004).

Wheatley, Margaret. *Turning to One Another: Simple Conversations to Restore Hope to the Future*. San Francisco: Berrett-Koehler Publishers, 2002.

White, Chris. "Why I Oppose the U.S. War on Terror: An Ex-Marine Sergeant Speaks Out." *The Grass Root* 3, no. 1 (Spring 2003). *The Grass Root* is the paper of the Kansas Greens (Kansas Green Party, Box 1482, Lawrence, KS 66044).

"Why Is Everybody Always Pickin' on Me?" Dispatches, *Outside Online*, July 1998. http://web.outsideonline.com/magazine/0798/9807disprod.html (accessed July 10, 2003).

Wikle, Thomas A. "Cellular Tower Proliferation in the United States." *The Geographical Review* 92, no. 1 (January 2002): 45–62.

Wilkinson, Bob. "Trained Killers." *Anderson Valley Advertiser*, April 30, 2003, 3.

Williams, Martyn. "UN Study: Think Upgrade Before Buying a New PC: New Report Finds 1.8 Tons of Material Are Used to Manufacture Desktop PC and Monitor." *Infoworld*, March 7, 2004. http://www.infoworld.com/article/04/03/07/hnunstudy_1.html (accessed March 12, 2004).

Wilson, Jim. "E-Bomb: In the Blink of an Eye, Electromagnetic Bombs Could Throw Civilization Back 200 Years. And Terrorists Can Build Them for $400." *Popular Mechanics*, September 2001. http://popularmechanics.com/science/military/2001/9/e-bomb/print.phtml (accessed August 22, 2003).

Wingate, Steve. "The OMEGA File—Concentration Camps: Federal Emergency Management Agency." http://www.posse-comitatus.org/govt/FEMA-Camp.html (accessed July 21, 2004).

"Witch Hunting and Population Policy." http://www.geocities.com/iconoclastes.geo/witches.html (accessed September 23, 2002).

The World Factbook, s.v. "Afghanistan." CIA. http://www.odci.gov/cia/publication/factbook/geos/af/html (accessed November 19, 2001).

Wyss, Jim. "Ecuador Free-for-All Threatens Tribes, Trees: Weak Government Lets Loggers Prevail." *San Francisco Chronicle*, September 3, 2004, W1.

Yergin, Daniel. *The Prize: The Epic Quest for Oil, Money & Power*. New York: Simon & Schuster, 1991.

Z Magazine, July/August 2000, 62.

About the Author

DERRICK JENSEN is the acclaimed author of *A Language Older Than Words* and *The Culture of Make Believe*, among many others. Author, teacher, activist, small farmer, and leading voice of uncompromising dissent, he regularly stirs auditoriums across the country with revolutionary spirit. Jensen lives in Northern California and organizes on issues surrounding deforestation, dam removal, restoration of habitat for salmon and other fish, habitat improvement for amphibians, the promotion of organic farms, and the preservation of family farms. He holds a degree in Creative Writing from Eastern Washington University, a degree in Mineral Engineering Physics from the Colorado School of Mines, and has taught at Eastern Washington University and Pelican Bay State Prison.